Currents in High-Energy Astrophysics

NATO ASI Series

Advanced Science Institutes Series

A Series presenting the results of activities sponsored by the NATO Science Committee, which aims at the dissemination of advanced scientific and technological knowledge, with a view to strengthening links between scientific communities.

The Series is published by an international board of publishers in conjunction with the NATO Scientific Affairs Division

A Life Sciences **B Physics**	Plenum Publishing Corporation London and New York
C Mathematical and Physical Sciences **D Behavioural and Social Sciences** **E Applied Sciences**	Kluwer Academic Publishers Dordrecht, Boston and London
F Computer and Systems Sciences **G Ecological Sciences** **H Cell Biology** **I Global Environmental Change**	Springer-Verlag Berlin, Heidelberg, New York, London, Paris and Tokyo

PARTNERSHIP SUB-SERIES

1. Disarmament Technologies	Kluwer Academic Publishers
2. Environment	Springer-Verlag
3. High Technology	Kluwer Academic Publishers
4. Science and Technology Policy	Kluwer Academic Publishers
5. Computer Networking	Kluwer Academic Publishers

The Partnership Sub-Series incorporates activities undertaken in collaboration with NATO's Cooperation Partners, the countries of the CIS and Central and Eastern Europe, in Priority Areas of concern to those countries.

NATO-PCO-DATA BASE

The electronic index to the NATO ASI Series provides full bibliographical references (with keywords and/or abstracts) to more than 30000 contributions from international scientists published in all sections of the NATO ASI Series.
Access to the NATO-PCO-DATA BASE is possible in two ways:

– via online FILE 128 (NATO-PCO-DATA BASE) hosted by ESRIN, Via Galileo Galilei, I-00044 Frascati, Italy.

– via CD-ROM "NATO-PCO-DATA BASE" with user-friendly retrieval software in English, French and German (© WTV GmbH and DATAWARE Technologies Inc. 1989).

The CD-ROM can be ordered through any member of the Board of Publishers or through NATO-PCO, Overijse, Belgium.

Series C: Mathematical and Physical Sciences – Vol. 458

Currents in High-Energy Astrophysics

edited by

Maurice M. Shapiro
Department of Physics and Astronomy,
University of Maryland,
College Park, MD, U.S.A.

Rein Silberberg
Universities Space Research Association,
Washington, D.C., U.S.A.

and

John P. Wefel
Department of Physics and Astronomy,
Louisiana State University,
Baton Rouge, LA, U.S.A.

Kluwer Academic Publishers

Dordrecht / Boston / London

Published in cooperation with NATO Scientific Affairs Division

Stc.

Proceedings of the NATO Advanced Study Institute
and Ninth Course of the International School of Cosmic Ray Astrophysics on
Currents in High-Energy Astrophysics
Ettore Majorana Centre, Erice, Sicily, Italy
7–18 May 1994

A C.I.P. Catalogue record for this book is available from the Library of Congress

ISBN 0-7923-3354-3

Published by Kluwer Academic Publishers,
P.O. Box 17, 3300 AA Dordrecht, The Netherlands.

Kluwer Academic Publishers incorporates the publishing programmes of
D. Reidel, Martinus Nijhoff, Dr W. Junk and MTP Press.

Sold and distributed in the U.S.A. and Canada
by Kluwer Academic Publishers,
101 Philip Drive, Norwell, MA 02061, U.S.A.

In all other countries, sold and distributed
by Kluwer Academic Publishers Group,
P.O. Box 322, 3300 AH Dordrecht, The Netherlands.

Printed on acid-free paper

Printed in the Netherlands

This volume is dedicated

in fond memory

to three giants of cosmic-ray physics:

Pierre Auger

Giuseppe Occhialini

Bruno Rossi

and to a promising young scientist

Kyoung Hye Moon.*

They are sorely missed by their colleagues and friends.

*See Preface

TABLE OF CONTENTS

III. PULSARS AND SUPERNOVA REMNANTS

IV. COSMOLOGY

PREFACE

This volume comprises selected lectures presented in the Ninth Course of the International School of Cosmic-Ray Astrophysics held at the Ettore Majorana Centre in Erice, Sicily, May 7-18, 1994. Director of the Centre is A. Zichichi, assisted by M. Zaini. Director of the School is M. M. Shapiro. J. P. Wefel was co-director of the Ninth Course, which was also a NATO Advanced Study Institute (ASI), and NATO support is gratefully acknowledged. Devoted to problems and prospects in high-energy astrophysics and cosmology, the major areas explored in this course were: gamma-ray, X-ray, and neutrino astronomies; cosmic rays; pulsars and supernova remnants; and cosmology, as well as cosmogony.

Among the principal developments in gamma-ray astrophysics were those generated by the Compton Gamma Ray Observatory. Cosmic neutrinos at MeV energies, i.e., those from the sun and from Supernova 1987a, were discussed, as well as neutrino masses in astrophysics. The source composition of cosmic rays, and extensive air shower experiments, received special attention. The early universe according to COBE data, and as viewed by theorists of cosmology, was reviewed. Finally, the connections with particle physics occasioned a timely description of the Standard Model of elementary particles.

The cosmic-ray community, the International School of Cosmic-Ray Astrophysics, and indeed the world of science have recently suffered grievous losses in the demise of Pierre Auger, Giuseppe Occhialini, and Bruno Rossi. These distinguished physicists had served for eighteen years on the scientific advisory committee of our School in Erice. Accordingly, a memorial session was included in our program. In the same session a farewell tribute was paid by M. M. Shapiro and A. E. Smith to the late Kyoung Hye Moon, an enthusiastic student of our School, who missed this Course when a rampant illness cut short her promising career.

If our Course succeeded in its aims, this was due largely to the following: Dr. Alberto Gabriele, Jerry Pilarski, and Pinola Savalli, and the staff of the Centre; our lecturers--not in the least for their stimulating interactions with younger participants; NATO's Division of Scientific Affairs, and the Director of its ASI program, Dr. L. V. da Cunha; the Italian Ministry of Education, the Sicilian Regional Government, the National Science Foundation, the International Science Foundation, Mrs. Shirley Ratner of Chevy Chase, MD; and scientists from Huntsville, AL and other locations who contributed to a fellowship fund in memory of Dr. K. H. Moon.

The scientific Advisory Committee consists of Professors W. A. Fowler, V. L. Ginzburg, Rein Silberberg, J. A. Simpson, J. A. Van Allen, A. Zichichi and the Director of the School. Their counsel helped to make this ASI a memorable experience for students and lecturers alike.

Maurice M. Shapiro
Rein Silberberg
John P. Wefel

The Sky as viewed from the Compton Gamma Ray Observatory

G. Kanbach
Max-Planck-Institut für Extraterrestrische Physik
Giessenbachstr. 1
85748 Garching, Germany

1. Introduction

This review is on observational gamma ray astronomy and on the exciting results that have been flowing from the instruments on the Compton Gamma Ray Observatory. Before we present the present status of the observations and the latest discoveries I'd like to shortly outline the scientific background, the reason for doing astronomy in the gamma ray energy range, the questions on which this discipline got started and why it took a long time before the present mature state as a member in the spectrum of multiwavelength astronomers was reached.
After it had been realized that the cosmic radiation is mainly composed of positively charged particles – the latitude dependence and the east-west assymmetry of the cosmic ray intensity due to the Earth's magnetic field being a most convincing argument – it was quite clear that a component of high energy photons must be associated with the cosmic radiation. Nuclear and particle physics as well as electrodynamics allowed predictions of what happens when high energy protons, ions and electrons get into contact with target matter, photons or electromagnetic fields, either in their places of origin or while they propagate through the interstellar medium – gamma radiation will inevitably be produced in one of the following reactions:

p, ions $\rightarrow$ material targets $\rightarrow$ secondary particles
$\rightarrow$ gamma ray lines
$\rightarrow \pi^0 \rightarrow 2\,\gamma$'s ($E_0 = 69.5$ MeV)
$\rightarrow \pi^{\pm} \rightarrow e^{\pm} \rightarrow$ Bremsstrahlung
$\rightarrow$ Kaons etc. $\rightarrow \gamma$'s

Electrons $\longrightarrow$ material targets $\longrightarrow$ Thermal or nonthermal Bremsstrahlung
$\longrightarrow$ photons $\longrightarrow$ Inverse Compton scattering (Energy transfer from electron to photon)
$\longrightarrow$ magnetic field $\longrightarrow$ Synchrotron radiation, Curvature radiation, e-m cascades
Matter – Antimatter annihilation: $e^+ - e^- \rightarrow 2\,\gamma$'s at 0.511 MeV or positronium (3 γ's)
$p - \bar{p} \rightarrow \pi^0\,,\ \pi^{\pm}$, etc.

All of these gamma ray production processes, with the possible exception of the last one, are realized in the universe and are being observed. For many cosmic high energy sites the gamma radiation provides the most direct view and information of the processes that take place at the highest energies known in nature. Many of the theoretical predictions for gamma ray astronomy were formulated in the late 1940's and 50's. Why has it taken about 20 years before the first real

M. M. Shapiro et al. (eds.), Currents in High-Energy Astrophysics, 1–20.

observational results became available? First of all gamma ray photons are a very rare species of cosmic radiation – at GeV energies only one in a thousand cosmic rays is a photon, which corresponds to a typical flux of $10^{-4}\ \gamma\,cm^{-2}\,s^{-1}\,sr^{-1}$. That means that not only have the detectors to be of sufficient mass to detect this penetrating radiation, but they also have to have typical areas of $10^4\,cm^2$ and they must be extremely selective to discriminate charged particles that enter the instrument. Furthermore the Earth's atmosphere not only shields even the highest mountain tops from the primary cosmic photons but also is a bright source of gamma ray background under the impact of the charged cosmic radiation. Gamma ray astronomy therefore has to depend on the technology and transportation systems that the space age provided since the 1960's although at first stratospheric balloons carried instrumentation to the edge of the atmosphere. The success of balloon borne telescopes was however limited because of the high background levels, rather small instrumentation and short observation times. Only with satellite instruments did this discipline of astronomy begin to produce convincing results. In 1967 a directional scintillation detector on the OSO–3 mission discovered the ~ 100 MeV emission from the Milky Way (Kraushaar et al., 1972). The first truly imaging spark chamber telescope SAS–2 measured the galactic distribution of diffuse gamma ray emission in much greater detail and discovered several point sources, i.e.the Crab and Vela pulsars, a mysterious unidentified source called Geminga and emission from the Cygnus region presumed to come from Cyg–X3 (Fichtel et al., 1975). From 1975 to 1982 the COS–B satellite, similar to SAS–2 but augmented with an energy calorimeter that provided improved spectral resolution, measured the galactic emissions in more detail and refined the observations of Crab, Vela and Geminga (Mayer-Hasselwander et al., 1982). Cyg–X3 was not confirmed although approximately 20 other galactic unidentified sources could be extracted from the COS–B data (Swanenburg et al., 1981). The first extragalactic source, the quasar 3C273 was also revealed as a gamma ray emitter and it became clear that the emission of gamma radiation would dominate the radiation budget of this cosmic object. At the same time the observations between hard X–rays (~ 100 keV) and low energy gamma rays ($\leq$ 10 MeV) made a lot of progress. Balloon instruments led to the discovery of cyclotron lines from X-ray binaries and high resolution Ge spectrometers detected the 0.511 MeV annihilation lines from the inner Galaxy. Spectrometers on OSO–7 and the Solar Maximum Mission provided the first detection of gamma ray line emission from solar flares and from the interstellar medium. The number and properties of hard X-ray and soft gamma ray sources was determined by satellite telescopes on HEAO–1 and –3. Many of these sources turned out to be active galaxies of Seyfert 1 type. Finally, the study of the mysterious gamma ray bursts with many detectors in interplanetary space achieved such results that the need for a new mission, dedicated to gamma ray astronomy over a wide energy range, became very urgent. At this point in about 1980, the plan for a satellite mission, covering with a set of instruments the range from about 20 keV to more than 30 GeV, took shape and the Gamma Ray Observatory was announced. Four instruments form the payload of what is now called the Compton Observatory (after Arthur H. Compton, 1892–1962, Nobelprice in 1927 for his research in electron interactions with photons).

2. CGRO Mission and Instrumentation

The Compton Gamma Ray Observatory (CGRO) carries four instruments which detect photons from the upper end of the X-ray spectrum at about 20 keV to more than 30 GeV: the Burst and Transient Source Experiment (BATSE), the Oriented Scintillation Spectrometer Experiment (OSSE), the Imaging Compton Telescope (COMPTEL), and the Energetic Gamma Ray Experiment Telescope (EGRET). The CGRO spacecraft is shown schematically in figure 1. Each instrument on CGRO is designed and optimized for its spectral band to detect the incident radiation via the principal interaction processes (photoabsorption, Compton scattering and pair creation) and to suppress the unwanted background as much as possible. The instrument properties are summarized in table 1:

Table 1: Characteristics of the Compton Gamma Ray Observatory Instruments

	BATSE		OSSE	COMPTEL	EGRET
	Large Area	Spectroscopy			
Energy Range (MeV)	0.03 – 1.9	0.015 – 110	0.06 – 10	0.8 – 30	30 – 30000
Effective Area (cm^2)	1800 ea. at 100 keV	127 ea. at 200 keV	1480 at 1.0 MeV	30 at 3 MeV	1600 at 500 MeV
Field of View	4π sr	4π sr	$3.8° \times 11.4°$	1 sr	0.6 sr
Position Localization	3°		10 arc min	0.5° – 1°	5 – 10 arc min
Source Sensitivity (2 weeks obs.)	3×10^{-8} erg cm^{-2} (1 sec burst)		$(3–8)\times10^{-5}$ $cm^{-2}s^{-1}$ (line)	1.6×10^{-4} $cm^{-2}s^{-1}$ (1–30 MeV)	7×10^{-8} $cm^{-2}s^{-1}$ (>100 MeV))

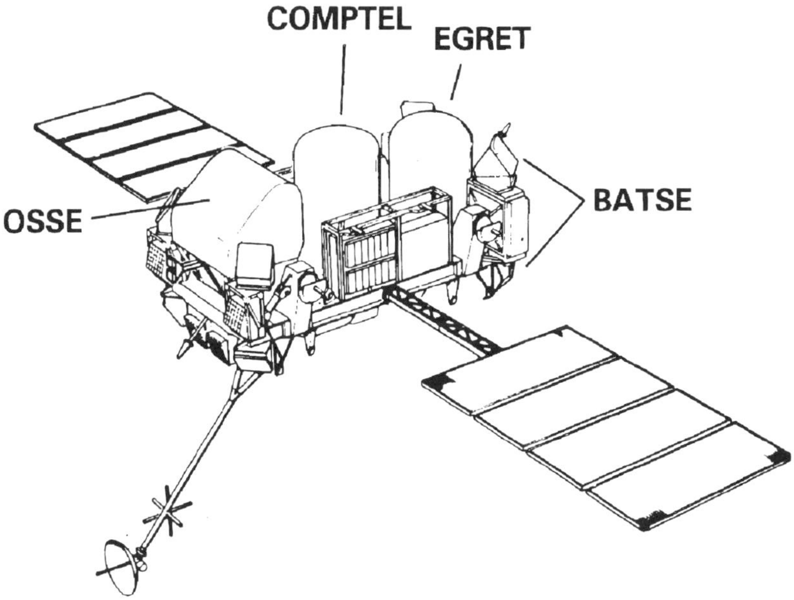

Figure 1: Schematic layout of the Compton Gamma Ray Observatory (CGRO)

BATSE is an all sky monitor for transient sources of low energy radiation. It consists of 8 modular large area scintillation counters that are placed at the corners of the satellite platform and are oriented towards the sky like the faces of a regular octahedron. Each module intersects an incoming flux of radiation with a different projected area which leads to characteristic counting rates in each detector which are used to locate the direction of sources. Newly developed analysis techniques based on the accurately predictable occultation of steady sources by the earth's limb enable BATSE also to monitor a wide range of strong hard X-ray sources (PI G.J. Fishman, NASA-MSFC, Huntsville, USA).

OSSE, the instrument primarily dedicated to spectroscopy of astrophysical objects, consists of four actively shielded NaI(Tl)-CsI(Na) phoswich detectors which are collimated to a field of view of 3.8° × 11.4° (FWHM). Each module can be oriented independently in the X–Z plane over a range of 192°. Usually OSSE operates in an on-source/off-source mode to determine the intrinsic detector background and the net source signal. OSSE can also react quickly to observe solar flares or transient sources when a trigger signal is generated by BATSE (PI J.D. Kurfess, NRL, Washington, USA).

COMPTEL is an imaging telescope for medium energy photons, which interact primarily in Compton scatterings off electrons. Two interactions are recorded in two detector layers, a scattering of the incoming and a subsequent absorption of the scattered photon. From the location of the interactions and the energy deposits one can, after a complete exposure, reconstruct an image of the sky in the large field of view. COMPTEL observations have been used to map the galactic emissions in spectral lines and in the continuum and in the search and measurement of point sources (PI V. Schönfelder, MPE, Garching, Germany).

The telescope for the highest energies, EGRET, detects photons via pair creation events, which are imaged in a digital multi layer spark chamber. The chamber is filled with a Ne/Ar gas mixture that needs to be flushed and refilled about once a year to maintain a good quality of track detection. EGRET had about 5 refills of spark chamber gas in high pressure tanks on board, which allows about a lifetime of 6 years of normal operations. The direction of the incoming photons is derived from the tracks of the created electrons and the energy of the released cascade is measured in a thick NaI calorimeter. The principal objectives for EGRET are the search for new gamma ray sources and the measurement of the galactic and extragalactic diffuse emissions. Like the other CGRO instruments EGRET has been designed to discriminate effectively against instrumental background radiation in order to achieve a sensitivity which is only limited by the size of the instrument and the incoming fluxes from cosmic sources (Co-PIs C.E. Fichtel, NASA-GSFC, Greenbelt, USA and K. Pinkau, IPP, Garching, Germany).

The Compton Observatory was launched on April 5, 1991 into a 450 km circular orbit inclined by 28.5° on the space shuttle Atlantis. With a total mass of 15.9 tons, of which about 6 tons are scientific instrumentation, CGRO is one of the largest scientific satellites ever launched. The first aim for Compton was a complete sky survey with the large field of view instruments COMPTEL and EGRET. This goal was achieved in about 50 individual pointings between May 16, 1991 and November 17, 1992. Since then an increasing portion of the mission observations are being proposed and analysed by guest investigators that are selected in approximately yearly intervals. The failure of the tape recorders on CGRO in March 1992 required that the mission is now operated in a real time telemetry mode through the TDRSS relay satellites. This mode however also recovers about 80% of the available data. During October/December 1993 the first orbital maintenance, i.e. a lift of the orbit altitude from 350 km to the original 450 km, was performed with the on-bord rocket engines. The essential technical systems of the CGRO platform and instruments are in excellent condition and several more years of observations are hoped for (Gehrels, et al., 1993).

3. The puzzle of Gamma Ray Bursts

Before the launch of the Compton Observatory gamma ray bursts (GRBs) had been observed with a wide variety of instruments in space and on balloons. The first reports of this unexpected cosmic phenomenon consisting of short duration (up to tens of seconds), high energy transients (typically 100s of keV but occasionally up to 100 MeV) came from instruments on the Vela system of surveillance satellites (Klebesadel et al, 1973). These measurements were soon followed up by detectors on soviet (e.g. Mazets et al., 1979) and other international missions. Most bursts were detected with no directional information; only in a few cases an interplanetary network of detectors enabled the arrival time triangulation of the burst locations within a few arcminutes. No convincing counterparts for the burst sources in other wavelengths could be found with the possible exception of GRB790305, a unique burst located within the supernova remnant N49, in the Large Magellanic Cloud. The other spatially located bursts seemed to be distributed isotropically on the sky and didn´t have any obvious counterparts at other wavelengths (Atteia et al, 1987). The limited sensitivity of the pre-CGRO instruments, i.e. the limited reach for fainter, more distant bursts, did however not allow firm conclusions on the nature and distribution of burst sources. As many researchers proposed galactic neutron stars as burst sites a correlation with the galactic disk would have been expected if the measurements would reach distances of the order of several kpc. These expectations were completely confounded after the much more sensitive BATSE instrument had collected burst data at a rate of about one event per day for several years (for an early report see Meegan et al., 1992).. BATSE burst number 1000 was recorded on May 27, 1994 and the distribution of these gamma ray bursts on a skymap in galactic coordinates is shown in figure 2 (Fishman et al., 1994). The distribution is completely isotropic and in particular no concentration to the galactic disk is visible. If one tentatively assumes that the sources are also distributed homogeneously over the volume that is accessible with BATSE the number of sources N that are brighter than S should scale like $S^{-3/2}$ due to the constant space density ($N \propto \rho V \propto \rho R^3$) and the inverse square law that determines the brightness ($S \propto R^{-2}$). The BATSE observations however indicated a clear deficit of dimmer bursts as shown in figure 3. This indicates that the observed bursts come from a limited volume and are not homogeneously distributed around us.

What type of astrophysical object could be distributed in a spherical volume around the Earth that only extends to a certain maximal radius? The mainstream of current explanations for gamma ray bursts holds that the sources are either distributed in a large galactic halo or that they are at cosmological distances of up to redshifts of about 1. In the first case typical luminosities of 10^{43} erg/s would be required while the cosmological source must supply about 10^{49} erg/s in the gamma ray range (isotropic emission assumed, see Harding, 1993). The high energy emission is produced with extremely short time scales and the time profiles of bursts are unique signatures of individual events. This suggests that the source regions are very compact. The degree of isotropy and the absence of localized excesses in the celestial distribution of the BATSE bursts already can put severe limits on a potential population located in the galactic halo. As Hakkila et al., 1994 have shown the BATSE data have already ruled out all but the largest galactic halo models because the anisotropy that results from the excentric location of the Sun in the Galaxy is not seen. On the other side a galactic halo of burst sources could not be too large because one would start to detect the halos around neighbouring galaxies, like the Andromeda nebula. These are also not detected.

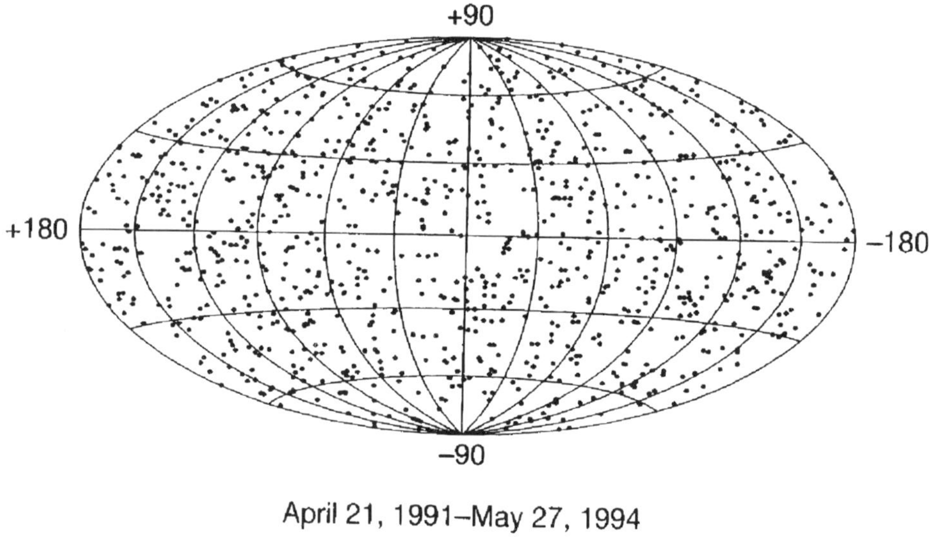

Figure 2: Distribution of 1000 gamma ray bursts detected by BATSE (galactic coordinates)

A widely discussed current hypothesis holds that gamma ray bursts are generated in normal galaxies when a binary stellar system made of two neutron stars merges due to the continous loss of orbital energy that is radiated away as gravitational radiation (Paczinski, 1986). Such a terminal cataclysm of two magnetized stars that can be considered as the collision of two macroscopic nuclei could well result in the generation of a fireball of high energy radiation. The observation in our own Galaxy of these compact binary systems that slowly spiral into each other allows to estimate that about one merger in about a million years occurs in a normal galaxy. This rate is sufficient to provide the observed number of gamma ray bursts. The hypothesis could also explain that no obviously repeating sources have been observed so far.

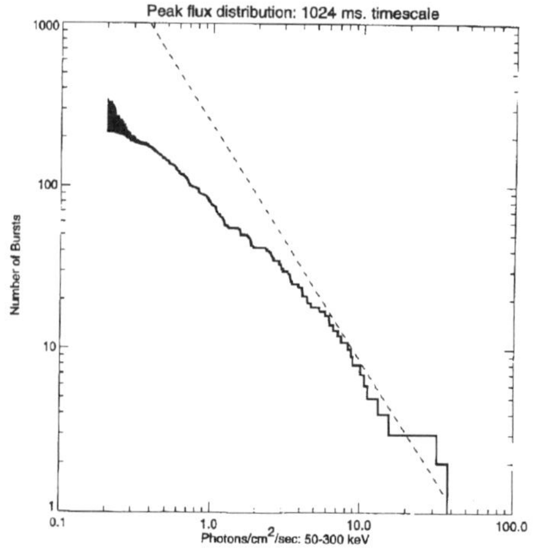

Figure 3: logN – logS diagram of BATSE bursts (Fishman et al., 1994)

If bursts occur within the field of view of EGRET or COMPTEL the possibility exists to directly image the burst sources. Comptel has imaged 18 bursts (up to Sept 21, 1994) with a positional accuracy of typically ±1.5°. EGRET detected the bursts of May 3, 1991 (Schneid et al, 1992), June 1, 1991 (Kwok et al., 1993), January 31, 1993 (Sommer et al, 1993) and March 1, 1994 (Schneid et al., 1994); they all show emission into the GeV range. The burst locations agree well with the positions derived from interplanetary triangulation and restrict the position to typically a radius of 1°. No obvious counterparts have been found in these error boxes. Recently the observation with EGRET of photons up to several tens of GeV related to a gamma ray burst on

February 17, 1994 (Hurley et al., 1994) has raised considerable attention. GRB940217 displayed convincingly for the first time that very high energy photons can be delayed by typically an hour after the peak of the low energy radiation. An explanation for this delayed emission could be, that the energetic event which spreads in the form of a blast wave reaches a shell of debris that had been ejected previous to the final cataclysm. EGRET has so far imaged a high energy component (>30 MeV) from 5 of the about 150 BATSE bursts, that occured in the field of view of the telescope. These 5 bursts are also the brightest (peak flux and/or fluence) BATSE events in the sample. One can therefore speculate whether a more sensitive detector would be able to detect high energy photons also from the weaker events. It is conceivable that the summed high energy contribution from these bursts will be detectable. The presence of photons above several MeV in gamma ray bursts leads to interesting considerations on the radiation pattern of these sources: a very compact source, as the time structure leads us to assume, would become optically thick against photon-photon pair creation if the photon density in the source is estimated from an isotropic emission model. A sharply beamed emission geometry, as it would be generated if the source material moves with relativistic velocities toward the observer, would alleviate the problem of intrinsic absorption. If the source is at typical galactic halo distances (20 kpc) a bulk Lorentz factor of ~ 14 is required, while for a cosmological source at 1 Gpc a factor of ~ 500 applies (Harding, 1993, see also a detailed recent review by Baring, 1994)).

In conclusion one can say that the observations of gamma ray bursts provided by the instuments on CGRO have given us a glimpse of perhaps the most energetic processes in the universe and allow us to describe accurately the distributions in space, time and energy of the burst phenomenon. The puzzle of the origin of bursts, including the question whether there are clearly distinct classes of bursts which have different origins, is however not solved and it seems that new and exciting surprises occur as the observations continue. The investigation of gamma ray bursts across the electromagnetic spectrum will certainly remain one of the most exciting fields of astrophysics for many years to come.

4. The Radioactive Universe

The observation of radioactive isotopes and their identification based on the characteristic gamma ray line radiation emitted is a certain sign that nucleosynthesis is occuring in astrophysical settings. While the fusion processes in normal stars are completely hidden deep inside the stars, an explosive, energetic event like a solar flare, a nova or supernova can generate and disperse enough radioactive material either on the surface of the parent body or into the interstellar medium. Especially the observed yield of radioactive isotopes in supernovae will be an important fact to verify models for these explosions which are thought to produce and distribute practically all higher elements in the universe. Radioactive nuclei are produced in a large variety of reaction chains depending on conditions and have widely different decay schemes and timescales. Only a limited number of isotopes is important for gamma ray observations - they are listed in table 2: The timeconstants already indicate for which type of astrophysical process the different nuclides play a role: the short lived nuclides ^{56}Ni and ^{57}Co are thought to be the main power sources in supernova ejecta; ^{22}Na has been searched for in novae which occur on typical time scales of years; ^{44}Ti traces the sites of historical supernovae while the longer lived isotopes ^{26}Al and ^{60}Fe accumulate over millions of years in the interstellar medium. COMPTEL and OSSE are the principal instruments on CGRO for spectroscopic observations. We review only three results from COMPTEL concerning the galactic emission of the ^{26}Al line at 1.8 MeV, emission from Cas A a historical supernova and from the Orion cloud complex.

The Galaxy imaged in the ^{26}Al line: A survey of the galactic disk, filtered to a 200 keV wide energy band around the 1.8 MeV line of ^{26}Al is shown in figure 4 (Diehl, 1994). The emission is concentrated on a ridge along the galactic equator with significant clumpy enhancements in Vela, Carina and Cygnus as well as regions of very low intensity, e.g. at longitudes 40°–70°, 130°–180°, and 200°–260°. The emission maximum is in the central region of the Galaxy although there is an offset from the center position. The overall map is asymmetric with respect to the Galaxy with a higher intensity of the ^{26}Al line visible from the third and fourth galactic quadrant. Previously models had proposed a fairly smooth distribution of ^{26}Al in the Galaxy due to the superposition of many novae. These models are not supported by the data – instead a mixture of strong, concentrated sources like the Vela supernova remnant and contributions from local star forming regions has to be assumed. The required mass of ^{26}Al in Vela is compatible with current calculations of the production of this isotope in a core collapse supernova if the Vela SNR is in the distance range 120–350 pc, which is entirely in agreement with recent estimates from X-ray observations (Oberlack et al., 1994).

Table 2: Radionuclides important for gamma-ray astronomy

Isotope	Half-Life (years)	Decay Mode	Photons E_γ (MeV)	Prob (%)
^{56}Ni	0.018	e^+, γ	0.847	100
			1.238	100
^{57}Co	0.745	EC, γ	0.014	10
			0.122	86
			0.136	11
^{22}Na	2.602	β^+ (90%), EC (10%)	0.511	Annihil.
			1.275	100
^{44}Ti	54	β^+, γ	0.068	100
			0.078	98
			1.157	100
^{26}Al	0.72×10^6	β^+, γ	1.809	
^{60}Fe	1.52×10^6	β^+, γ	0.058	100
			1.332	100
			1.173	100

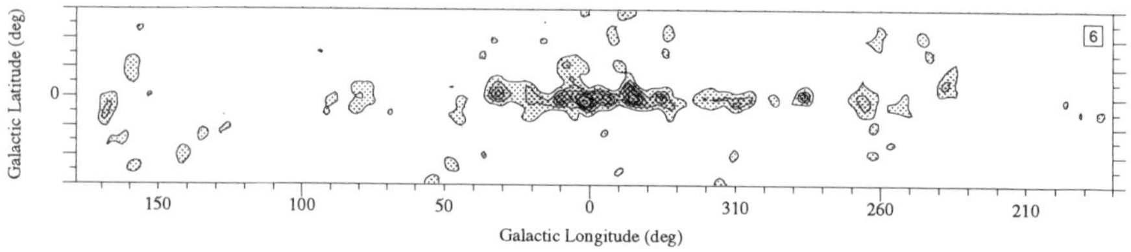

Figure 4. Maximum entropy flux map of the Galaxy in the light of the ^{26}Al line at 1.8 MeV

^{44}Ti in Cas A: Cas A is a young supernova remnant with an age of about 300 years that has been mainly detected in the radio range because of severe interstellar absorption. COMPTEL exposures of the region have produced evidence for a source of 1.16 MeV line radiation concentrated on the SNR (Iyudin et al., 1994). The observed line flux of $(7.0 \pm 1.7)\times10^{-5}\ \gamma / cm^2$ s, the distance of 2.8 kpc and the age of the radioactive source yield an estimate of the ejected mass of ^{44}Ti of $(2.5 \pm 1.3) \times 10^{-4}\ M_0$. This mass is consistent with the yield expected in the supernova explosion of a 20 M_0 · Wolf Rayet progenitor star. Presently the Comptel data are being searched for other localized ^{44}Ti sources which could indicate sites of young SNRs that are shrouded by the clumpy interstellar medium.

$^{16}O^*$ and $^{12}C^*$ lines from Orion: Energetic protons and ions (typically ≥ 10 MeV/nucleon) interacting with matter of normal composition excite prompt gamma ray lines in many of the participating nuclei. These processes occur frequently in solar flares and complex spectra consisting of broad and narrow lines and nuclear continua in the energy range from 1 to 8 MeV are typical for high energy flares. In the gamma ray energy range from 3–7 MeV the most prominent lines are from the abundant elements $^{12}C^*$ (4.4 MeV) and $^{16}O^*$ (6.13 MeV). COMPTEL observations of the giant molecular clouds in Orion have now, for the first time from an object out of the solar system, produced evidence for emission of broadened gamma ray lines at 4.4 and 6.1 MeV. The flux from 3–7 MeV is localized on the maximum of the cloud complexes (Bloemen et al., 1994). The maximum likelihood image in the gamma ray energy range from Orion overlaid on the radio map tracing the CO molecule is shown in figure 5. With a flux of $(1.0 \pm 0.15) \times 10^{-4}$ γ(3–7 MeV) / cm^2 s the Orion region is much brighter than had been anticipated from the interaction of cosmic rays of local density with the cloud material – actually the cosmic ray density would have to be raised by about a factor of 500 to account for the line flux. Such a density of ionizing radiation is however completely ruled out inside a molecular cloud. Only a flux of low energy (~ 10 MeV/nucleon) medium mass nuclei, enhanced by a factor of about 30 with respect to the normal abundances, which interacts with the stationary cloud gas, could generate the required slightly broadened line flux. A recent proposal by Ramaty, 1994 suggests that supernova ejecta, enriched in the medium mass nuclei provide such a flux of low energy cosmic rays. The required energy to excite the observed gamma ray lines could be provided by the kinetic energy of the supernova remnant expansion

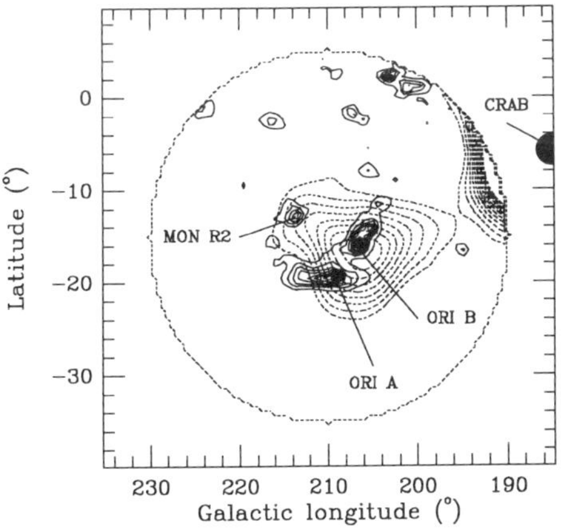

Figure 5: COMPTEL maximum likelihood map of the Orion region in the range 3–7 MeV (dashed). The molecular clouds are traced by the CO contours (solid lines) from the Columbia survey.

5. EGRET high energy sources

5.1. THE FIRST EGRET CATALOGUE OF POINT SOURCES

From the EGRET data of the all sky survey a catalogue of high energy gamma-ray sources was derived by Fichtel et al., 1994. A graphical display of the catalogue sources is given in figure 6 where the size of the symbols is a measure of the source intensity. In this catalogue a wide range of objects is represented: clearly identified sources like solar flares, pulsars, active galactic nuclei such as radio loud quasars and BL Lac objects as well as unidentified galactic and extragalactic sources and detections of gamma ray burst events up to energies of several tens of GeV. Table 3 shows an overview of the EGRET sources detected during the sky survey: Most of the listed sources have been analysed at energies above 100 MeV which not only is a conventional reference energy but also affords a compromise between the better counting statistics at lower energies and the better angular resolution at higher energies. Essential for the analysis of point sources embedded in a strong background of galactic and extragalactic background radiation is a good model of this background. Based on radio surveys of the galactic atomic and molecular gas and on assumptions of the cosmic ray distribution in the Galaxy the gamma ray emissivity of the diffuse interstellar medium was modelled by Bertsch et al., 1993. The catalogue sources are seen in excess to this background model.

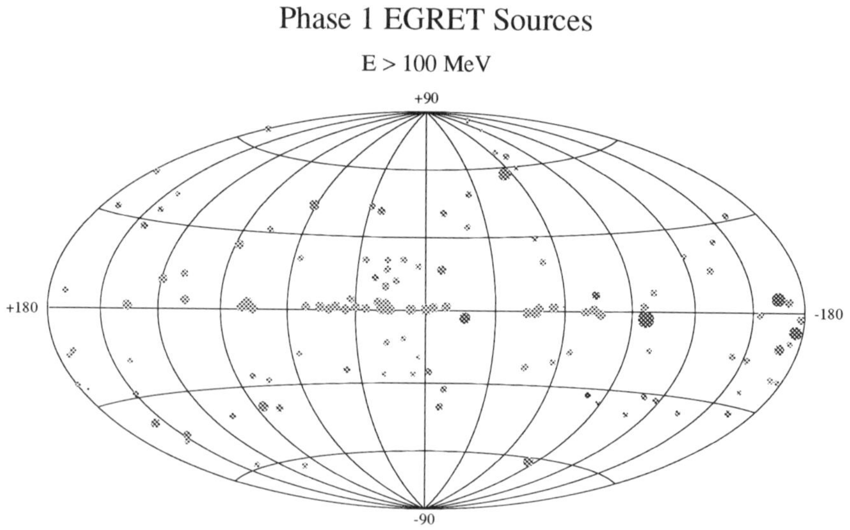

Figure 6: Graphical representation of the > 100 MeV gamma ray sources listed in the first EGRET catalogue (Fichtel et al., 1994)

Table3. Summary of the first EGRET catalogue

Source Type	Nr. of Detections	Comment
Solar Flares	2 in spark chamber 6 in TASC calorimeter	June 9, 11, 1991
Pulsars	5	Crab,Vela, PSR1706-44, PSR1055-52, Geminga
Unidentified sources at $\|b\| < 10°$	10 with significance $> 6\sigma$ 25 5–6 σ	
Normal Galaxies	1 detection 4 upper limits	LMC SMC, M31, M82, NGC253
Active Galactic Nuclei (QSOs and BL Lac Objects)	33 with significance $> 5\sigma$ 11 4–5 σ	most AGN sources are variable
Unidentified sources at $\|b\| > 10°$	8 with significance $> 5\sigma$ 35 4–5 σ	
Gamma Ray Bursts GRB	3 in spark chamber 9 in TASC calorimeter	910503, 930131, 940217

5.2. ACTIVE GALACTIC NUCLEI

Active galactic nuclei (AGN) are the most powerful localized emitters of radiation known in the universe. From a volume of space that is about the size of the solar system, radiation across the electromagnetic spectrum is produced with a luminosity that exceeds the total output of a normal galaxy by orders of magnitude. These extraordinary objects were first discovered as quasistellar radio sources, i.e. unresolved pointlike objects, but it was soon realized that quasars have blue star-like optical counterparts that show emission lines at large red shifts (Schmidt, 1963). The implied cosmological distances of the order of Gigaparsecs and the subsequent discovery of many quasars up to X-ray energies indicated that extremely powerful thermal and nonthermal processes occur in these sources. Only the release of gravitational energy around a massive black hole, which accretes matter from a surrounding galaxy, can account for the required luminosity from a small volume that is compatible with the rapid variability observed from many quasars. Since quasars were found to emit roughly equal amounts of energy in every band of the electromagnetic spectrum that became available through the development of sensitive detectors, it was expected that these objects should also be visible at high gamma ray energies.

The first extragalactic high energy gamma ray source, the quasar 3C273, was detected with the COS-B telescope (Swanenburg et al., 1978). Although this detection was at the limit of COS-B's sensitivity it was clear that the gamma ray range above 50 MeV contains a major fraction of the quasar's bolometric luminosity. Scaling from known quasar fluxes in the hard X-ray range it was expected that several more AGN sources would be detectable with CGRO. The predictions were borne out beyond the most optimistic expectations: after the first years of CGRO more than 40 radioloud quasars and BL Lac objects have been detected above 100 MeV at high galactic latitudes in the EGRET data (Fichtel et al., 1994, von Montigny et al., 1995). Table 4 gives a current list of these identifications together with the principal observational results. One of the sources, the BL Lac object Mkn 421, has subsequently been observed at TeV energies with a ground based Cherenkov telescope (Punch et al., 1992).

Considering the characteristics of the detected gamma ray quasars in the radio and optical bands one finds several common traits: nearly all (>90%) are radio loud and show a flat or complex radio spectrum; many (30–40%) are known to display violent optical variability on short time scales and emit polarized light which indicates the nonthermal nature of the emission processes; many of the sources display superluminal motions (15–30%) in jet like structures that extend from the unresolved core of the source. The statistics of the optically violent variable, polarized and superluminal sources is of course incomplete since not all EGRET sources have been investigated with optical monitoring and VLBI techniques, i.e. the quoted percentages are lower limits for these source characteristics. AGN that have these characteristics are grouped in the class of 'blazars' (BL Lac objects, optically violent variable and polarized quasars). The available EGRET measurements of blazars can be summarized as follows:

- the luminosity in the gamma ray energy band in many of the sources is the dominant emission, often exceeding the radiation at all lower energies.
- sources for which redshifts are available are distributed over the wide range z=0.03 to 2.28. The z distribution of gamma ray quasars is compatible with the distribution of radio loud, flat spectrum radio quasars, indicating that a complete sample of these objects is detected.
- the luminosity calculated under the assumption of isotropic emission (which is almost certainly not applicable) ranges from about 10^{44} erg/s to 10^{49} erg/s, with most values around 10^{48} erg/s. A more realistic luminosity can be estimated in a model with a relativistic beam that points at the observer: the intrinsic energy requirement could be lowered by up to 4 orders of magnitude if the emission is concentrated in such a beam.

Table 4: List of Active Galaxies detected with EGRET (significance >4σ) during phase 1 and 2 (May 1991–September 1993) (von Montigny et al., 1995)

Name	l	b	$^{*}\mathcal{F}$	S. I.	z	$^{+}\mathcal{L}$	OVV	Pol	BLLac	SLM
0202+149 (4C+15.05)	17.93	–44.04	0.26	2.4 ± 0.2	–	–		•		
0208-512	276.10	–61.78	1.1	1.7 ± 0.1	1.003	2.1		•		
0234+285 (4C 28.07)	149.47	–28.53	0.16	2.4 ± 0.3	1.213	0.3		•		•
0235+164 (OD+160)	156.77	–39.11	0.82	2.0 ± 0.2	0.94	1.2		•	•	?
0420-014 (OA 129)	195.29	–33.14	0.45	1.9 ± 0.3	0.92	0.7	•	•		
0446+112	187.43	–20.74	1.04	1.8 ± 0.3	1.207	2.8				
0454-234	223.71	–34.90	0.14	–	1.009	0.2		•		
0454-463	251.97	–38.81	0.29	1.9 ± 0.4	0.86	0.4				
0458-020	201.45	–25.30	0.28	–	2.286	3.0	•			
0506-612	270.55	–36.07	0.06	–	1.093	0.1				
0521-365	240.61	–32.72	0.20	–	0.055	.0008		•		
0528+134	191.37	–11.01	1.6	2.6 ± 0.1	2.06	16	•			
0537-441	250.08	–31.09	0.32	2.0 ± 0.2	0.894	0.4	•	•	•	
0716+714	143.98	+28.02	0.50	2.0 ± 0.3	–	–		•	•	?
0804+499	169.06	+32.56	0.29	2.5 ± 0.2	1.43	1.1		•		
0827+243	200.02	+31.87	0.21	2.2 ± 0.4	2.046	1.6	•			
0829+046	220.69	+24.33	0.14	–	0.18	0.006	•		•	
0836+710 (4C+71.07)	143.54	+34.43	0.34	2.4 ± 0.2	2.17	3.5	•	•		•
0954+658	145.75	+43.13	0.21	1.7 ± 0.2	0.368	0.05		•	•	
1101+384 (Mrk 421)	179.83	+65.03	0.14	1.7 ± 0.2	0.031	.0002			•	?
1156+295 (4C+29.45)	199.41	+78.37	0.63	1.8 ± 0.4	0.729	–	•	•		•
1219+285 (ON 231)	201.74	+83.29	0.17	1.4 ± 0.4	0.102	0.004	•	•	•	
1222+216 (4C 21.35)	255.07	+81.66	0.17	2.4 ± 0.2	0.435	0.04				
1226+023 (3C273)	289.95	+64.36	0.21	2.4 ± 0.1	0.158	0.005	•	•		•
1229-02	293.16	+60.10	0.12	–	1.045	0.2				
1253-055 (3C279)	305.10	+57.06	2.7	2.0 ± 0.1	0.538	1.2	•	•		•
1313-333	308.80	+28.94	1.3	2.0 ± 0.2	1.21	0.3	•			
1406-076	333.88	+50.28	0.41	1.9 ± 0.1	1.494	1.7				
1510-089	351.29	+40.14	0.23	2.6 ± 0.4	0.361	0.03	?	•		
1604+159	29.38	+43.41	0.17	–	–	–		•	•	
1606+106 (4C+10.45)	23.03	+40.79	0.53	2.5 ± 0.2	1.23	1.4				
1611+343	55.15	+46.38	0.33	2.1 ± 0.3	1.40	1.2	•			
1622-253	352.14	+16.32	0.47	2.0 ± 0.1	–	–				
1633+382 (4C+38.41)	61.09	+42.34	1.0	1.9 ± 0.1	1.81	6.3	•			•
1739+522 (4C+51.37)	79.56	+31.75	0.36	1.9 ± 0.2	1.38	1.3				
1741-038	21.59	+13.13	0.34	3.0 ± 0.4	1.054	0.6		•		
1933-400	359.16	–25.37	0.88	–	0.966	0.3-1.4				
2005-489	350.37	–32.60	0.18	–	0.071	0.001	•		•	
2022-077	36.90	–24.38	0.63	1.5 ± 0.2	–	–				
2052-474	352.59	–40.38	0.28	2.4 ± 0.4	1.489	1.1				
2209+236	82.24	–26.09	0.15	–	–	–				
2230+114 (CTA 102)	77.44	–38.58	0.46	2.6 ± 0.2	1.037	0.4		•		?
2251+158 (3C454.3)	86.11	–38.18	1.35	2.2 ± 0.1	0.859	0.9	•	•		•
2356+196	106.38	–41.25	0.29	–	1.066	0.6				

$^{*}\mathcal{F}$: max.observ. Flux; units:(E>100MeV) ($10^{-6}\ \gamma\,cm^{-2}\,s^{-1}$); S.I.:spectral index of power law

$^{+}\mathcal{L}$: Luminosity: (E>100MeV) units: f×10^{48} erg/s; f = beaming factor;

OVV: optically violent variable; Pol: >3% polarization; SLM: superluminal motion

- many sources show time variability of their gamma ray fluxes with factors of 5 – 10 over time scales from days to months.
- the photon spectra can generally be described with power laws over energy. The indices range for different sources from −1.6 to −2.6.
- many AGN, that would have been expected as gamma ray emitters (sources at small distances, noteable blazars and other superluminal objects) have not (yet?) been detected.

As an illustration and example we consider the observations of 3C279, a noteable nearby quasar (z=0.54) which has been observed several times during phase 1 of the CGRO mission. Figures 7 and 8 show the time variation of 3C279 and its multiwavelength spectrum (Kniffen et al, 1993, von

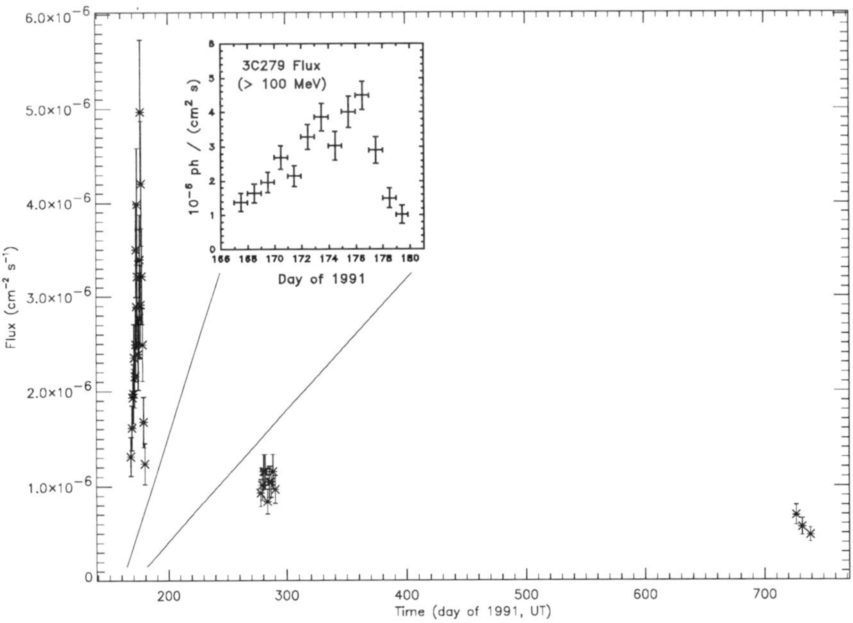

Figure 7: The time variability of 3C279 from June 1991 to April, 1992. The inset shows an enlarged intensity profile for the flare during the time interval June 16 – 2, 1991.

Montigny et al., 1995).
In June 1991 3C279 displayed an unprecedented outburst of >100 MeV radiation with a steady increase (factor of ≈ 4) in intensity over a week and a subsequent rapid fall-off with a 50% decrease within about half a day. Applying the standard light crossing time argument to derive the size of the emission region it is clear that this gamma ray flare must have come from a very compact source. The density and directional distribution of high energy photons in the source can be estimated from the observed fluxes if an emission geometry is assumed. The optical depth with respect to γ–γ collisions in the source is a function of this photon density and it must be small enough to enable gamma ray photons to emerge from the source. The apparent compactness of

many of the variable AGN gamma ray sources is found to be far in excess of this requirement on optical depth if an isotropic emission is assumed.

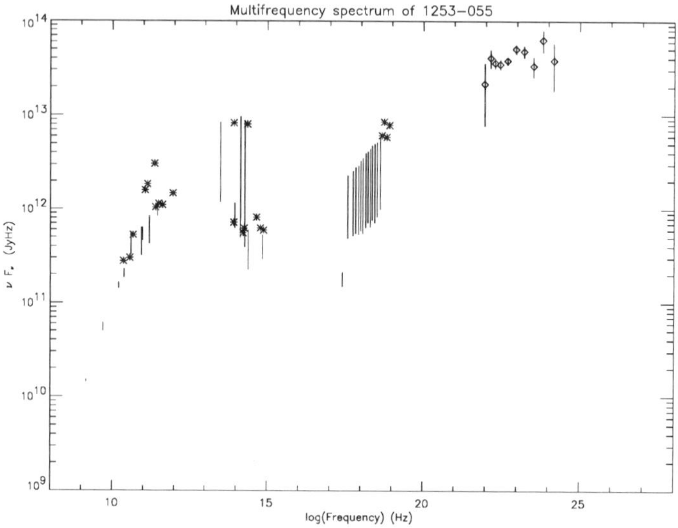

Figure 8: Multifrequency spectrum of 3C279 (von Montigny et al., 1995)

This is one of the most compelling arguments in favor of a beamed emission model for the gamma radiation. Models in which a relativistic beam is directed close to the line of sight to the observer have been widely discussed in the literature (Blandford and Königl, 1979, Begelman, Blandford and Rees, 1984). The jet is assumed to contain a population of energetic particles that generates gamma ray photons through inverse Compton processes (electrons on low energy photons) or through proton–photon interactions with production of pions and electrons depending on the species that has been energized by the central accelerator in an AGN. The radiation from the relativistic jet is then strongly collimated (opening angles of less than about 10° are typical) and the apparent brightness in the jet direction is amplified as well by the blue shift of a presumed power law emission as by the relativistic beaming. The intrinsic luminosities of the AGN could therefore be up to a factor of 10^4 smaller than the 'naive' luminosity calculated under the assumption of isotropy. Although we now know much more about the total power budget of an active galactic nucleus and we have viable models for the generation of radiation throughout the observed spectrum there are still major questions that need to be addressed through observational and theoretical research. Let me mention a few of the 'hot topics' that are presently discussed:

- what is the nature of the central engine in a quasar and how is the relativistic jet formed?
- how and which species of relativistic particles are accelerated inside the jet?
- how are the different regions of the multiwavelength spectrum related to each other? Can we separate the emissions from the central accretion disk and the jet? What are the temporal and intensity relations of different components in the spectrum?
- what is the cause for the large variability observed at high energies?
- what is the integrated contribution of quasars to the extragalactic gamma ray background?

5.3. PULSARS

The first clear identifications of galactic gamma ray sources with pulsars were achieved on the basis of the periodicities seen in radio and gamma ray emissions simultaneously. The Crab (PSR B0531+21) and Vela (PSR B0833–45) pulsars were identified nearly 20 years ago with the telescopes on SAS-2 and COS-B. Up to now, the instruments on GRO have raised the number of gamma-ray pulsars to six. They are listed in table 5 with their relevant parameters and shown as lightcurves in all observed wavelength ranges in figure 9.

Table 5: Characteristics of gamma ray pulsars detected with CGRO instruments

Pulsar	l	b	P (ms)	τ	D (kpc)	Flux	S.I.	ΔE	L	η
Crab	184.6	–5.8	33	1.30	2.0	1.8 ± 0.1	–2.15	40 keV – 4 GeV	272	6.1×10^{-4}
1509–58	320.3	–1.2	150	1.55	4.2	–	–1.68	50 keV – 5 MeV	89	5.0×10^{-3}
Vela	263.6	–2.8	89	11	0.3	7.8 ± 1.0	–1.70	3 MeV– 3 GeV	6.2	9.0×10^{-4}
1706-44	343.1	–2.7	102	17	1.82	1.0 ± 0.2	–1.72	30 MeV–10 GeV	36	1.1×10^{-2}
Geminga	195.1	+4.3	237	340	0.25	2.9 ± 0.1	–1.50	30 MeV–2 GeV	1.4	4.3×10^{-2}
1055–52	286.0	+6.6	197	530	1.53	0.24± 04	–1.18	300 MeV–4 GeV	9.7	3.2×10^{-1}

Notes:

τ : rotational age (= $P/2\dot{P}$) in units of 10^3 years

D: distance in kpc; generally from radio dispersion measurements. For Geminga and Vela the distances are based on X- ray data

Flux: >100 MeV; units 10^{-6} $cm^{-2}s^{-1}$ when available

S.I.: spectral index of power law fit over energy range ΔE

L: luminosity over ΔE in units of10^{33} erg/(s sr); Note: beaming into 1 sr is assumed

η: Efficiency for conversion of rotational energy loss into gamma radiation $\eta = L_\gamma / \dot{E}_{rot}$

where $\dot{E}_{rot} = 10^{45} \cdot 4\pi^2 \cdot \nu\dot{\nu}$ erg/s

In addition to Crab and Vela, which are also detected to energies below a MeV, EGRET detected PSR B1706–44 and PSR B1055–52 above 100 MeV. Between 60 keV and 2 MeV PSR B1509–58 has only been detected by BATSE, OSSE and COMPTEL. It is noteworthy that EGRET places a significant upper limit on the flux of PSR B1509-58 in the 100 MeV range, which indicates that its spectrum must turn over in the MeV range. The long time mysterious gamma ray source 'Geminga' (PSR J0633+1746) in the galactic anticenter was finally identified as a pulsar in 1992 after a deep X-ray observation revealed a 237 ms periodicity, which was immediately found also in the gamma ray data. So far the Geminga pulsar has not been detected at gamma ray energies below about 30 MeV and there is no radio signal from this source.

Spectra of the total pulsed emission above 30 MeV were derived for Crab (Nolan et al., 1993), Vela (Kanbach et al., 1994), PSRs 1706–44 and 1055–52 (Thompson et al., 1992, Fierro et al., 1993), and Geminga (Mayer–Hasselwander et al., 1994). Power laws provide a good representation for these spectra over most of the EGRET energy range although spectral cut-offs are observed for several of the more intense pulsars. The spectral indices vary in a characteristic way with the apparent age of the pulsars, i.e. the spectra become harder with increasing age.

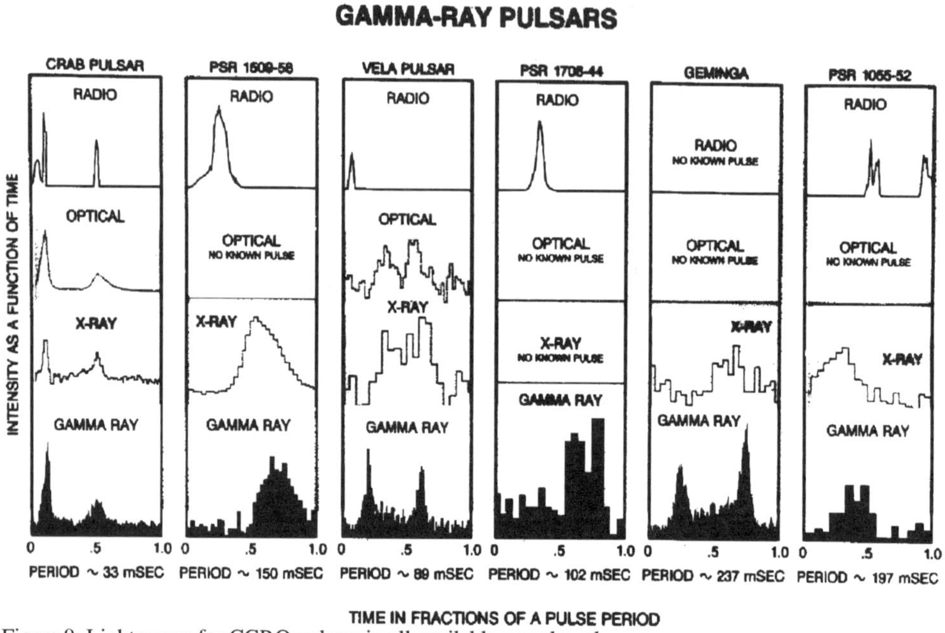

Figure 9: Lightcurves for CGRO pulsars in all available wavelength ranges.

We observe for the five pulsars detected by EGRET that the high energy spectral form and the gamma ray efficiency vary systematically with apparent age. The power law indices become flatter and the efficiency increases as the pulsar ages. Thompson, 1993 derived empirical fits to these trends:

$$\text{Power Law Index} = 0.308 \times \log(\tau) - 3.04$$

$$\log \eta = 1.79 \times \log(\tau) - 8.64 - 1.3 \times \log\left(\dot{P}_{Crab}/\dot{P}\right)$$

where the scaling with the period derivative is based on a polar cap emission model by Daugherty and Harding, 1982.

It should however be noted that upper limits on some pulsars, like PSR B0656–14 or B1929–10 , indicate a gamma ray efficiency that falls short of the expected empirical value by at least an order of magnitude. This might indicate that individual effects, e.g. differing beaming geometries, play an important rôle in determining the gamma ray brightness of a pulsar.

From the brightest pulsars Crab, Geminga and Vela the EGRET telescope has accumulated sufficient statistics to perform detailed analyses of the emission spectra as a function of rotation phase. The lightcurves of these pulsars are generally similar: they show emission concentrated in two sharp peaks with phase separation of 0.40 ± 0.02 (Crab), 0.424 ± 0.002 (Vela) and 0.50 ± 0.03 (Geminga) and lower emission in the interval regions of the lightcurves. Significant spectral variations have been found for all 3 objects as the neutron star turns: the peaks display a softer spectrum than the interval regions. The exception to this rule is the second peak of Geminga which has a very hard spectrum, fitted with an $E^{-1.22}$ power law.

As an example the lightcurve of Vela with definitions of 8 phase intervals is shown in figure 10. The spectral form, expressed as a power law index, for these phase intervals varies as shown in figure 11 (Kanbach et al., 1994).

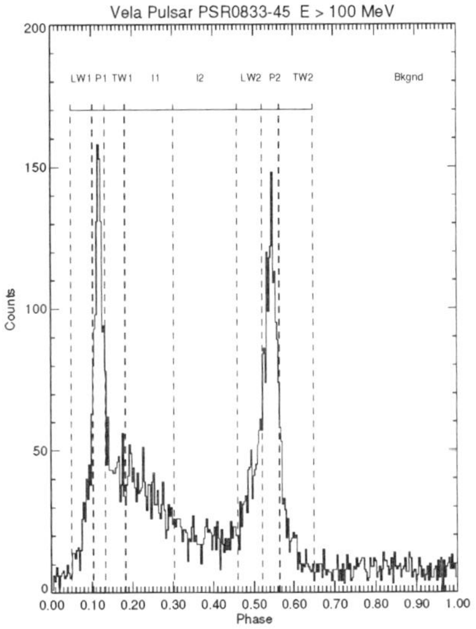

Figure 10: Lightcurve of PSR B0833-45 (Vela) for E > 100 MeV. Phase intervals as defined by Kanbach et al., 1994 are shown.

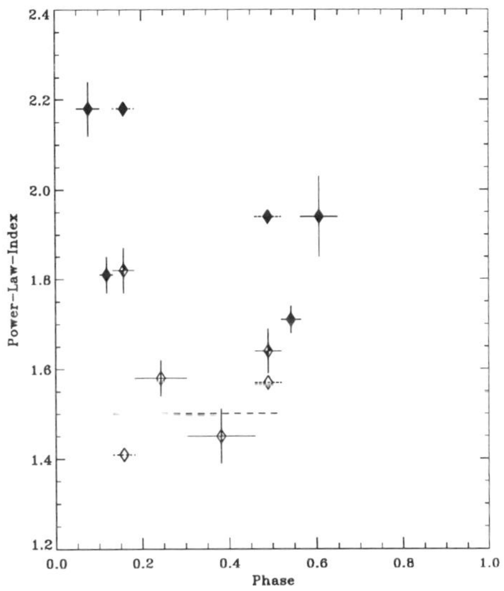

Figure 11. Power law indices for the spectra as a function of phase for PSR B0833-45 (Vela)

The total pulsed emission of Vela is shown in figure 12. Data from OSSE (0.06–9 MeV, Strickman et al., 1993), Comptel (1–30 MeV, Hermsen et al., 1993) and EGRET (30 MeV–10 GeV, Kanbach et al., 1994) are shown together with current models by Ho, 1993 (outer gap model) and Harding and Daugherty, 1993 (polar cap model). Both models can account for the general form of the pulsar spectrum with a power law approximation between about 1 MeV and 2 GeV. The low and high energy cutoffs together with the phase dependent spectra should be decisive parameters for future theoretical work on pulsar emission models.

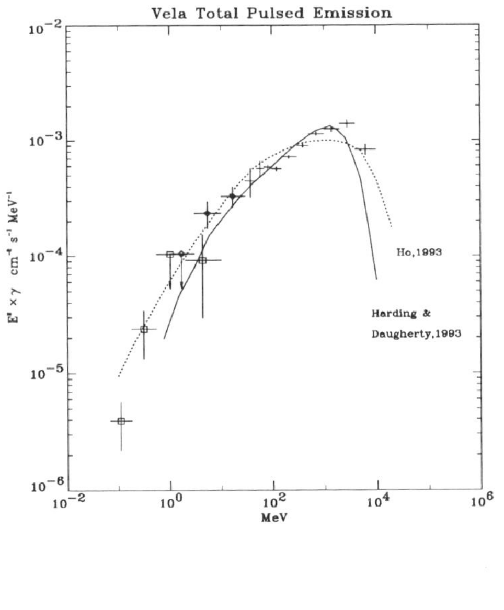

Figure 12. Total pulsed emission spectrum for PSR B0833-45 (Vela) as observed by EGRET (no symbol), Comptel (diamond) and OSSE (square). The curves are theoretical spectra for a polar cap model (Harding and Daugherty, 1993) and an outer gap model (Ho, 1993)

5.4. UNIDENTIFIED GALACTIC SOURCES

The first EGRET catalogue listed 35 unidentified sources with greater than 5σ significance for the latitude interval |b| < 10°. The brighter sources have fluxes above 100 MeV around 10^{-6} cm^{-2} s^{-1} and the weaker detections range to several times 10^{-7} cm^{-2} s^{-1}. If these fluxes are emitted isotropically the typical source luminosities for E > 100 MeV must be in the range 10^{34} – 10^{35} erg/s × $(d/1\ kpc)^2$. In table 6 the luminosities of the five identified pulsars for E > 100 MeV are listed under the assumption that their flux is emitted by an isotropic source into 4 π sr.

Table 6. PSR luminosities for E > 100 MeV

Pulsar	4 π Luminosity (erg/s)
Crab	7.0×10^{35}
B 1706–44	3.1×10^{35}
Vela	5.7×10^{34}
B 1055–52	2.6×10^{35}
Geminga	1.5×10^{34}

It is quite evident that old pulsars like Geminga can explain only those unknown sources that are not much more distant than about a kpc. Such a close population of galactic objects would be expected to be distributed to large galactic latitudes. As we show in the latitude distribution in figure 13 the distribution of the catalogue sources is concentrated quite markedly towards the galactic equator. Assuming a minimum scale height of 40 pc for young galactic populations the apparent concentration of gamma ray sources, which are believed to belong to that category, demands that they are at distances more than about 2 kpc. On the other hand the longitude profile in figure 5 does not indicate a strong excess of sources towards the inner galaxy. Although the bias on source recognition toward the inner galaxy, with its intense diffuse emission regions, has to be realized, we argue that at the chosen significance threshold the catalogue is nearly complete. The absence of a strong peak in source numbers toward the inner galaxy can therefore be taken to indicate a distance for the detected source not in excess of about 2/3 to galactic center, i.e. about 6 kpc. With these estimates of the distance range the luminosities of the typical gamma ray source (> 100 MeV) can be limited to 4×10^{34} – 4×10^{36} erg/s. Several possibilities can be mentioned as counterparts for these sources:

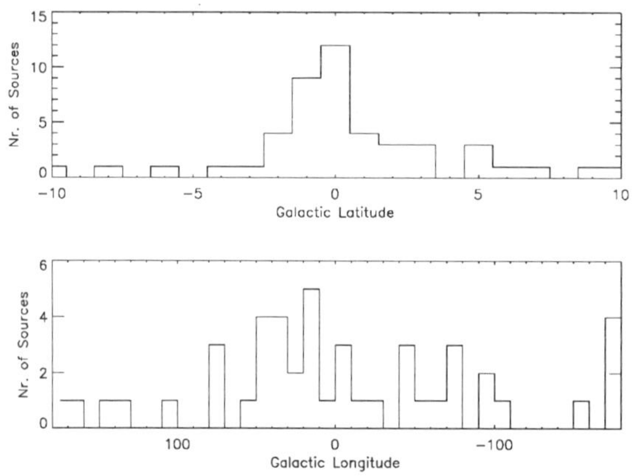

Figure 13. Distribution of EGRET catalogue sources in latitude and longitude

– Young pulsars that are not yet detected at other wavelength ranges (radio or X-ray). The case of PSR B1706–44 is illustrative of this possibility: the COS-B catalogue listed a significant gamma ray source 2CG 342–02 and the EGRET observations readily confirmed this source in 1991. Only during a dedicated search for new radio pulsars was the periodicity discovered in 1991 and the identification could be achieved. Similarily the case of Geminga could only be solved after the discovery of pulsations at X-ray energies.

– Other compact objects that generate gamma radiation in a way analogous to the mechanisms discussed for active galactic nuclei, although on a much reduced scale. Ingredients for such a scenario are a compact object surrounded by an accretion disk which generates a highly relativistic jet containing relativistic particles. Gamma ray photons would be generated in the jet in inverse Compton processes between the relativistic particles and the ambient photon fields and would be emitted in a highly collimated beam along the jet axis. Examples for such objects in the Galaxy are SS433 (Margon et al., 1988) the sources in the galactic center region 1E1740.7–2942 and GRS 1758–258 (Mirabel and Rodriguez, 1994a) where the energetic jets are seen from the side and, recently discovered, the first superluminal galactic source GRS1915+105 (Mirabel and Rodriguez, 1994b). If the jet axis of such objects is aligned to the observer and relativistic particles are contained in the jet high energy emission might be detectable. One could call such a source a galactic 'mini-quasar'.

– Sites of enhanced cosmic ray density and/or enhanced density in the interstellar medium (clouds). Galactic cosmic rays in the GeV energy range are thought to be produced locally in our Galaxy. An energy source sufficient for the observed galactic cosmic ray density can be provided by the cumulated power of supernovae explosions in the Galaxy. These energetic events can generate at least a sizeable fraction of the cosmic rays. Several processes how the blast energy of a supernova is converted into high energy particles have been proposed: Diffusive shock acceleration processes on the shocks travelling with the remnant into the interstellar medium (e.g. Blandford and Ostriker, 1980) or pulsar generated cosmic rays in the center of the remnant (Harding et al., 1991) or the interaction of the SNR with young, active stars or stellar clusters which provide suprathermal particles for the acceleration processes (Cassé & Goret, 1978, Montmerle, 1979). If such an object with enhanced cosmic ray content is in contact with sufficient target material a steady gamma ray source will be created.

Acknowledgement: I thank my colleagues in the EGRET and COMPTEL teams, who have supplied much of the material used in this review and have, through discussions and explanations, helped me to understand a wider range of gamma ray astronomy.

6. References

Atteia, J.-L., et al., 1987, *ApJ Suppl.*, **64**, 305
Baring, M., 1994, Proceedings of the Erice NATO-ASI, May 1994,
eds. .Wefel, . Silberberg, and. Shapiro, in press
Begelman, M.C., Blandford, R.D., and Rees, M.J., 1984, *Rev.Mod.Phys.*, **56**, 255
Bertsch D.L., et al., 1993, *ApJ*, **416**, 587
Blandford, R.D. and Königl, A., 1979, *ApJ*, **232**, 34
Bloemen H., et al., 1994, *A&A*, **281**, L5
Daugherty, J.K. and Harding, A.K., 1982, *ApJ*, **252**, 337
Diehl, R., 1994, Proceedings of the Les Houches school, Jan. 1994, in press NATO-ASI
Diehl, R., 1994, Proceedings of the 30th COSPAR meeting, July 1994, Hamburg
Dingus, B.L., et al., 1994, Huntsville Gamma Ray Burst Workshop, October 1993
Fichtel, C.E., et al., 1975, *ApJ*, **198**, 163
Fichtel, C.E., et al., 1994, *ApJ Suppl.*, in press
Fierro, J.M., et al., 1993, *ApJ*, **413**, L27
Fishman, G.J.1994, private communication
Gehrels N., E. Chipman, and D.A. Kniffen, 1993, Second Compton Symposium, College Park, Md., AIP Conf. Proc. 304, eds. Fichtel, Gehrels and Norris, p.3
Hakkila, J., et al., 1994, *ApJ*, **422**, 659
Harding, A.K., 1993, Second Compton Symposium, College Park, Md., AIP Conf. Proc. 304,

eds. Fichtel, Gehrels and Norris, p.30
Harding, A.K. and Daugherty, J.K., 1993, *Proceedings of the Los Alamos Workshop on 'Isolated Pulsars', eds Van Riper, Epstein, and Ho*, Cambridge Univ. Press, p. 279
Hermsen W., et al., 1993, *Proc. of the Integral Workshop*, Feb. 1993, Les Diablerets, to be published in *ApJ Suppl.*
Ho, C., 1993, *Proceedings of the Los Alamos Workshop on 'Isolated Pulsars', eds.. Van Riper,. Epstein, and Ho*, Cambridge Univ. Press, p. 271
Hurley, K., et al., 1994, submitted to *Nature*
Iyudin A.F. et al., 1994, *A&A Lett.,* **284**, L1
Kanbach, G., et al., 1994, *A&A*, in press
Klebesadel, R., et al., 1973, *ApJ Lett.,* **182**, L85
Kniffen, D.A., et al., 1993, *Ap.J.*, **411**, 133
Kraushaar, W.L., et al., 1972, *Ap.J.*, **177**, 341
Kwok, P.W., et al., 1993, *AIP Conf.Proc. 280, eds. Friedlander M., Gehrels, N., and Macomb D.J.*, p. 855
Meegan, C.A., et al., 1992, *Nature,* **355**, 143
Margon, B.A., 1988, *Rev Astr Astrophys,* **22**, 507
Mayer-Hasselwander, H.A., et al., 1982, A*&A,* **105**, 164
Mayer-Hasselwander, H.A., et al., 1994, *ApJ*, **421**, 276
Mazets, E., et al., 1979, *Nature*, **282**, 587
Mirabel, F. and Rodriguez, L.F., 1993, AIP Conf. Proc. 304, eds.Fichtel, Gehrels and Norris, p. 413
Mirabel, F. and Rodriguez, L.F., 1994, *Nature,* **371**, 46
Montigny, C. von, et al., 1995, *ApJ,* in press
Nolan, P.L., et al., 1993, *ApJ*, **409**, 697
Oberlack, U., et al., 1994, *ApJ Suppl.*, **92**, 433
Paczynski, B., 1986, *ApJ Lett.*, 308, L43
Punch, M., et al., 1992, *Nature,* **358**, 477
Ramaty, R., 1994, Proceedings of the Les Houches school, Jan. 1994, in press NATO-ASI
Schneid, E.J., et al., 1992, *A.&A,* **225**, L13
Schneid, E.J., et al., 1994, in preparation
Schmidt, M., 1963, *Nature,* **197**, 1040
Sommer et al., 1994, *ApJ* 422, L63
Strickman, M.S., et al., 1993, *AIP Conf.Proc. 280, eds. Friedlander M., Gehrels, N., and Macomb D.J.*, p. 209
Swanenburg, B.N., et al., 1981, *ApJ*, **243**, L69
Swanenburg, B.N., et al., 1978, *Nature,* **197**, 1040
Thompson, D.J., et al., 1992, *Nature* , **359**, 615
Thompson, D.J., et al., 1993, *ApJ Suppl.* , **86**, 629
Thompson, D.J., 1993, Second Compton Symposium, College Park, Md., AIP Conf. Proc. 304, eds. Fichtel, Gehrels and Norris, p.57

COSMIC GAMMA-RAY BURSTS

Matthew G. Baring*
NASA Goddard Space Flight Center,
Lab. for High Energy Astrophysics, Code 665,
Greenbelt, MD 20771, U.S.A.

ABSTRACT. Gamma-ray bursts are an exotic class of astronomical sources that are currently extremely topical within the astrophysical community; their enigmatic nature has provoked considerable controversy over the years. This paper briefly reviews the generic properties of these transient sources, obtained from observations accumulated over the last two decades, and then discusses some of the latest observations, focusing on recent interpretations of the data, and theoretical approaches that may help unravel the mysteries of gamma-ray bursts.

1. An Introduction to Gamma-ray Bursts

Gamma-ray bursts (GRBs) have intrigued observers and confounded theorists ever since their discovery just over twenty years ago.[1] The results obtained by the Compton Gamma-Ray Observatory (CGRO) have perpetuated the confused picture we have of these enigmatic objects. Yet the GRB field is alive and more dynamic than ever, perhaps due to the exotic nature of these objects and the fascinating physics that may be needed to describe and understand them. The enigmatic nature of GRBs is largely because they are transient phenomena. They are bursts of gamma-rays, usually somewhere between about 10 keV and 10 MeV in energy (e.g. see ref. 2), and last for as little as milliseconds or as long as several minutes.[2–6] Typical burst time histories are displayed in Fig. 1 and also in the First BATSE GRB Catalogue;[5] an exceptionally good burst spectrum is presented in Fig. 2. While other transient astronomical sources exist, bursts are unique in that they are the only source population that emits only in the gamma-ray range of the spectrum; associated emission in other wavebands has never been observed. In fact, searches (e.g. see papers in ref. 7) for convincing transient or quiescent steady-state counterparts to classical GRBs in radio, optical and X-ray bands have proved negative. Note that such counterparts have been found for the soft-gamma repeaters, a separate class of objects that will not be considered here. This null result has prolonged the mystery of GRBs, since the discovery of associated sources would provide distances to GRBs and thereby tell astronomers their origin.

---

* NAS–NRC Research Associate

M. M. Shapiro et al. (eds.), Currents in High-Energy Astrophysics, 21–30.

Gamma-ray bursts have rapid time-variability with intensity fluctuations on timescales t_{var} sometimes less than a millisecond,[8] motivating the premise that they originate in compact environments of size $c\,t_{var} \sim 10^7$ cm. This naturally lead to the popularity of neutron stars as a site for GRBs, an idea that was bolstered by the realization that their extremely strong magnetic fields could contain the emitting plasma in the source for typical burst durations. The time-histories of GRBs, as illustrated in ref. 5 and Fig. 1, display a diversity of structure, with single or multiple peaks, subpeaks and lengthy gaps between peaks appearing in various sources. There is fairly strong evidence for a bimodality in the distribution of GRB durations,[6] which may not necessarily be an indication of two classes of bursts. Typical burst spectra $f(\varepsilon)$ are such that the power $\varepsilon^2 f(\varepsilon)$ (e.g. see Fig. 2) peaks in the range 50 keV – 5 MeV, a rare feat for astronomical sources; even more remarkable is the fact that less than 2% of the GRB flux appears in the range[9] 2–10 keV. This paucity of X-rays renders bursts unique among classes of cosmic gamma-ray sources, and severely constrains theoretical GRB models. The observed GRB fluxes at earth imply source luminosities of $L \sim 10^{38}$ erg/sec if they are 1 kpc distant, suggesting a neutron star origin, or $L \sim 10^{50}$ erg/sec if 1 Gpc away, reminiscent of supernova events.

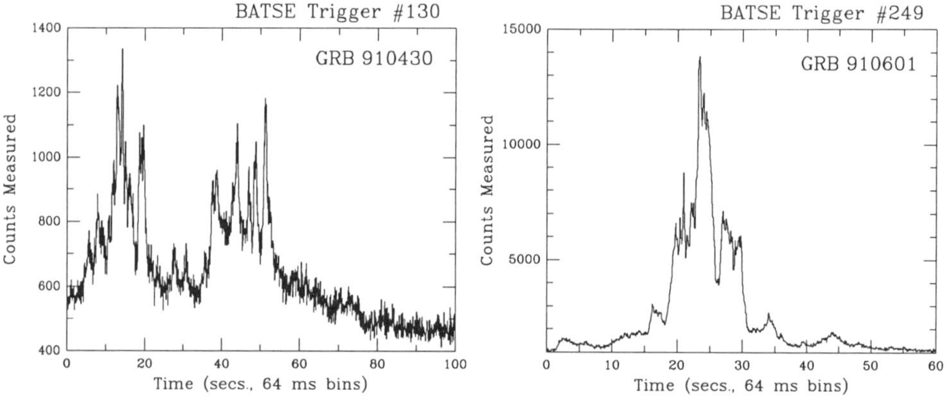

Fig. 1 The time-histories of two gamma-ray bursts, GRB910430 and GRB910601, taken from the First BATSE GRB Catalogue,[5] revealing complex temporal structures that are common in bursts.

The early observations of gamma-ray bursts yielded spectra that were mostly quasi-thermal in nature (e.g. see the KONUS data in ref. 10), however the results[11] obtained by instruments aboard the Solar Maximum Mission (SMM) suggested that many if not most continuum spectra were non-thermal and extended well above 1 MeV, with a quasi-power-law shape. This GRB property has been largely confirmed by the CGRO detectors, with COMPTEL and EGRET seeing several bursts with high energy emission[12,13] up to around 1 GeV (e.g. GRB930131). These energetic phenomena have recently been augmented by the startling observation[14] of a 25 GeV photon from GRB940217, which indicates that it must be closer than about a redshift of two in order to avoid absorption in collisions with infra-red background photons. In addition, this photon and others from GRB940217 were emitted over a period of an hour after the peak in the BATSE count rate, indicating a much longer burst timescale than had previously been known: this may have profound implications for theoretical models of bursts. The detection rate of high energy emission from GRBs is statistically significant, suggesting many (or most) emit at these energies. The earlier quasi-thermal observations may reflect intrinsic properties of those bursts, or that they may have been influenced by threshold effects at high energies in less sensitive detectors like KONUS.

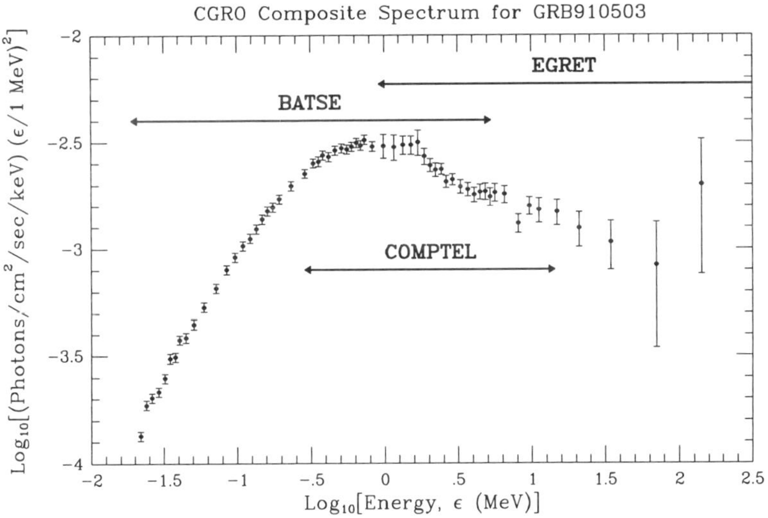

Fig. 2 The composite time-integrated spectrum[15] for the burst GRB910503 (detected May 3, 1991) as obtained by the BATSE, COMPTEL and EGRET instruments on the Compton Gamma-Ray Observatory (CGRO), multiplied by the square of the photon energy ε to depict the emitted power per logarithmic energy bandwidth. The various detector responses have been accounted for in producing this source spectrum, and the energy ranges over which the three instruments are sensitive are indicated.

Prior to the launch of CGRO, many observations of spectral lines from bursts were reported. About 20% of sources[10] showed single absorption dips in their spectra between 20 and 60 keV, which would yield magnetic field strengths of $2 - 5 \times 10^{12}$ Gauss in their emission regions if these broad features were interpreted as cyclotron absorption lines. Further, around 5% of bursts exhibited broad emission lines[10] at around 400 keV, which were naturally interpreted as electron-positron (pair) annihilation lines redshifted from 511 keV by the gravitational field of neutron stars. Therefore, around 1988, the weight of evidence in favour of a galactic neutron star association for bursts was substantial, and the stunning observation of double narrow absorption dips in two burst spectra by the GINGA X-ray satellite[16] (interpreted as cyclotron harmonics) all but clinched the case for proponents of galactic burst sources. Since then, the CGRO instruments, in particular BATSE, have observed no lines of either type in over 900 burst detections, spawning a controversy over the existence of GRB line features. The consistency of the apparently conflicting BATSE and GINGA results concerning absorption lines is hotly debated, and presently it is uncertain whether or not the discrepancy can be explained by the different instrumental line-detection capabilities.

The launch of the Compton Gamma-Ray Observatory created an even bigger upheaval to the "conventional wisdom" of the galactic neutron star hypothesis when the BATSE experiment observed that the burst population was isotropic but inhomogeneous with a comparative deficiency of fainter sources[17] (i.e. presumably at greater distances). For the first 743 BATSE detections, the dipole and quadrupole moments[18] of the population (in galactic coordinates) were $\langle\cos\theta\rangle = 0.031 \pm 0.021$ and $\langle\sin^2 b\rangle = 0.326 \pm 0.011$, respectively; i.e. close to the values of 0 and 1/3 expected for isotropy. Although this result was suggested by earlier experiments that observed significant numbers of GRBs, BATSE is a more sensitive detector and has been able to observe more bursts with a broader range of source fluxes. As such it was intended to observe to the edge of a galactic disk population and therefore detect an anisotropy in the spatial distribution of bursts; the contrary result has indicated that most (or all) bursts may in fact be cosmological in origin.[19] This is currently the new conventional wisdom, and while the observed isotropy and inhomogeneity may be explained by GRBs originating in a large spherical galactic halo, the expectation that halo bursts from M31, the nearest large galaxy (Andromeda), should be detected and upset the observed isotropy is now severely constraining[20] a possible halo population.

A good review of the various population hypotheses is presented in ref. 21. It is now extremely unlikely that all bursts could be confined within the galaxy, though a significant minority ($\lesssim 30\%$) disk population is still quite feasible. Recent work[22] on average burst temporal profiles as functions of burst brightness has suggested a time dilation effect for fainter bursts that is consistent with, but not conclusively implying, a cosmological origin of bursts; the statistics on this result need to be improved in order to make more definitive deductions. There have also been recent controversial claims of anisotropic sub-populations of bursts and repetitions of bursts in the BATSE observations; these are discussed in ref. 23.

2. Spectral Constraints on Emission Mechanisms

Information about individual bursts is still best deduced from their spectra, which can provide much insight concerning the physical processes occurring in GRB emission regions. The natural example is the interpretation of low energy (20-60 keV) absorption features seen in spectra obtained by the KONUS[10] and GINGA[16] experiments, referred to earlier, as being due to absorption of radiation at the cyclotron energy in the strong magnetic fields of neutron star magnetospheres. Since the BATSE instrument on CGRO has made no definitive observation of any spectral line features the GRB continuum must now provide the clues for theorists. Many such continuum spectra $f(\varepsilon)$ have a general appearance similar[24] to Fig. 2 [which depicts $\varepsilon^2 f(\varepsilon)$], with a very flat spectrum at the lower energies (i.e. below 50 keV) "turning over" or "breaking" to a steep spectrum above 1 MeV. The energy ε_B of the turnover varies from burst to burst,[25] and a significant fraction of bright bursts[26] detected by BATSE have $\varepsilon_B > 100$ keV. The break energy ε_B and the spectral index $\alpha = -d[f(\varepsilon)]/d[\log_e \varepsilon]$ at energies above and below ε_B provide important constraints on theoretical models.

The most sophisticated model that generates turnovers in γ-ray continuum spectra is the $B = 0$ pair cascade model[27] that has been comprehensively developed for applications to active galactic nuclei (AGNs). This model used isotropic ultrarelativistic electrons, generated by some unspecified particle energization mechanism such as shock acceleration,[28] to produce a gamma-ray continuum by the inverse Compton scattering of UV photons, and then the two-photon pair production process $\gamma\gamma \to e^+e^-$ to absorb the γ-rays and initiate a cascade of several or many generations of pair production. The attenuation of the GRB continuum by $\gamma\gamma \to e^+e^-$ is an obvious candidate for a means to create the observed "MeV" turnovers. When this model is adapted[29] to try to fit the X-ray portion of the GRB continuum (flatter than for AGNs), multiple inverse Compton scattering of the seed low energy photons occurs, spawning a broad quasi-thermal bump due to pair annihilation and thermal bremsstrahlung around 1 MeV, contrary to observations. It now seems unlikely[30] that such field-free pair cascades can create the MeV breaks observed in bursts. Therefore it appears that the BATSE sources with breaks observed in the MeV energy range[26] require a mechanism other than $\gamma\gamma \to e^+e^-$ to generate their turnovers.[30]

For neutron star models of GRBs, MeV breaks can be generated using synchrotron pair cascades, where magnetic pair production $\gamma \to e^+e^-$ acts to attenuate the continuum,[31,32] since in the strong fields of neutron stars ($B \gtrsim 10^{12}$ Gauss) this process dominates $\gamma\gamma \to e^+e^-$ as a means of absorbing gamma-rays. In these cascades, ultrarelativistic electrons are injected into the emission region with significant pitch angles θ so that a synchrotron continuum is radiated, with the radiation being beamed close to the direction of the electrons' momenta. The dramatic increase[2,31,33] in the pair production optical depth above the effective threshold of $\varepsilon = 2m_e c^2/\sin\theta$ efficiently truncates the continuum above this energy,[31] subsequently generating a pair cascade. Since this truncation energy is a function of pitch angle, spectra integrated over θ can generate broken power-laws given suitable angular distributions of electrons.[34] The spectral breaks in this scenario can be substantial or non-existent, an attractive versatility, though the model suffers from the drawback that the spectrum below the break is never flatter than $\alpha = 3/2$, i.e. it cannot model many flat spectrum sources like GRB 910503 (see Fig. 2).

Another mechanism for generating MeV breaks in magnetized environments is resonant inverse Compton scattering of X-rays from the surface of a neutron star by ultrarelativistic electrons. When $B \neq 0$, the Compton cross-section has a resonance at the cyclotron energy $\varepsilon_c \propto B$ in the electron rest frame, which dominates the scattering process, easily generating spectral breaks[35–37] at MeV energies. The observed spectra are strongly dependent on the angle of photon emission with respect to the field lines. Resonant scattering has no problem dealing with those GRB spectra that are flat below 100 keV, though it has difficulty describing sources with small breaks such as GRB930131; possibly[30] a synchrotron self-Compton hybrid cascade model might succeed in eliminating the deficiencies of the two neutron star models discussed here. The few cosmological models for GRB spectra that exist, including the fireball scenarios discussed below and AGN-like models,[38] are considerably less sophisticated than the neutron star variety and need much more development for serious comparison with burst continuum observations.

3. Relativistic Motion in Bursts

The indication earlier that $\gamma\gamma \to e^+e^-$ is not responsible for MeV turnovers in gamma-ray bursts implies that it is probably not operating to attenuate the gamma-ray continuum within the range of observed burst energies. An important consequence of this is that relativistic bulk motion may be common in burst sources. Two-photon pair production was first considered as a potential mechanism for attenuating GRB spectra by Schmidt,[39] who proposed that it could be used to provide upper limits to the distance to burst sources that revealed no spectral breaks above 1 MeV. Schmidt's application focussed on the case of isotropic photons, and concluded that the bursts seen before then (at energies less than a few MeV) must be galactic, since they had no breaks, in direct conflict with the recent BATSE results' implications for burst populations. If the majority of bursts are at cosmological distances (~ 1 Gpc), then the enormous luminosities expected ($L \sim 10^{50}$ erg/sec), combined with the compact source size implied by the observed rapid time variability, give photon densities high enough to make GRBs optically thick to $\gamma\gamma \to e^+e^-$ by many orders of magnitude; i.e. the free escape of gamma-rays from burst sources would be impossible. Such a situation is incompatible with the observed spectra of GRBs detected by EGRET, which show no attenuation at energies in some cases as high as 1 GeV (see refs. 12, 13).

One way around this problem is to assume beaming of the radiation, which raises the pair production turnover to energies above those observed. Relativistic motion of the source is a natural way to achieve beaming of the emission and avoid attenuation in GRBs. The radiation from a source that is emitting isotropically in a reference frame that moves relativistically with respect to an observer on earth will be beamed roughly within an angle $1/\Gamma$ in the observer's frame,[30,40] where Γ is the bulk Lorentz factor. Such *relativistic beaming* of radiation naturally suppresses the pair production rate, since the $\gamma\gamma \to e^+e^-$ mechanism has a threshold energy $2(m_ec^2)^2/[\varepsilon(1 - \cos\Theta)]$ for target photons of energy ε; this is strongly dependent[30,40] on the angle Θ between the photon directions. Relativistic beaming of radiation in bursts with declining spectra ($\alpha = -d[f(\varepsilon)]/d[\log_e \varepsilon] > 0$: true for all EGRET sources) will consequently "blueshift" the $\gamma\gamma \to e^+e^-$ turnover up in energy by a factor of Γ (since then $\Theta \sim 1/\Gamma$); that this moves above the EGRET energy range therefore provides a determination of a lower bound to Γ.

When the source opening angle is of order $1/\Gamma$, the pair production optical depth $\tau_{\gamma\gamma}$ is reduced by a factor $\Gamma^{-(1+2\alpha)}$ below the optical depth for isotropic radiation.[30,40] The condition that no spectral attenuation occurs is that $\tau_{\gamma\gamma} < 1$ up to the highest energies observed (the optical depth is effectively the probability that $\gamma\gamma \to e^+e^-$ occurs in the source). If $t_{var} = R_v/c$ is the variability timescale of the burst, then for a source at distance $d_{\rm kpc} \equiv d/1\,{\rm kpc}$ with flux at 1 MeV of

F(1 MeV), this condition translates to[41,42]

$$\Gamma^{1+2\alpha} \gtrsim \frac{(3.83)^\alpha}{\alpha^{5/3}(4/3+\alpha)^{27/11}} d_{\rm kpc}^2 \left[\frac{\varepsilon_{\rm max}}{1\ {\rm MeV}}\right]^{\alpha-1} {\rm F(1\ MeV)} \left(\frac{R_v}{10^7 {\rm cm}}\right)^{-1} . \quad (1)$$

The bulk Lorentz factors Γ consequently required for the bright "superbowl" burst GRB930131, detected by EGRET[13] up to an energy of 1 GeV, are $\Gamma \gtrsim 10^3$ if it is 1 Gpc distant and $\Gamma \gtrsim 10$ at 30 kpc (ref. 41). Similar estimates of bulk relativistic motion are obtained for the other EGRET sources,[42] as the results in Table 1 reveal, indicating that this phenomenon is common in bursts. An advantage of relativistic beaming is that a smaller source luminosity is required because the observed flux is enhanced by a solid angle factor Γ^2 (ref. 40). However, then the number of sources must really be a factor Γ^2 higher than detected in order to account for the observed rate of gamma-ray bursts. In the case of cosmological GRBs, this factor could be as high as 10^6 for the typical Lorentz factors $\Gamma \sim 10^3$, which is unacceptably large for many models. Source geometry therefore has significant impact on the viability of models. Expanding the opening angle to become much larger than $1/\Gamma$ could ease this problem[43] if the high energy photons were able to escape; in fact estimates of the minimum Γ for completely spherical expanding sources are qualitatively similar to the highly beamed case.[43]

TABLE 1

LOWER BOUNDS FOR Γ AND PARAMETERS FOR ENERGETIC BURSTS

GRB	Detecting Instrument	$\varepsilon_{\rm MAX}$ (MeV)	α	f(1 MeV) (/cm^2 secMeV)	Γ 30 kpc	Γ 1 Gpc
910503	EGRET	170	2.2 ± 0.2	8.71 ± 0.49	13	565
910601	EGRET	3.5	3.7 ± 0.2	0.98 ± 0.08	2.9	36
910601	COMPTEL	9.4	2.7 ± 0.2	0.40 ± 0.07	3.6	89
910814	EGRET	60	2.8 ± 0.2	13.5 ± 0.85	10	240
930131	EGRET	1000	2.0 ± 0.3	1.90 ± 0.26	13	867

The estimated minimum bulk Lorentz factor Γ for four of the GRBs detected by EGRET, assuming that two-photon pair production does not influence the spectrum below the maximum detected energy $\varepsilon_{\rm MAX}$. The observed spectral index α below $\varepsilon_{\rm MAX}$, as well as the source power-law flux F(1 MeV) at 1 MeV are also given; the EGRET power-law fits were extrapolated down to 1 MeV.

4. Fireball Models

While relativistic motion in GRBs that have $\tau_{\gamma\gamma} < 1$ may be common, it is also probable in bursts of high optical depth to $\gamma\gamma \to e^+e^-$. Such situations lead to many generations of pair production and naturally arise out of several cosmological models for bursts. Among these is the merger scenario, where two neutron stars or a neutron star and a solar mass black hole coalesce[44–46] due to the loss of their orbital energy via the radiation of gravitational waves. Such mergers, which are envisaged as forming a cosmological population of GRBs, may occur at the rate of about one per galaxy per million years since there are four binary neutron star systems observed in our own galaxy,[47] and can release a large fraction of a solar mass of energy (about 10^{54} ergs) in an extremely small volume, typically about $R \sim 10^8$ cm in radius. The energy density is therefore enormous, as is $\tau_{\gamma\gamma}$, and thermal equilibrium is established rapidly, generating a pair plasma at relativistic temperatures[44] $[kT/(m_e c^2) \sim (k/m_e c^2)[M_\odot c^3/(4\sigma R^3)]^{1/4} \sim 20]$. This energy naturally must disperse adiabatically,[44,48] and the resulting expansion of pairs is called[49] a *fireball*. The early stages of such a dynamic fireball may involve rapid neutrino production (and $\nu\bar{\nu}$ annihilation to produce pairs with 0.1% efficiency) so that enormous numbers of neutrinos are emitted[45] ($E/c^2 \sim M_\odot \Rightarrow 10^{60}\nu$); unfortunately this leads to about 10^3 neutrinos per square cm on earth, a signal that is many orders of magnitude too small for detection.

Other models where fireballs arise include failed type 1b supernovae,[50] in which a star with a massive iron core fails to expel its outer layers as a supernova, allowing the would-be ejecta to form a transient accretion disk and then collapse onto the condensed core (a newly-formed neutron star) and explosively depositing energy to form a fireball. Copious production of neutrinos would be expected in this model. It has also been suggested[51] that rapidly spinning neutron stars in distant galaxies could form with extremely high magnetic fields, 10^{15} Gauss, much stronger than conventional pulsar fields. These so-called *magnetars* would result from the collapse of accreting white dwarfs with unusually high magnetic fields (about 10^9 Gauss), and would rapidly lose angular momentum through gravitational and magnetic-dipole radiation. A predominantly electromagnetic fireball would then be initiated, comprised mostly of electron-positron pairs and photons. Both models are envisaged as defining cosmological burst populations.

Non-relativistic temperatures (typically ~ 20 keV) are usually achieved[44,48] in the adiabatic cooling of fireballs, which naturally imply that the bulk velocities attained by the expansions when they become optically thin to $\gamma\gamma \rightarrow e^+e^-$ are ultrarelativistic (typically $\Gamma \sim 10^3$); this is required in order to satisfy energy conservation. These properties are confirmed by hydrodynamic analyses of fireball evolution.[52] Rudimentary calculations[48] of the output spectrum of this fireball is a cool (narrow) Planck continuum blueshifted to MeV energies and distorted by bulk motion of the plasma. This is unlike the observed non-thermal spectra of bursts, leading immediately to a problem with pure fireball models. More detailed computations[53] of radiative transfer effects in late stages of fireball evolution indicate that high energy power-law tails may be possible, though refinement of these calculations is needed.

The fireball phenomenon is not as simple as an adiabatically expanding plasma of pairs and photons; realistic models must include some form of baryonic pollution swept up from the environment of the progenitor. This contamination depends strongly on the geometry in the progenitor. It was realized[54] that the presence of baryonic matter in the fireball must rob it of its radiative efficiency since the baryons can dominate the kinetic energy of the expansion. Effectively the initial energy of the fireball is deposited as kinetic energy of the baryons rather than emitted as radiation. Simple computations[54] reveal that the final bulk Lorentz factor

Γ attained by the fireball is less than the ratio of pair to baryon energy in its initial stages, and that the fraction of the available energy that appears as radiation is generally much less than unity, posing a severe problem for fireball models. A natural way of avoiding these problems is to allow the fireball to impact on the surrounding interstellar medium,[55] generate one or several shocks (much like the propagation of supernova ejecta), and accelerate particles in the shock environs, i.e. convert the kinetic energy of the expansion to non-thermal energy. The resulting ultrarelativistic particles radiate efficiently via synchrotron radiation or inverse Compton scattering[56] guaranteeing that a large fraction of the initial fireball energy is extracted in the form of radiation. Sub-parsec scale clumpiness of the interstellar medium can provide the GRB time variability as the fireball *apparently* sweeps up matter on a timescale of order seconds.[55] The resulting spectra[56] can have small spectral breaks in the GRB energy range, however there is no preferred energy for the breaks in this scenario. In addition, it is presently unclear that the model can fit the variety of spectral slopes α present in GRB data.

5. Conclusion

This concludes a brief summary of gamma-ray burst properties, some recent observations and interesting theoretical work. Clearly the dilemma continues: bursts remain as enigmatic as ever. It appears unlikely that any stunning new observations may appear on the scene from CGRO, which will largely be devoted to refinement of existing results by increasing the database. However, positive results from searches[57] for gravitationally-lensed GRBs (by distant galaxies) would provide the smoking gun for cosmological pundits, and any observation of 20–60 keV absorption features by BATSE would excite proponents of the galactic neutron star hypothesis. Neither discovery would conclusively prove that *all* bursts belonged to either population. Observations of photoelectric absorption of the very low energy X-ray spectra of bursts could flip the cosmological/galactic coin either way. For theorists, who depend on the observations, the GRB question will almost certainly be resolved by the observations; however development of various models, particularly in regard to spectral issues, will aid in reducing the plethora of viable models.

6. References

1. Klebesadel, R. W., Strong, I. B. and Olson, R. A.: Ap. J. **182**, L85 (1973).
2. Harding, A. K.: Phys. Rep. **206**, 327 (1991).
3. Fishman, G. J. et al.: A&AS **97**, 17 (1993).
4. Higdon, J. C. and Lingenfelter, R. E.: Ann. Rev. Astron. Astr. **28**, 401 (1990).
5. Fishman, G. J. et al.: Ap. J. Supp. **92**, 229 (1994).
6. Kouveliotou, C. et al.: Ap. J. **413**, L101 (1993).
7. Paciesas, W. S., Fishman, G. J., eds.: Gamma-Ray Bursts, (AIP 265, NY, 1991).
8. Laros, J.: Nature **318**, 448 (1985).
9. Laros, J. et al.: in High Energy Transients in Astrophysics, ed. S. E. Woosley (AIP 115, NY, 1984), p. 378.
10. Mazets, E. P. et al.: Ap. Space Sci. **80**, 3,85,119 (1981).
11. Matz, S. M. et al.: Ap. J. **288**, L37 (1985).
12. Schneid, E. J. et al.: A&A **255**, L13 (1992).
13. Sommer, M. et al.: Ap. J. **422**, L63 (1994).

14. Hurley, K. et al.: to be submitted to Ap. J. Letters, (1994).
15. Schaefer, B. E. et al.: in Gamma-Ray Bursts, ed. G. Fishman et al. p. 280 (AIP 307, NY, 1994).
16. Murakami, T. et al.: Nature **335**, 234 (1988).
17. Meegan, C. et al.: Nature **355**, 143 (1992).
18. Meegan, C. et al.: in Gamma-Ray Bursts, ed. G. Fishman et al. p. 3 (AIP 307, NY, 1994).
19. Mao, S. and Paczyński, B.: Ap. J. **388**, L45 (1992).
20. Hakkila, J. et al.: Ap. J. **422**, 659 (1994).
21. Hartmann, D.: in The Lives of Neutron Stars, eds. A. Alpar and J. van Paradijs (Kluwer Academic, Dordrecht, 1994).
22. Norris, J. et al.: Ap. J. in press, (1994).
23. Baring, M. G.: in Proc. XIVth Moriond Workshop (Editions Frontières, 1994).
24. Schaefer, B. E. et al.: Ap. J. Supp. **92**, 285 (1994).
25. Band, D. et al.: Ap. J. **413**, 281 (1993).
26. Schaefer, B. E. et al.: Ap. J. **393**, L51 (1992).
27. Lightman, A. P. and Zdziarski, A. A.: Ap. J. **391**, 643 (1987).
28. Dermer, C. D.: in Proc. XIVth Moriond Workshop (Editions Frontières, 1994).
29. Zdziarski, A. A., Coppi, P. S. and Lamb, D. Q.: Ap. J. **357**, 149 (1990).
30. Baring, M. G.: Ap. J. **418**, 391 (1993).
31. Baring, M. G.: A&A **225**, 260 (1989).
32. Preece, R. and Harding, A. K.: Ap. J. **347**, 1128 (1989).
33. Daugherty, J. K. and Harding, A. K.: Ap. J. **273**, 761 (1983).
34. Baring, M. G.: MNRAS **244**, 49 (1990).
35. Dermer, C. D.: Ap. J. **347**, L13 (1989).
36. Dermer, C. D.: Ap. J. **360**, 197 (1990).
37. Baring, M. G.: in GRBs, ed. G. Fishman et al. p. 572 (AIP 307, NY, 1994).
38. Dermer, C. D. and Schlickeiser, R.: in Gamma-Ray Bursts, ed. G. Fishman et al. p. 510 (AIP 307, NY, 1994).
39. Schmidt, W. K. H.: Nature **271**, 525 (1978).
40. Krolik, J. H. and Pier, E. A.: Ap. J. **373**, 277 (1991).
41. Harding, A. K.: in Proc. 2nd Compton Symp., ed. C. Fichtel et al., p. 30 (AIP 304, NY, 1994).
42. Baring, M. G. and Harding, A. K.: in Proc. 23rd ICRC **1**, 53 (1993).
43. Harding, A. K. and Baring, M. G.: in Gamma-Ray Bursts, ed. G. Fishman et al. p. 520 (AIP 307, NY, 1994).
44. Paczyński, B.: Ap. J. **308**, L43 (1986).
45. Eichler, D. et al.: Nature **340**, 126 (1989).
46. Mészáros, P. and Rees, M. J.: Ap. J. **397**, 570 (1992).
47. Narayan, R., Piran, T. and Shemi, A.: Ap. J. **379**, L17 (1991).
48. Goodman, J.: Ap. J. **308**, L47 (1986).
49. Cavallo, G. and Rees, M. J.: MNRAS **183**, 359 (1978).
50. Woosley, S. E.: Ap. J. **405**, 273 (1993).
51. Usov, V.: Nature **357**, 472 (1992).
52. Piran, T., Shemi, A. and Narayan, R.: MNRAS **263**, 861 (1994).
53. Carrigan, B. J. and Katz, J. I.: Ap. J. **399**, 100 (1992).
54. Shemi, A. and Piran, T.: Ap. J. **365**, L55 (1990).
55. Mészáros, P. and Rees, M. J.: Ap. J. **405**, 278 (1993).
56. Mészáros, P. and Rees, M. J.: Ap. J. **418**, L59 (1993).
57. Nemiroff, R. J. et al.: in Gamma-Ray Bursts, ed. G. Fishman et al. p. 150 (AIP 307, NY, 1994).

GAMMA-RAY PRODUCTION IN THE WINDS OF WOLF-RAYET STARS

P.J. MEINTJES, G. KANBACH
Max Planck Institute for Extraterrestrial Physics
Postfach 1603, Garching bei Munich
D-85740
F.R.G

1. Introduction

The highly successful missions of the American SAS II (30-200 MeV) and European Cos-B (70 MeV - 5 GeV) orbiting γ-ray detectors showed that the diffuse gamma-ray sky correlates with the galactic disk structure (Mayer-Hasselwander et al. 1980), on which there are a number of sources superimposed with typical error boxes of 1° and typical flux of $F_\gamma \sim 10^{-6} ph.cm^{-2}\, s^{-1}$. On a large scale these sources show a correlation with the galactic arms (Caraveo & Paul 1979), and their latitude distribution shows that their average distance from the sun is between 2 and 5 kpc, with corresponding luminosities $L_\gamma \sim 10^{34} - 10^{35}\, erg\, s^{-1}$ (Swanenburg et al. 1981). It was shown that up to 50% of the detected sources could be linked to active star forming regions, which includes the ρ Oph dark cloud complex (D $\sim$ 160 pc) (Cassé & Paul 1980), the Orion complex (D $\sim$ 500 pc) (Caraveo et al. 1980) and possibly also supernova remnants physically linked with OB associations (Montmerle 1979). Cassé & Paul (1980) showed that these objects show a substantial enhancement in cosmic ray (CR) intensity relative to the Solar neighborhood and they suggested that most of the observed cosmic rays in these objects may be accelerated in the expanding atmospheres of mass-losing stars, rather than by supernova (SNR) shock waves, which has been advocated by Montmerle (1979) to explain these γ-ray sources.

Chaotic stellar winds as an accelerating site for charged particles were investigated by (White 1985; Chen & White 1991a), and these authors showed that shocks imbedded in the stellar wind may accelerate a small fraction of the charged particles like electrons and ions by first order Fermi acceleration to relativistic energies, which explained the non-thermal radio emission observed from many hot luminous stars, particularly Wolf Rayet stars (Abbot et al. 1986). Chen & White (1991b) progressed along this line and showed that the stellar winds of luminous young O-type stars can be a new class of discrete low energy γ-ray sources, with the γ-ray luminosity $L_\gamma \propto \frac{\dot{M}^2}{V_\infty^2 R_*}$, where $\dot{M}$, V_∞ and R_* represent the stellar mass-loss rate, stellar wind terminal velocity and stellar radius respectively. So stars with higher mass-loss rate should in principle be strong γ-ray emitters. Pollock (1987a) pointed out that electrons accelerated by shocks in the dense winds of high mass-losing Wolf Rayet stars may be responsible for the non-thermal radio emission observed from some of these

M. M. Shapiro et al. (eds.), Currents in High-Energy Astrophysics, 31–38.

stars, and that it may probably also generate high levels of inverse-Compton X-rays and bremsstrahlung GeV γ-rays. However, the bremsstrahlung γ-ray emission seems unlikely since the inverse Compton losses in the atmospheres of hot young stars will dominate and it may not be detectable (White & Chen 1992). However, relativistic protons accelerated in the wind- generated shocks will encounter mainly energy loss via particle-particle collisions, which will result in the γ-ray emission through the decay of π° mesons produced in the collision process. Therefore, in this Paper we consider the stellar winds of Wolf Rayet stars as possible cosmic ray and γ-ray production sites, since these stars show profound mass-loss which may be an order of magnitude more than the maximum mass-loss observed in young O-type stars. Another fact which might indicate that Wolf Rayet stars are cosmic ray production sites, is the high ratio of $^{22}Ne/^{20}Ne$ as well as the abundance of ^{4}He, ^{14}N and ^{16}O in the observed cosmic rays. Up to $\sim$ 33% of the observed cosmic rays consist of these elements, which are produced in the CNO cycles of massive hot stars, especially Wolf Rayet stars which are the exposed helium-rich cores of evolved O stars (Cassé & Paul 1980).

The detectability of these sources was hampered by the poor resolution of both previous generations of γ-ray detectors, but with the arrival of the **Compton Gamma Ray Observatory** (CGRO), γ-ray astronomy entered a new era, which enabled us to study potential γ-ray emitting objects at much greater distances than was possible with previous generations of γ-ray detectors. In Section 2 of this Paper we will outline some basic details about shock acceleration in stellar winds and some aspects of the available energy budget for particle acceleration in the wind generated shocks, and in Section 3 we will show the typical expected γ-ray flux for some nearby Wolf Rayet stars. In Section 4 we will summarize our results and discuss their implications.

2. Wind Generated Shocks and Particle Acceleration

Early type stars (O, early B and Wolf-Rayet stars) occupy the upper left corner of the Hertzsprung-Russel (H-R) diagram. These stars are all very hot ($T_{eff} >$ 20 000 K) and with masses that exceeds 20 $M_{\odot}$ (progenitor mass for Wolf Rayet (WR) stars) and very luminous ($L_{bol} \sim 10^5 - 10^6\, L_{\odot}$). The enormous radiation from a massive star provides enough UV photons to ionize surrounding HII regions, and the extreme radiation pressure in its atmosphere drives the outer atmosphere to form a stellar wind, which may reach terminal velocities which can exceed 3000 $km.s^{-1}$ (Cassinelli 1979). In the case of WR stars, these winds are carrying mass away which is orders of magnitude greater than the radiation pressure and there is still no mechanism which successfully explains this phenomenon. It was shown (Owocki & Rybicki 1984) that the line driven winds of high mass-losing stars are very unstable and that shocks accompanied by large density inhomogeneities do form in the wind. Numerical simulations (Owocki, Castor & Rybicki 1988) show that instabilities in the wind can grow exponentially into strong multiple shocks. The existence of these shocks in the wind has long been used to explain the non-thermal radio emission (White 1985) from massive stars, as well as the EINSTEIN X-ray emission (Lucy & White 1980). The success of shock acceleration to account for the non-thermal emissions from massive young stars has prompted many authors to speculate that this mechanism may provide a natural platform for a unified theory of non-thermal emission from hot stars (White 1985;

Pollock 1987a; Montmerle 1990; Chen & White 1991b).

We consider the acceleration of particles via the first-order Fermi mechanism (see Bell 1978; Blandford & Ostriker 1978; and Blandford & Eichler 1987 for a review) in the winds of Wolf Rayet stars. High energy particles will be scattered across a shock front, usually by irregularities in the magnetic field imbedded in the stellar wind. The scattering of particles will increase the average particle energy slightly while the incoming gas is decelerated. For effective particle acceleration the shocks must be collisionless, otherwise particle-particle collisions will hamper effective acceleration. The criteria for collisionless shocks is that the collisional mean free path of particles in the wind, given by

$$l = 4 \times 10^9 \left(\frac{T}{10^8\,K}\right)^2 \left(\frac{\dot{M}}{10^{-5}\,M_\odot\,yr^{-1}}\right)^{-1} \left(\frac{V_\infty}{2500\,km\,s^{-1}}\right) \left(\frac{r}{R_*}\right)^2 cm \quad (1)$$

must be larger than the gyroradius of an accelerated particle in the wind, given by

$$R_g = 3.13 \times 10^6 \left(\frac{\gamma}{10^3}\right) \left(\frac{B_*}{1000\,G}\right)^{-1} \left(\frac{r}{R_*}\right)^2 cm \quad (2)$$

for $r \leq 10\,R_*$ and

$$R_g = 3.13 \times 10^8 \left(\frac{\gamma}{10^3}\right) \left(\frac{B_p}{10G}\right)^{-1} \left(\frac{r}{R}\right) cm \quad (3)$$

for $r \geq 10\,R_*$ if we consider a Parker-type wind.

The maximum energy a particle can reach via shock acceleration in the wind is limited by several energy-loss mechanisms, like diffusion from the shock, particle-particle collisions, and in the case of the atmospheres of luminous stars, particle-photon collisions. The maximum energy to which a particle can be accelerated is estimated by balancing the acceleration time with the energy-loss time scales (Achterberg 1992)

$$\left(\frac{d\gamma}{dt}\right)_{acc} = \left(\frac{d\gamma}{dt}\right)_{diff} + \left(\frac{d\gamma}{dt}\right)_{p-p} + \left(\frac{d\gamma}{dt}\right)_{p-ph} \quad (4)$$

where $(\frac{d\gamma}{dt})_{acc} = \frac{\beta V_\infty^2}{c R_o}$ where R_o is the Larmor radius of a $\gamma = 1$ particle ($\beta \sim 0.15$). The energy loss terms are respectively $(\frac{d\gamma}{dt})_{diff} = \frac{\gamma V_\infty}{L}$, $(\frac{d\gamma}{dt})_{p-p} = \gamma\, n_p\, \sigma_{p-p}\, c$ and $(\frac{d\gamma}{dt})_{p-ph} = \gamma\, n_{ph}\, \sigma_{p-ph}\, c$, where L is the Bohm diffusion coefficient and where the particle and photon densities in the wind are

$$N_p = \frac{\dot{M}}{4\pi m_i r^2 V_\infty} \sim 3 \times 10^{11} \left(\frac{\dot{M}}{10^{-4}\,M_\odot\,yr^{-1}}\right) \left(\frac{V_\infty}{3000\,km\,s^{-1}}\right)^{-1} \left(\frac{R_*}{r}\right)^2 cm^{-3} \quad (5)$$

$$N_{ph} = \frac{L_{bol}}{4\,\pi r^2 c E_{ph}} \sim 2 \times 10^{14}\, L_{B,39} \left(\frac{R_*}{r}\right)^2 \left(\frac{E_{ph}}{4\,eV}\right)^{-1} cm^{-3} \quad (6)$$

respectively. By adopting a wind speed of $V_\infty \sim 3000\,km\,s^{-1}$, a surface magnetic field strength of 1000 Gauss, the maximum energy to which a proton or ion can be accelerated is $E_{p,max} \sim \gamma_{max}\, m_p\, c^2 \sim 30$ TeV for regions $\leq 100\, R_*$ where the energy-losses are most severe. Further out in the wind this may be a lower limit to the maximum energy.

The net result of this process is a power-law distribution in particle energy (Chen 1992)

$$\frac{dN(E)}{dE} = (\beta - 1)\, N_p\, E_o^{\beta - 1}\, E^{-\beta}\, cm^{-3}\, erg^{-1} \qquad (7)$$

where N_p is the particle density in the vicinity of the shock and $\beta = \frac{(r+2)}{(r-1)}$, with $r = \frac{u_1}{u_2}$ the compression ratio of the shock, which is determined by the Mach number of the outward moving shocks in the stellar wind. For strong adiabatic shocks r = 4, and the resulting particle spectrum in the vicinity of the shock is then

$$\frac{dN(E)}{dE} = N_p\, (k_B\, T_s)\, E^{-2}\, cm^{-3}\, erg^{-1} \qquad (8)$$

where $E_o \sim (k_B\, T_s)$ is the thermal energy of the particles in the post-shock region. In this Paper we assume that 20% of the thermal particles will be scattered up to relativistic energies, which correlates with the value at the Earths bowshock and from non-thermal X-ray observations of the Orion stars (Cassinelli & Swank 1983).

3. Wolf Rayet Winds and Gamma-Ray Production

Wolf Rayet stars show the highest fluxes of mass outflow of any stellar type (Abbot & Conti 1987), with the mass-loss rates $\dot{M}_{WR} \leq 10^{-4}\, M_\odot\, yr^{-1}$ and wind terminal velocities $V_{WR}^\infty \leq 5000\, km\, s^{-1}$ respectively, in comparison to O-type stars which have $\dot{M}_O \leq 10^{-5}\, M_\odot\, yr^{-1}$ and $V_O^\infty \leq 3000\, km\, s^{-1}$ (Garmany & Conti 1984). This implies that the stellar winds of WR stars drives a wind luminosity $L_{WR} \sim \frac{1}{2}\dot{M}_{WR} V_{WR}^2 \leq 8 \times 10^{38}\, erg\, s^{-1}$, in comparison to O-type stars which have wind luminosities of order $L_O \leq 3 \times 10^{37}\, erg\, s^{-1}$. It is clear that WR stars have a much higher power output than O-type stars and therefore potentially more energy is available for the acceleration of charged particles. In the Lucy (1982) model 1% of the wind power can be converted into shocks, but if we assume that 0.5% of the wind energy goes into shocks, the total shock power is $L_s \leq 4 \times 10^{36}\, erg\, s^{-1}$. If the efficiency of shocks is 20% for accelerating particles, then the power into relativistic ions can also be of the order of $L_{ions} \leq 8 \times 10^{35}\, erg\, s^{-1}$. This power in relativistic ions may be sufficient to result in γ-ray luminosities of order $L_\gamma \leq 4 \times 10^{34}\, erg\, s^{-1}$ if 5% of these particles are converted into γ-rays through π^o decay. Through this crude estimate it is clear that the winds of WR stars, especially those with the highest mass-loss, provide enough power for γ-ray generation in the range observed by COS-B (Swanenburg et al. 1981). These γ-ray luminosities imply that the sources should be visible above the EGRET sensitivity threshold ($E_{th} \sim 100$ MeV) of $F_{th,G-C} \sim 3 \times 10^{-7}\, ph.\, cm^{-2}\, s^{-1}$ and $F_{th,A-C} \sim 1.7 \times 10^{-7}\, ph.\, cm^{-2}\, s^{-1}$ for the galactic center and anti-center respectively (Radecke & Kanbach 1992) if they are at distances $D_{kpc} \leq 2.6$ (galactic center) and $D_{kpc} \leq 3.5$ (anti-center) respectively. For an effective area of 1000 cm^2 at 100 MeV (Thompson et al. 1993) and a typical observation period of two weeks these fluxes imply source- induced γ-ray counts of $N_s \sim 363$ (Gal. Center) and $N_s \sim 206$ (Anti-Center) respectively. The galactic background emission at 100 MeV, convolved with the EGRET point spread function (Bertsch et al. 1993), is $F_{\gamma,G-C} \sim 8 \times 10^{-6}\, cm^{-2}\, s^{-1}$ (Gal. Center) and $F_{\gamma,A-C} \sim 2 \times 10^{-6}\, cm^{-2}\, s^{-1}$ respectively, implying two week background counts of $N_b \sim 9677$ (Gal. Center) and $N_b \sim 2419$ (Anti-Center) respectively. The significance of these limiting fluxes is ($\sigma \sim \frac{N_{s+b} - N_b}{\sqrt{N_{tot}}}$) 2.6$\sigma$ (Gal. Center) and 2.9σ (Anti-Center) respectively, and is barely distinguishable from the background emission.

The galactic distribution of WR stars is given by Hidayat et al. (1982), and there are a substantial number ($\sim$ 32) of WR stars with distances around 2 kpc from the solar system. Most of these stars lie in the line of sight of the galactic center, where the background γ-ray radiation is the highest. Unfortunately, the galactic anti-center region which has the lowest background γ-ray radiation is poorly represented in WR stars, and there seems to be a death zone between $250° > l^{II} > 150°$ where there are no Wolf Rayet stars at all. With this in mind, we selected a sample of the nearest and highest mass-losing Wolf Rayet stars, with distances $\leq$ 2 kpc from the Solar system. The expected γ-ray flux is estimated by

$$F_\gamma \sim \frac{2 \int \frac{dN(E)}{dE}\, dE \int dV \int d\,N_p \langle \sigma_{p-H} \rangle\, c\, P_{esc}}{D^2}\ ph.\, cm^{-2}\, s^{-1} \tag{9}$$

where N (E) is the input particle spectrum, $\int dV$ is the volume of the source determined by 1000 stellar radii around the star, $\int d\,N_p$ is the particle density in the wind and $P_{esc} \sim 1 - \int_r^\infty N_{p,ph}\, \sigma\, dr$ is the probability that these γ-rays may escape without absorbtion due to photon-photon and photon-particle collisions. The effect of photon-particle collisions is negligible in a wind density of approximately $n_p \sim 10^{11}\, (\frac{R_*}{r})^2\, cm^{-3}$, with cross section $\sigma \sim 10^{-27}\, cm^2$ and is $P_{esc} >$ 99.99%. The effect of photon-photon collisions is determined by $(E_\gamma\, E_{ph})(1 - cos(\theta)) = 4(m_e\, c^2)^2$. From the average bolometric magnitudes of WR stars, which is $M_v \sim$ -8.5, average blackbody surface temperatures of 44000 K are deduced, resulting in peak blackbody emission at photon energies of $E_{ph} \sim 4$ eV (UV). This implies that γ-rays with energies of $\sim$ 100 GeV will be the most absorbed by photon-photon collisions. The magnitude of the effect is estimated as follows: The photon density is derived from the bolometric luminosity and in $n_{ph} \sim 2 \times 10^{14}\, (\frac{R_*}{r})^2\, cm^{-3}$. This implies optical depths $\tau \sim n_{ph}\, \sigma\, r \sim 56$ for a cross section of $\sigma \sim 10^{-25}\, cm^2$, integrated over 100 stellar radii. However, TeV γ-rays produced out > 100 stellar radii away from the star will have a $P_{esc} >$ 50% chance of escaping. For lower energy γ - rays, higher energy photons are required for pair creation, for example 1 GeV γ-rays will create pairs in collisions with soft X-rays of 500 eV. These X-ray luminosities are however $< 10^{-4}$ times the bolometric luminosity, implying an escape probability of $P_{esc} >$ 99.99%. The resulting gamma-ray flux above 100 MeV is then

$$F_\gamma \sim 5 \times 10^{-7}\, (\frac{\dot{M}}{1 \times 10^{-4}})^2\, D_{kpc}^{-2}\, (\frac{P_{esc}}{100\%})\, (\frac{E}{100\, MeV})^{-1}\ ph.\, cm^{-2}\, s^{-1} \tag{10}$$

The closest Wolf Rayet stars $D_{kpc} \leq 2$ with the highest mass-loss were taken, and their γ-ray fluxes above 100 MeV were calculated (see Table 1).

Table 1: Estimated γ-ray fluxes above 100 MeV of the nearest high mass-losing Wolf Rayet stars.

WR	$\dot{M}_{-4}$	D_{kpc}	F_γ	$N_{2w}{}^*$	$N_{4w}{}^*$	σ_{2w}	σ_{4w}
6	0.76	1.56	0.7	85	170	1.2	1.7
11	0.85	0.5	3.60	436	874	4.94	7
79	0.9	2.0	0.5	61	122	-	-
86	0.7	1.95	0.34	41	82	-	-
90	0.7	1.97	0.33	41	82	-	-
93	0.74	1.74	0.5	61	122	-	-
110	0.96	1.89	0.64	77	154	-	-
134	0.74	2.0	0.34	41	82	-	-
136	0.90	1.82	0.61	74	148	-	-
140	0.91	1.30	1.22	148	296	2.1	3.0

Mass-losses obtained from Abbot et al. (1986)
Distances from van der Hucht et al. (1988)
Gamma-ray flux in units of $10^{-7}\ cm^{-2}\ s^{-1}$
* *represents the expected 2 and 4 week γ-ray counts*

It is clear that two WR stars, i.e. WR 11 = γ^2 Vel and WR 140 (HD 193793) might show detectable γ-ray emission above the EGRET detection threshold. However, WR 140 might only be detectable after 4 weeks of observations of the Cygnus region of the sky. The significances in columns 7 and 8 are calculated by comparing the Wolf Rayet counts with the galactic background counts in that particular region of the sky (Bertsch et al 1993). Pollock et al. (1985) pointed out that HD 193793 was 2.3° from the center of the error box of the COS-B γ-ray source 083+03 (Swanenburg et al. 1981). This WR star is also a strong non-thermal radio emitter which might indicate that particle acceleration occurs in the wind.

A recent catalogue of galactic γ-ray emitters which represents detections from the Phase I period of EGRET (Fichtel et al. 1994) shows several as yet unidentified γ-ray excesses. The γ-ray source GRO J 1111-60 ($l^{II} = 290.99°\,; b^{II} = -0.23°$) corresponds positionally with Wolf Rayet stars WR37, WR38, WR39 and WR42 in the WR list of van der Hucht et al. (1988). The γ-ray source GRO J 2021+37 ($l^{II} = 75.52°\,; b^{II} = 0.28°$), which coincides with the COS-B γ-ray source 2CG75+00 (Swanenburg et al. 1981) lies only 11.4 minutes in longitude, and 13.2 minutes in latitude away from WR142 and may also correlate with WR141 (van der Hucht et al. 1988). Both these WR stars are part of open clusters, i.e. Berkley 87 and Cyg OB1 respectively. Associated with Cyg OB1 is the HII region S 109, and Be 87 also contains approximately 100 massive young stars and belongs to the star formation region ON 2 with several compact HII regions and CO molecular clouds (Polcaro et al. 1991). The open cluster Be 87 was identified recently (Polcaro et al. 1994) as the counterpart for the γ-ray source GRO J 2021+37. The γ-ray source GRO J 2032+40 ($l^{II} = 79.77°\,; b^{II} = 0.70°$) also coincides with the Wolf Rayet binary WR145 (AS 422), which is part of the association Cyg OB2. However, none of these stars can individually account for the observed γ-ray fluxes, but their winds, in association with other high mass-losing stars in the cluster to which they belong, may boost the local cosmic ray density. These cosmic rays may leak out to the surrounding cloud complexes associated with these HII regions and produce the γ-ray emission through particle collisions. A similar case may be the open cluster IC 1805 (= Mel 15) ($l^{II} = 134.7°\,; b^{II} = 1°$) which is similar to Berkeley 87. This cluster also contains a peculiar O4f mass losing star HD 15570 ($V_\infty \sim 3800\, km.s^{-1}$), and it is also imbedded in the star forming region Cas OB6, which is very similar to ON 2. This region also contains the compact HII region G133.9+1.1 (W3-OH) (see Polcaro et al. 1991 and references therein), and also corresponds positionally with the COS-B γ-ray source CG 135+4 (see e.g. Montmerle 1979). Incidently, the phase I catalogue of γ-ray sources detected by EGRET (Fichtel et al. 1994) also show a γ-ray source which is positionally not far away, i.e. GRO J 239+61, which are situated at $l^{II} = 135.53°\,; b^{II} = 1.25°$.

4. Discussion

In this Paper we did not attempt to give a detailed theoretical account of possible γ-ray emission in the expanding envelopes of WR stars, but we made an attempt to make optimistic estimates of the possible γ-ray production of individual close-by WR stars. Our estimates showed that 2 WR stars may (WR11 and WR140) gives γ-ray fluxes which may be detectable by EGRET. However, our best candidate WR11 (γ Vel) is positionally to close to the Velar pulsar to be resolved by EGRET. The other candidate, WR140 (HD 193793), also shows strong X-ray emission and non-thermal radio synchrotron emission, which may indicate that shocks exist in the wind which accelerate charged particles to relativistic energies. Its flux is however too low to qualify even as a marginal detection (5 - 6 σ) in the Fichtel et al. (1994) EGRET source catalogue. Some of the other Wolf Rayet stars might show detectable γ-ray emission if the observation period is increased to longer than 4 weeks.

However, although the results of this study are somewhat disappointing concerning possible EGRET detections of WR stars, one important fact forthcoming from our investigation is that these sources might be detectable with ground based air Cherenkov detectors ($E_\gamma >$ 1 TeV). Recent improvements in sensitivities of Cherenkov detectors, especially imaging Cherenkov telescopes (Akerlof et al. 1989), enable these detectors to detect TeV γ-rays at flux levels of $10^{-12}\,ph.\,cm^{-2}\,s^{-1}$. Our best candidate, WR11 (γ Vel) may show a flux $F_\gamma(\geq\, 1\,TeV) \sim 9 \times 10^{-12}\,ph.\,cm^{-2}\,s^{-1}$. Its declination of $\sim\, -47°$ makes it suitable for TeV γ-ray observations from Nooitgedacht (South-Africa) and Narrabri (Australia). Since the field of view of air Cherenkov detectors is typically between 1° and 2°, this star should not be influenced by emission from the Vela Pulsar. The other candidate, WR140 may give a flux upper limit $F_\gamma(\geq\, 1\,TeV) \sim 2.3 \times 10^{-12}\,ph.\,cm^{-2}\,s^{-1}$, and is at high northern declination ($\sim\, +44°$). These fluxes are estimated by integrating over that part of the wind where the TeV γ-rays have a > 50 % escape probability. A thorough TeV γ - ray study of these systems may provide essential clues as to what energies particles can be accelerated in the winds of high mass losing stars.

References

Abbot D.C. et al. 1986, ApJ, 303, 239
Abbot, D.C. & Conti, P.S. 1987, ARA&A, 25, 113
Achterberg, A. 1992 in Lecture Notes in Physics no 418, 1
Akerlof, C.W. et al. 1989, Harvard-Smithsonian Preprint No. 2945
Bell, A.R. 1978, MNRAS, 182, 147
Blandford, R. & Ostriker, J.P. 1978, ApJ, 221, L29
Blandford, R. & Eichler, D. 1987 Phys. Rept., 154, 1
Caraveo, P. & Paul, J. 1979, A&A, 75, 340
Caraveo, P. et. al. 1981, A&A, 91, L3
Cassé, M., Paul, J.A. 1980, ApJ, 237, 236
Cassinelli, J.P. 1979, Ann. Rev. A.Ap, 17, 275
Cassinelli, J.P. & Swank, J.H. 1983, ApJ, 271, 681
Cavallo, G. & Gould, R.J. 1971, Nuovo Cimento, 2B, 77
Chen, W. 1992, Ph.D. thesis (unpublished), Johns Hopkins Univ.
Chen, W. & White, R.L. 1991a, ApJ, 366, 512
Chen, W. & White, R.L. 1991b, ApJ, 381, L63
Dermer, C. 1986, A&A, 157, 223
Eichler, D. ApJ, 229, 419
Fichtel, C.E. et. al. 1994, in Press
Garmany, C.D. & Conti, P.S. 1984, ApJ, 284, 705
Hidayat, B. et. al. 1982, in Wolf Rayet Stars: Observations, Physics, Evolution, ed. C.W.H. de Loore & Willis, A.J., 27
van der Hucht, K.A. et al. 1988, A&A, 199, 217
Lucy, L.B. 1982, ApJ, 255, 286
Lucy, L.B. & White, R.L. 1980, ApJ, 241, 300
Montmerle, T. 1979, ApJ, 231, 95
Montmerle, T. 1990, in IAU Symp. 143, eds. K.A. van der Hucht & B. Hidayat (Dordrecht: Kluwer), 397
Owocki, S.P. & Rybicki, G.B. 1984, ApJ, 284, 337
Owocki, S.P. et al. 1988, ApJ, 335, 914
Polcaro et al. 1991, A&A, 252, 590
Polcaro et al. 1994, MNRAS, in press
Pollock, A.M.T. 1978, A&A, 171, 135
Pollock, A.M.T et al. 1985, A&A, 146, 352
Radecke, H.-D., Kanbach, G. 1992, in Data Analysis in Astronomy IV, ed. V. Di Gesu et al. Plenum Press, New York, 300
Swanenburg, B.N. et. al. 1981, ApJ, 234, L69
Thompson, D.J. et al. 1993, ApJ S, 86, 629
Torres, A.V., Conti, P.S., Massey, P. 1986, ApJ, 300, 379
van der Hucht, K.A. et al. 1981, Space Sci. Rev., 28, 227
White, R.L. 1985, ApJ, 289, 698
White, R.L. & Chen, W. 1992, ApJ, 387, L81
White, R.L. 1992, in Nonisotropic and Variable Outflows from Stars, ASP Conf. Ser., Vol 22, eds. L. Drissen, C. Leitherer & A. Nota, 274

ORIGIN OF HIGH–ENERGY X–RAY EMISSION FROM NGC253

OLEG GOLDSHMIDT
School of Physics and Astronomy
Tel Aviv University
69978, Tel Aviv, Israel

ABSTRACT. Compton emission from the disk and radio halo of the starburst galaxy NGC253 is calculated. It is shown that the emission can account for the high energy X-ray emission recently detected by *OSSE* (Bhattacharya *et al.* 1994). The galactic disk provides less than a third of the total flux, with the bulk originating in the large halo. If this explanation of the detected flux is viable, the *OSSE* result will be the first detection of a truly diffuse extragalactic Compton emission, underlining the significance of the escape of cosmic ray electrons from galaxies, and highlighting the interest in starburst galaxies for the study of this and related phenomena.

1 Introduction

NGC253 is a prototypical starburst galaxy that has been extensively studied in many regions of the electromagnetic spectrum: in the radio (e. g. Klein *et al.* 1983), the infrared (e. g. Telesco & Harper 1980, Rice *et al.* 1988), and in the X-ray (Boller *et al.* 1992, Bhattacharya *et al.* 1994) ranges. Recently Carilli *et al.* (1992) reported the discovery of an extended radio halo around the galaxy. Synchrotron emitting galactic halos had been clearly detected before only in three edge-on spiral galaxies: NGC4631 (Ekers & Sancisi 1977), NGC891 (Allen, Baldwin, & Sancisi 1978) and M82 (Seaquist & Odegard 1991). These halos are thought to be due to synchrotron emission from cosmic ray electrons that escape from the galactic disks by diffusion and convection in supernova-driven outflows of gas. Such outflows have been seen in line emission (McCarthy, Heckman, & van Breugel 1987), and in particular are characteristic of all the powerful far-infrared (FIR) galaxies (Heckman, Armus, & Miley 1987).

The discovery of an extended radio halo around NGC253 by Carilli *et al.* (1992) is thus of great interest, especially in connection with the recent detection of high energy X-ray emission from the galaxy by the *OSSE* detector on board the *Compton Gamma-Ray Observatory* (Bhattacharya *et al.* 1994). The emission would be most naturally explained as due to Compton scattering of the FIR radiation off the radio emitting electrons, but Bhattacharya *et al.* (1994) argued that this mechanism can produce only a small part of the observed flux. However, they took into account only the electrons in the disk, while with sufficiently high convection most of the Compton flux would come from the halo. The halo is indeed large and rather strong in the radio (see section 2.1 below). Presumably the magnetic field in the halo is weak compared to that in the disk, implying that the halo

M. M. Shapiro et al. (eds.), Currents in High-Energy Astrophysics, 39–44.

contains a large number of relativistic electrons. Although both the electron density and the FIR radiation field decrease with distance from the galaxy, the halo must be regarded as a potentially strong source of high-energy X-ray emission. This is the primary motivation for the calculation presented below.

The basic outline of the calculation is as follows: Analyzing the synchrotron emission from the disk of NGC253, one can derive the energy distribution of cosmic ray electrons in the disk. Then one can employ a particular model for electrons' escape from the galaxy into the halo, incorporating convection, diffusion, radiative (Compton and synchrotron), and excitation losses. The model used here is based on the work of Lerche & Schlickeiser (1980; hereafter LS); it represents a reasonable compromise between preserving the basic physics and keeping the treatment tractable (as shown below). Knowing the FIR radiation field, one can calculate the Compton emission of all the electrons in the disk *and* the halo.

The relevant observations of NGC253 are described in Section 2; an overview of the LS model is given in Section 3. The calculation and results are presented in Section 4, followed by a summary. The distance to NGC253 is taken to be 4.17 Mpc.

2 Observations of NGC253

2.1 RADIO

Radio observations of the disk of NGC253 were summarized by Klein *et al.* (1983). The non-thermal emission spectrum exhibits a break around 400 MHz: the spectral index changes from 0.46 at low frequencies to 0.72 at high frequencies. The equipartition magnetic field value is $9\,\mu$G, and I shall use the 10.7 GHz flux (corrected for thermal emission) of 1.556 Jy. The radio disk is thicker than the stellar one; a thickness of 0.5 kpc is adopted here. Klein *et al.* (1983) also present the distribution of the brightness temperature of the radio emission in the plane of the galaxy. I shall make use of the FWHM of this distribution, equal to $\simeq 1.0$ kpc.

The radio halo around NGC253 was detected by Carilli *et al.* (1992) at two frequencies, 330 MHz and 1.4 GHz. Its observed diameter is $16'$ (corresponding to 19.4 kpc at the distance of 4.17 Mpc). The halo provides about 25% of the total radio emission at 330 MHz, and the spectral index between the two frequencies is approximately 1.0.

2.2 INFRARED

NGC253 was first detected in the infrared range by Harper & Low (1973), and was subsequently observed by many researchers (see e. g. Telesco & Harper 1980). It is also a strong *IRAS* source, with a luminosity of $3.04 \times 10^{10}\,L_\odot$ between 42.5 and 122.5 μm (Rice *et al.* 1988). Rice *et al.* interpreted the *IRAS* data as emission from two dust components: cool ($T_{\rm cool} \approx 36.8$ K) and warm ($T_{\rm warm} \approx 172.8$ K). I shall adopt their model, which assumes that the dust absorption efficiency is proportional to frequency in both these components. Also, the combined modified Planck spectra provide a total FIR luminosity of $5.58 \times 10^{10}\,L_\odot$, with $4.07 \times 10^{10}\,L_\odot$ coming from the cool dust, and $1.51 \times 10^{10}\,L_\odot$ — from the warm dust. Note that the modification of the Planck spectrum due to the frequency dependence of the dust absorption efficiency is equivalent to ascribing a finite optical depth to the FIR source (Dwek & Arendt 1992).

Clearly, almost all FIR emission comes from the very compact nucleus of the galaxy. Harper & Low (1973) estimated the radius of the FIR source as 150 pc. Telesco & Harper

(1980) estimated the FIR source diameter between 120 and 600 pc. Observations at 1300 μm (Krügel *et al.* 1990) resolved a source with FWHM of $16'' \times 11''$, corresponding to mean radius of 136.5 pc. It is this last value that I shall use in the calculations, assuming that emissivity is constant within this radius and negligible outside.

2.3 HIGH ENERGY X-RAYS

Bhattacharya *et al.* (1994) reported *OSSE* detection of NGC253 in the high-energy X-ray range of 50–300 keV. The X-ray flux in this range is $\sim 10^{-3}$ photons cm^{-2} s^{-1} MeV^{-1}. The total flux between 50 and 165 keV is $\sim 3 \times 10^{-11}\,\mathrm{erg\,cm^{-2}\,s^{-1}}$, corresponding to a luminosity of $6.25 \times 10^{40}\,\mathrm{erg\,s^{-1}}$. An estimate of the Compton flux from the synchrotron-emitting electrons in the disk can yield at most 20–25% of the observed emission. This might actually be an overestimate, because Bhattacharya *et al.* (1994) assumed a low value for the magnetic field, $B = 1\,\mu$G. Bhattacharya *et al.* (1994) also estimated that free-free emission from the interstellar gas cannot provide more than 3% of the observed flux. They proposed several types of discrete sources (such as low mass X-ray binaries and supernova remnants), but one would need a prohibitively large number of such sources to obtain the necessary luminosity. Nor is continuum emission from an obscured supernova at peak brightness a viable explanation, because in that case *OSSE* should have also detected lines from the ^{56}Ni radioactive decay chain.

3 The Lerche – Schlickeiser Model

This model was decribed in detail and applied to NGC4631 and NGC891 by LS. It assumes that the cosmic ray electrons are injected into a thin galactic disk with a power-law spectrum with index p, and that a steady state is maintained. The electrons propagate in directions orthogonal to the disk by convection and diffusion, losing energy by electronic excitations and by Compton and synchrotron processes. I shall assume that the convection velocity, V, the diffusion coefficient, $D(E) = D_0 E^{\mu}$, and the radiative energy loss rate, $b_0 E^2$, do not depend on the spatial coordinates. In their analysis of NGC4631 and NGC891, LS concluded that this simplification allows one to obtain a reasonably complete picture of the relevant physical processes, while keeping the calculations relatively simple.

The radiative loss coefficient b_0 is $b_0 = 0.04[B^2/(8\pi) + W_{FIR} + W_{CMB}]\,\mathrm{erg^{-1}\,s^{-1}}$, where W_{FIR} and W_{CMB} are the energy densities of the FIR and cosmic microwave background (CMB) radiation in $\mathrm{erg\,cm^{-3}}$. Realistically, both B and W_{FIR} vary in space. I shall take this into account when I calculate the Compton flux, but *not* when I calculate the electron distribution. If anything, this underestimates the importance of halo electrons relative to those of the disk. At lower energies excitation losses dominate. They equal $9 \times 10^{-19}\,\mathrm{erg\,s^{-1}}$, assuming a free electron density of $1\,\mathrm{cm^{-3}}$.

In order to understand the effects of convection, one can compare the distances travelled by an electron with energy E by convection, $\sim V\tau(E)$, and by diffusion, $\sim 2\,[D(E)\tau(E)]^{1/2}$, during the energy loss time $\tau(E) = (b_0 E)^{-1}$. It is easy to see that low energy electrons travel mainly by convection, whereas diffusion is the dominant propagation mode of high energy electrons; the boundary energy is $E_c = [(1-\mu)V^2/(4b_0 D_0)]^{1/(1+\mu)}$. The electron energy distribution exhibits a break at $E = E_c$, with a corresponding break in the radio spectrum (LS). I interpret the observed break in the spectrum of NGC253 (see section 2.1 above) as a manifestation of the interplay between convection and diffusion.

4 Compton Flux from NGC253

4.1 OUTLINE OF THE CALCULATION

The calculation is based on properties of the disk radio emission, for which there are good spectral data. For the magnetic field in the disk I shall use the equipartition value of $B = 9\,\mu$G. The field is not necessarily in equipartition, so one should keep in mind that this value may be substantially uncertain. In particular, if the field is actually lower, the predicted Compton flux will be higher.

The distribution of cosmic ray electrons in the plane of the galaxy can only be assessed through the distribution of the brightness temperature of the radio emission. In order to obtain only a semi-quantitative estimate of the Compton emission — which is our goal here — we can adopt a "top hat" function, constant out to a certain radius, and zero beyond it. I shall take this effective radius to be equal to half the FWHM of the brightness temperature distribution, $\simeq 1.0$ kpc.

As described above, the radio spectrum of the disk of NGC253 breaks at $\sim 400\,$MHz. With the assumed value of B, $E_c \approx 1.4\,$GeV. Then, from B and the FIR luminosity ($L_{FIR} = 5.58 \times 10^{10}\, L_\odot$; see section 2.2) I estimate that $b_0 \approx 4.75 \times 10^{-13}\,\mathrm{erg}^{-1}\,\mathrm{s}^{-1}$, averaged in the galactic plane. From the spectral index map of Carilli *et al.* (1992) one can estimate the off-plane distance at which the spectral index steepens to its mean halo value of ~ 1.0. This distance equals roughly $V/(b_0 E_c)$ (LS), or about 2 kpc. This yields $V \approx 66\,\mathrm{km/\,s}$, in rough agreement with the gas outflow velocity (McCarthy *et al.* 1987).

The broken power-law spectrum is consistent with $p \approx 1.92$ and $\mu \approx 0.04$ (see LS). Note that at high frequencies the radio spectral index of the halo turns out close to 1.0, in accordance with the observations of Carilli *et al.* (1992). Finally, from the known E_c, b_0, and V one can also estimate the diffusion coefficient: $D(E) = D_0 E^{0.04}$, $D_0 \approx 1.2 \times 10^{28}\,\mathrm{cm}^2\,\mathrm{s}^{-1}$. These data are all one needs to calculate the electron distribution using the LS model, except normalization.

The normalization can be set by using the radio flux at 10.7 GHz (at this frequency the high energy approximation, $E \gg E_c$, is valid), which is 1.556 Jy. This accomplished, the only remaining task is to calculate the Compton photon flux, integrating the electron distribution over the whole volume (both the disk and the halo), and using the FIR radiation field according to the model of Rice *et al.* (1988), the CMB spectrum, and the basic formulae of Blumenthal & Gould (1970). However, one must take into account that here the FIR radiation field is not uniform over all the volume occupied by the electrons, but weakens with distance from the small central source. Therefore, one has to calculate the so-called radiation dilution factor, which is a measure of anisotropy of the radiation field; outside of the source this factor is simply the fractional solid angle subtended by the source. For a meaningful estimate, the dilution factor must be weighed by the electron distribution.

4.2 RESULTS

For direct comparison with observations, I present the results of this calculation together with the "binned" data of Bhattacharya *et al.* (1994) in Fig. 1. It seems clear from the figure that the Compton flux from the disk is too small (the disk thickness is assumed to be 300 pc, as in Bhattacharya *et al.*), while the total calculated flux can account for the observed emission in the *OSSE* band. However, one must be cautious in drawing conclusions because the model used here is too simple to produce accurate results; it can only serve for the purpose of a rough estimate. Let us briefly discuss the most important sources of

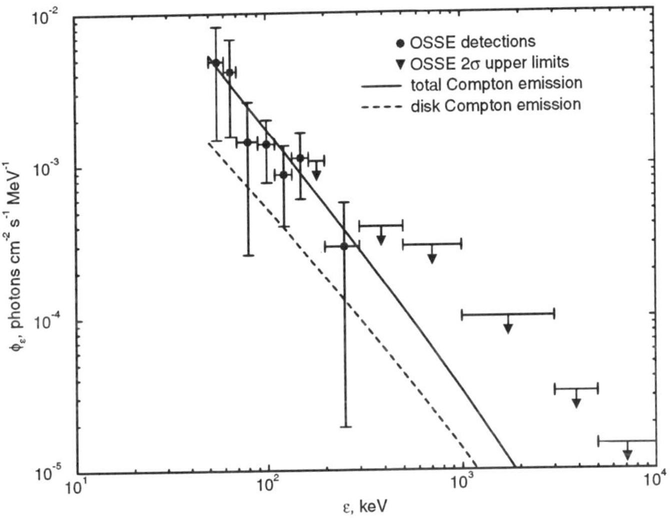

Fig. 1. The Compton flux from NGC253. The filled circles are the data in the energy channels of Bhattacharya et al. (1994) with error bars. The arrows indicate *OSSE* 2σ upper limits. The solid line is the calculated total (disk and halo) emission, and the dashed line is the emission from the disk only.

uncertainties.

As emphasized above, the magnetic field can be different from the adopted "equipartition" value. The dependence of the Compton flux on the field is rather strong, so a large uncertainty is introduced. It should be pointed out that the equipartition value is the only meaningful estimate available, and that the actual field is more likely to be weaker rather than stronger, thus strengthening the conclusion that all of the *OSSE* flux can be accounted for.

The simplifications of the LS model also introduce a substantial uncertainty. It is worth noting, for instance, that the typical "starburst" activity period, which is estimated to be $10^7 - 10^8$ years, is of the order of the energy loss time for electrons of relevant energies, $\tau(E_c) \approx 3 \times 10^7$ yr, so the steady state assumption has to be justified. However, as noted by LS in connection with NGC4631 and NGC891, the model describes the relevant physics reasonably well, and none of the assumptions made is specifically directed at increasing the resultant Compton flux.

The simplified geometry and the dilution factor also lead to uncertainties, especially recalling that the latter is weighed by the LS distribution, which is a simplification in itself. The result may deviate from the real situation in any direction: towards larger or smaller Compton emission.

Finally, the model of Rice *et al.* (1988) for FIR emission is essentially a fit to the *IRAS* data, and is but a rough approximation. Using this particular model can be justified only by the fact that it is the most recent detailed one, based on relatively good data.

Nevertheless, the goal was not to build a detailed model, or to obtain a precise fit to the observations. The available data of Bhattacharya *et al.* (1994) are not of sufficient quality to warrant an attempt to model the Compton emission in a much more realistic (and thus complicated) way. Although all of the above-mentioned uncertainties may be substantial, the calculation clearly shows that, contrary to the initial assertion of Bhattacharya *et al.* (1994), Compton emission from cosmic ray electrons in NGC253 *and its halo* can in principle account for the high-energy X-ray emission detected by *OSSE* . Since no other proposed emission mechanism can provide the necessary luminosity in the *OSSE* range, one can conclude that a Compton origin is the most viable interpretation of the observed emission.

5 Summary

Compton emission from NGC253, and its recently discovered radio halo, is calculated based on the radio emission from the disk and the Lerche & Schlickeiser (1980) model for cosmic ray electrons' escape from the galaxy into the halo. It is shown that the resultant flux can account for the observed high-energy X-ray emission in the *OSSE* range, while emission from the galactic disk comprises only about 30% of the total radiation. The calculated flux is quite uncertain, due to many simplifying assumptions. However, there is no indication that any of these assumptions lead to overestimation, rather than underestimation, of the emission.

If indeed this is the correct explanation of the observed flux from NGC253, then the *OSSE* result is the first detection of truly diffuse (non-AGN) extragalactic Compton emission. It also emphasizes the significance of the escape of cosmic ray electrons from galactic disks, that leads to formation of radio and X-ray halos around galaxies.

This short paper is based on work which will be reported in more detail by Goldshmidt & Rephaeli (1994, in preparation).

References

Allen, R. J., Baldwin, J. E., & Sancisi R.: 1978, *A&A*, **62**, 397

Bhattacharya et al. : 1994, *ApJ*, submitted

Blumenthal, G. R., & Gould, R. J.: 1970, *Rev. Mod. Phys.*, **42**, 237

Boller, Th., Meurs, E. J. A., Brinkman, W., Fink, H., Zimmermann, U., Adorf, H.-M.: 1992, *A&A*, **261**, 57

Carilli, C. L., Holdaway, M. A., Ho, P. T. P., & De Pree, C. G.: 1992, *ApJ*, **399**, L59

Dwek, E., & Arendt, R. G.: 1992, *ARA&A*, **30**, 11

Ekers, R. D., & Sancisi, R.: 1977, *A&A*, **54**, 973

Harper, D. A., & Low, F. J.: 1973, *ApJ*, **182**, L89

Heckman, T. H., Armus, L., & Miley, G. K.: 1987, *AJ*, **93**, 276

Klein, U., Urbanik, M., Beck, R., & Wielebinski, R.: 1983, *A&A*, **127**, 177

Krügel, E., Chini, R., Klein, U., Lemke, R., Wielebinski, R., & Zylka, R.: 1990, *A&A*, **240**, 232

Lerche, I., & Schlickeiser, R.: 1980, *ApJ*, **239**, 1089

MacCarthy, P. J., Heckman, T., & van Breugel, W.: 1987, *AJ*, **93**, 264

Rice W., Lonsdale, C. J., Soifer, B. T., Neugebauer, G., Kopan, E. L., Lloyd, L. A., de Jong, T., Habing, H. J.: 1988, *ApJS*, **68**, 91

Seaquist, E. R, & Odegard, N.: 1991, *ApJ*, **369**, 320

Telesco, C. M., & Harper, D. A.: 1980, *ApJ*, **235**, 392

SPECTRAL AND TEMPORAL VARIABILITY IN LOW-MASS X-RAY BINARIES

NIKOLAOS D. KYLAFIS
kylafis@iesl.forth.gr
University of Crete
Physics Department
714 09 Heraklion, Crete, Greece

ABSTRACT. The Low-mass X-ray binaries (LMXB) are reviewed. Some basic concepts about compact X-ray binaries are presented, the distinction between High-mass and Low-mass X-ray binaries is made and the phenomena observed in LMXB are introduced. An explanation is given as to why the LMXB exhibit Quasi Periodic Oscillations but not strictly periodic ones. The typical spectrum of the LMXB is discussed and the variation of this spectrum with time is given. A unified model for the LMXB is presented and the still unsolved problems are discussed. The recently discovered class of Super Soft X-ray sources is introduced and a possible connection with the LMXB is suggested.

1. Introduction

Taking into account the fact that a fraction of the participants of this school may not be familiar with the subject of compact X-ray sources, I will start with some basic concepts. Also, in view of the limited space, my review cannot be comprehensive. I will therefore refer the reader to detailed reviews of the specific subjects involved, which have exhaustive lists of references.

1.1. BASIC CONCEPTS

Compact X-ray sources are binaries with a normal star and a compact one (i.e., white dwarf, neutron star, or black hole). Here we will concentrate mainly on X-ray sources containing neutron stars.

The X-rays are produced when matter from the evolving companion star falls onto the neutron star. The luminosity resulting from the accreting matter is equal to the gravitational potential at the surface of the neutron star times the mass accretion rate $\dot{M}$ and is given by

$$L = \frac{GM}{R}\dot{M} \approx 10^{38} \left(\frac{M}{M_\odot}\right) \left(\frac{10^6 \mathrm{cm}}{R}\right) \left(\frac{\dot{M}}{10^{18} \mathrm{g/s}}\right) \mathrm{erg/s}, \quad (1)$$

where M is the mass of the neutron star, $M_\odot$ is the mass of the Sun, R is the radius of the neutron star and G is the gravitational constant.

M. M. Shapiro et al. (eds.), Currents in High-Energy Astrophysics, 45–54.

Due to the large gravitational potential of the neutron star the accreting matter reaches very high speeds when it hits the neutron star surface. For a particle of mass m, equating the potential energy at infinity to the kinetic energy at the surface of the neutron star

$$\frac{1}{2}mv^2 = \frac{GM}{R}m, \tag{2}$$

we obtain $v \approx 0.5c$, where c is the speed of light. Thus, effects of Special Relativity are important.

Assuming that the kinetic energy of the accreting matter is absorbed by the neutron star and re-emitted from its surface as blackbody radiation, we have

$$\frac{L}{4\pi R^2} = \sigma T_{BB}^4, \tag{3}$$

where σ is the Stefan-Boltzmann constant and the blackbody temperature of the emitted radiation is

$$T_{BB} \approx 2 \times 10^7 \left(\frac{L}{10^{38}\mathrm{erg/s}}\right)^{1/4} \left(\frac{10^6\mathrm{cm}}{R}\right)^{1/2} \mathrm{K}. \tag{4}$$

This is *not* to say that the spectra of the compact X-ray sources are of blackbody nature. On the contrary, as it will be discussed below. The point that I want to make is that the *minimum* temperature at the surface of an accreting neutron star is of order 10^7 K and therefore accretion onto a neutron star results in emission of radiation in the X-ray (and possibly γ-ray) region of the spectrum.

As the last of these basic concepts I would like to discuss the limiting luminosity from spherically accreting neutron stars. Consider matter falling radially onto a neutron star. At every point of this infall there are two forces acting on the matter: The inward gravitational force and the outward radiation pressure one. At low luminosities the gravitational force dominates. However, there is a luminosity, called Eddington luminosity L_E, at which the gravitational force is balanced by the radiation pressure one. This occurs when

$$\frac{1}{c}\sigma_T \frac{L_E}{4\pi r^2} = \frac{GM}{r^2}m_p, \tag{5}$$

from which one obtains

$$L_E = \frac{4\pi c m_p GM}{\sigma_T} \approx 1.3 \times 10^{38} \left(\frac{M}{M_\odot}\right) \mathrm{erg/s}, \tag{6}$$

where m_p is the mass of the proton and σ_T is the Thomson cross section. It is worth keeping in mind this value of the luminosity when we discuss the LMXB and the Super Soft Sources (SSS) below.

1.2. HIGH-MASS VERSUS LOW-MASS X-RAY BINARIES

Neutron star X-ray sources are divided into two broad categories: The High-mass X-ray Binaries (HMXB) and the Low-mass X-ray Binaries (LMXB). Below we compare their characteristics.

In the HMXB the companion star is typically of high-mass ($M \gtrsim 10M_\odot$) while in the LMXB it is typically of low-mass ($M \lesssim M_\odot$).

The age of the HMXB is consequently short ($\lesssim 10^7$ years) while the age of the LMXB is relatively long ($\sim 10^9$ years).

Most of the HMXB exhibit *periodic* oscillations (PO) in their intensity of frequency ν_* typically less than 1 Hz. Most of the LMXB exhibit *quasi periodic* oscillations (QPO), i.e., in the Fourier transform of the intensity one sees a broad peak, rather than a narrow one that is the signature of periodic oscillations. The frequency ν_Q of the QPO is typically greater than 5 Hz.

The magnetic field at the surface of the neutron star is large (typically $10^{12} - 10^{13}$ gauss) in the HMXB. In the LMXB the magnetic field has not been measured. It is thought to be relatively low (typically $10^8 - 10^9$ gauss).

In the HMXB the optical light of the system is dominated by the companion star, while in the LMXB it is dominated by the accretion disk.

The model of the HMXB is now well accepted (for a recent review see, e.g., Lamb 1989). The companion star, as it evolves, overflows its Roche lobe and spills matter over to the neutron star. Due to its angular momentum, the matter forms an accretion disk around the neutron star. In the disk the particles follow more or less Keplerian orbits because their radial velocities are small. The disk does not extend all the way to the neutron star surface. At some radius the magnetic field disrupts the disk and from there inward the accreting matter moves on field lines and ends up on the magnetic poles. There the kinetic energy is converted into X-rays which are emitted in a beam. As the neutron star rotates with frequency ν_*, and if the magnetic axis is not aligned with the rotation axis, the observed X-ray intensity exhibits *periodic* oscillations with frequency ν_*.

The model of the LMXB has many similarities with that of the HMXB and it will be discussed in detail below.

2. Characteristics of LMXB

The characteristic signatures of LMXB are described below. For detailed reviews see van der Klis (1989, 1991, 1994) and Lamb (1991).

2.1. LUMINOSITY AND SPECTRUM

The typical luminosity of LMXB ranges from moderate ($\sim 3 \times 10^{37}$ erg/s) to high ($\sim 10^{38}$ erg/s).

The typical photon number spectrum of LMXB in the energy range $2 - 30$ keV is described by the function (White *et al.* 1988; Schultz *et al.* 1989)

$$\frac{dN}{dE} = \left[A\left(\frac{E}{E_a}\right)^{-p} \exp(-E/E_a) + \frac{B(E/E_b)^2}{\exp(E/E_b)-1} \right] \exp[-N_H\sigma(E)], \qquad (7)$$

where, N_H is the equivalent column density of Hydrogen to the source and $\sigma(E)$ is the interstellar absorption cross section. The second term on the r.h.s. of equation (7) represents a blackbody component and it is absent in moderate-luminosity LMXB. In high-luminosity LMXB, it contributes at most 10 - 30% of the total flux and E_b is typically ~ 1 keV. The power p in the power-law term is typically 1 - 2, while the high energy cutoff E_a is 10 - 30 keV in the moderate-luminosity sources and 4 - 7 keV in the high-luminosity ones.

The X-ray spectra of LMXB do not depend strongly on the inclination of the system. Furthermore, the hard photons ($E \gtrsim 10$ keV) lag the softer photons ($E \lesssim 3$ keV) by a few msec. Both of these facts suggest that the neutron star is surrounded by a central, spherical, scattering corona.

2.2. VARIABILITY OF INTENSITY

The *quasi-periodic* variability in the intensity of LMXB came as a surprise. A number of people did not expect the bright LMXB to be variable and those that did expect variability, thought that it should be *periodic* with frequency ν_*, the rotation frequency of the neutron star. Thus, it was a search for PO that led to discovery of QPO (van der Klis *et al.* 1985). The QPO frequency ν_Q ranged from ~ 20 to ~ 50 Hz and it was proportional to approximately the square of the intensity. QPO frequencies around 6 Hz and in the interval 10 - 20 Hz have also been observed, but they will be discussed in §2.4.

Soon after the discovery of the QPO, the so called "beat-frequency model" was proposed to explain them (Alpar & Shaham 1985). They made the assumption that the model of the HMXB holds for the LMXB as well. The only difference is that, due to the significantly lower magnetic field in the LMXB, the inner edge of the accretion disk is now closer to the neutron star. They interpreted the frequency of the QPO as the difference between the Keplerian frequency ν_K of the inner edge of the accretion disk and the rotation frequency ν_* of the neutron star, with $\nu_K > \nu_*$. In other words, the accretion rate (and therefore the intensity) onto the surface of the neutron star varies quasi-periodically with frequency ν_Q. As the accretion rate increases, the inner edge of the accretion disk is pushed closer to the neutron star, the frequency ν_K increases and therefore $\nu_Q \equiv \nu_K - \nu_*$ increases. This explains the correlation between QPO frequency and intensity.

If the above picture is correct, then two questions arise. Since the same model is invoked for both HMXB and LMXB, why the HMXB do not show QPO? It took a while, but in the end they were found (Angelini, Stella, & Parmar 1989). This strengthened the belief in the beat frequency model. The second question is the opposite of the first, namely why the LMXB do not show PO at frequency ν_*? There are several ways to explain the lack of PO. For example, the magnetic field near the neutron star surface is dominated by components higher than the dipole. Thus the beaming expected from the dipole magnetic field is significantly reduced. Similarly a reduction in the beaming is expected if the accreting matter falls on a large fraction of the neutron star surface. Here we will investigate the effect of electron scattering on the oscillations of an X-ray source.

2.3. SMEARING OF OSCILLATIONS

Electron scattering has two effects on the oscillations of an X-ray source. It causes time delays in the scattered photons and it tends to isotropize the photons in a beam. Let us investigate each effect separately.

Consider a point source at the center of a spherical cloud of plasma of radius R and optical depth to electron scattering $\tau \gtrsim 3$. The source intensity is varying sinusoidally with period P. This is a simulation of quasi-periodic accretion onto a neutron star with a scattering cloud around it. Since the photons travel through the cloud in a diffusive way, the *mean travel time* is

$$\bar{t} \approx \bar{n}_{sc}\frac{l}{c} \approx \tau^2 \frac{R/\tau}{c} = \tau \frac{R}{c}, \tag{8}$$

where $\bar{n}_{sc} \approx \tau^2$ is the mean number of scatterings and $l \equiv R/\tau$ is the mean free path. If $\bar{t} \ll P$, the time delay of the photons in the cloud is insignificant and the oscillations of the intensity remain essentially unaffected. However, if $\bar{t} \gg P$, the oscillations are essentially completely smeared out. For the general case, if A_0 and A_∞ are the amplitudes of oscillation at the source and at infinity, we have (Kylafis & Klimis 1987; Kylafis & Phinney 1989)

$$\frac{A_\infty}{A_0} = \frac{\sqrt{2}x}{\sqrt{\sinh^2 x \cos^2 x + \cosh^2 x \sin^2 x}} + e^{-\tau} \approx 2\sqrt{2}xe^{-x} + e^{-\tau}, \tag{9}$$

where $x \equiv \sqrt{3\pi\bar{t}/P}$ and the approximate equality is better than 6% for $x > 1.3$.

Consider now at the center of the above cloud a point source which emits a rotating beam in the form of a δ-function of constant intensity I_0. This simulates a rotating neutron star emitting a beam from its magnetic pole and being surrounded by a scattering plasma. If $\hat{n}$ is the direction of the beam and $\hat{n}'$ is the direction of the observer, then the intensity seen by the observer is (Brainerd & Lamb 1987)

$$\frac{I_\infty}{I_0} = \frac{1}{4\pi}\left(1 + \frac{2}{1+\tau}\hat{n}\cdot\hat{n}'\right). \tag{10}$$

Thus, if we consider a beam rotating with period P and the observer is in the plane swept up by the beam, the observed amplitude of oscillations is

$$A_\infty = \frac{2}{1+\tau}. \tag{11}$$

Taking into account the light travel time of the photons in the cloud, the observed amplitude is the product of the expressions given by equations (9) and (11).

It is evident from equations (9) and (11) that, for reasonable values of the parameters (Kylafis & Klimis 1987), electron scattering can smear out the oscillations due to beaming (PO) but leave more or less unaffected the oscillations due to variable accretion rate (QPO).

2.4. VARIABILITY OF SPECTRUM

In §2.2 we examined the time variability of the LMXB without paying attention to their spectrum (i.e., integrating over the spectrum). It would be very nice to know how the spectrum varies with time. Unfortunately, since we are looking for millisecond variability, it is not possible to know the spectrum because only a few photons reach the detector every millisecond. Thus, instead of the spectrum, the photons are recorded in three channels: 1 - 3 keV, 3 - 6 keV and 6 - 17 keV. In a way similar to color-color diagrams in optical astronomy, the ratios of the intensities in the three channels are plotted in Figure 1 for Cyg X-2 (Schulz, Hasinger & Trümper 1989).

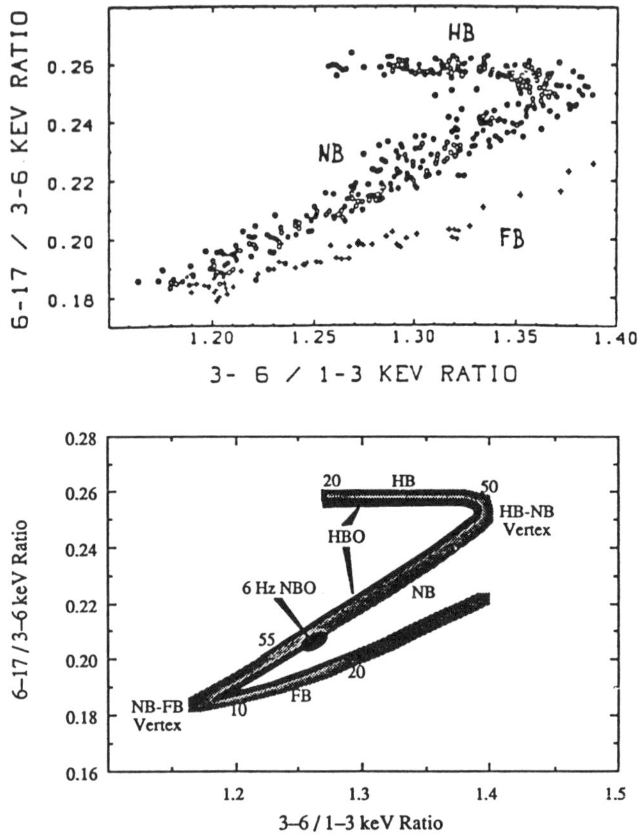

Figure 1.—X-ray color-color diagrams. Top: Cyg X-2 data from Schulz, Hasinger & Trümper (1989) showing the horizontal branch (HB), normal branch (NB) and flaring branch (FB). Bottom: A schematic Z-track from Lamb (1991) showing the connection between X-ray spectral state and QPO frequency.

The X-ray color-color diagram in Figure 1 exhibits a Z-shaped curve with three branches, the horizontal branch (HB), the normal branch (NB) and the flaring branch (FB). Similar curves have been observed from other LMXB.

No source has been seen to jump from one spectral branch to another. All of them move *along* the Z-track. Motion of a source from left to right across the HB corresponds to an increase in the size of the blackbody term in expression (7) with the other parameters remaining approximately constant. Motion down the NB corresponds to a decrease in the cutoff energy E_a with the other parameters approximately constant. Finally, motion up the FB can be reproduced by a monotonic increase in E_a, B and E_b. There are indications that the accretion rate and the bolometric luminosity increase as the sources progresses from the HB, to the NB and then to the FB.

In the HB the QPO frequency increases from 20 - 50 Hz while its strength decreases. In the NB, approximately half way down, the QPO is weaker and its frequency has increased to 55 Hz. In addition, a second QPO frequency of about 6 Hz is evident in the power spectrum. In the bottom half of the NB, the 55 Hz QPO has disappeared while the 6 Hz one has increased its frequency to about 10 Hz. A further increase to about 20 Hz is observed in the FB.

2.5. UNIFIED MODEL FOR LMXB

A unified model for the LMXB has been proposed and is currently widely accepted (Lamb 1991, 1994). In this model, the X-ray source is a neutron star with a relatively weak magnetic field ($B \sim 10^8 - 10^9$ G) accreting matter from a disk fed by the companion star. The disk is disrupted by the magnetic field of the neutron star at $\sim 15 - 20$ km. The energy to power the X-ray emission comes from release of the gravitational potential energy of the accreting matter as it approaches the neutron star.

The accretion flow pattern depends on the luminosity. At moderate luminosities ($\sim 0.5 - 0.9 L_E$), the outward momentum of the radiation escaping from the inner disk and neutron star causes the inner disk to thicken, forming a hot *central corona* around the small magnetosphere of the neutron star (see top panel of Figure 2). The electron scattering optical depth of the central corona is of order 5. As the luminosity increases from $0.5 L_E$ to $0.9 L_E$, the source moves from left to right on the HB (Fig. 1). The HB QPO with frequency $20 \lesssim \nu_Q \lesssim 50$ Hz is explained by the "beat frequency model".

At luminosities $\gtrsim 0.9 L_E$, the pressure of radiation escaping from the inner disk drives some plasma into a more extensive *inner disk corona* surrounding the central corona (see middle panel of Figure 2). Radiation drag causes plasma in the inner disk corona near the neutron star to lose its angular and vertical momentum and to fall approximately radially toward the neutron star. As the luminosity increases from $0.9 L_E$, the source traces down the NB (Fig. 1). The 55 Hz QPO is still explained by the "beat frequency model". However, the $\sim$ 6 Hz QPO is of a completely different nature. As the luminosity approaches L_E, the radiation pressure slows down the radially accreting matter. At constant luminosity the radial flow profile is constant. However, if the luminosity is, say, within a few % of L_E and increases slightly, the effect of the radiation pressure on the infalling matter is large. The inflow velocity is significantly reduced and the rate at which the matter reaches the neutron star surface is also reduced. This has as a consequence a decrease in the luminosity, which in turn implies a reduced radiation pressure and an enhanced accretion rate and luminosity. The cycle repeats and it gives rise to the $\sim$ 6 Hz QPO ($\nu_Q \approx 2/t_f \approx 6$ Hz), where t_f is the flow timescale.

When the luminosity becomes larger than L_E, non-axisymmetric photohydrodynamic modes develop in the slow radial flow near the magnetosphere. Growth of these modes segregates the accreting plasma and outflowing radiation as shown in the bottom panel of Figure 2. As the luminosity increases beyond L_E the source moves up on the FB (Fig. 1).

A theoretical reproduction of the Z curve in Figure 1 has not been accomplished yet. When it does, our confidence to the unified model described above will be significantly increased.

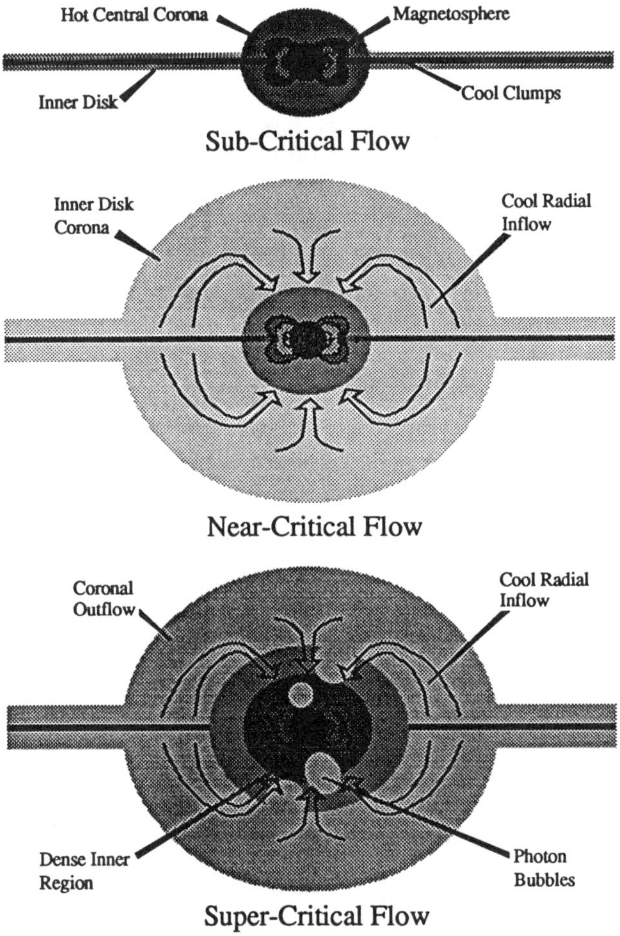

Figure 2.—Schematic side views of the accretion flows from Lamb (1991). The top panel represents the accretion flow at luminosities $\lesssim 0.9L_E$. *The middle panel represents the flow at luminosities near* L_E *and the bottom one at luminosities greater than* L_E.

3. Super Soft X-ray Sources

A new class of X-ray sources was recently discovered and from their extremely soft spectra they were named Super Soft Sources (SSS) (for a recent review see Hasinger 1994). These objects have luminosities around 10^{38} erg/s and spectra fitted with blackbody temperatures of order 30 eV.

A white dwarf model was proposed for their explanation (van den Heuvel *et al.* 1992). In this model, the supersoft spectrum is arising from *steady* nuclear burning of hydrogen accreting onto massive white dwarfs. In view of the fact that among the known SSS there are at least four systems in which the compact object is believed to be a white dwarf, there is no doubt that this model explains at least some of the SSS.

The orbital periods of some SSS, their observed aperiodic variability, and the similarity of their optical spectra with those of the LMXB, led us (Kylafis & Xilouris 1993, 1994) to investigate the possibility of a neutron-star model for the SSS. The requirement is that the photosphere of the accreting neutron star must be at a few thousand neutron star radii so that the blackbody temperature is in the tens of eV range.

We have performed detailed numerical calculations of sub-Eddington spherical accretion onto neutron stars and computed the emergent spectra. We have found that the supersoft X-ray spectra can be explained as the result of near-Eddington accretion onto neutron stars. Our model is consistent with the unified model for the LMXB described above. What determines whether the source appears as a "canonical" LMXB or as a SSS is the extent of the radial flow and its character (supersonic or subsonic). If the luminosity of the source is within 10% of L_E and the extend of the flow is at least $3 \times 10^3 R$, one can always find flows for which the emergent spectrum is a blackbody of temperature in the tens of eV range. If the extent of the radial flow is at most a few hundred neutron star radii, the source behaves as a "canonical" LMXB.

The discovery (Hughes 1994) of a very luminous transient X-ray pulsar in the Small Magellanic Cloud with a supersoft X-ray spectrum proves beyond any doubt that at least some of the SSS are accreting neutron stars.

6. References

Alpar, M. A. & Shaham, J. 1985, Nature, 316, 239

Angelini, L., Stella, L., & Parmar, A. N. 1989, ApJ, 346, 906

Brainerd, J. & Lamb, F. K. 1987, ApJ, 317, L33

Hasinger, G. 1994, in *The Evolution of X-ray Binaries*, 4th Annual Maryland Conference, October 1993, in press

Hughes, J. P. 1994, ApJ, 427, L25

Kylafis, N. D. & Klimis, G. S. 1987, ApJ, 323, 678

Kylafis, N. D. & Phinney, E. S. 1989, in *Timing Neutron Stars*, ed. H. Ögelman and E. P. J. van den Heuvel (Dordrecht: Kluwer), p. 731

Kylafis, N. D. & Xilouris E. M., 1993, A&A, 278, L43

——— . 1994, in *The Evolution of X-ray Binaries*, 4th Annual Maryland Conference, October 1993, in press

Lamb, F. K. 1989, in *Timing Neutron Stars*, ed. H. Ögelman and E. P. J. van den Heuvel (Dordrecht: Kluwer), p. 649

——— . 1991, in *Neutron Stars: Theory and Observation*, ed. J. Ventura and D. Pines (Dordrecht: Kluwer), p. 445

——— . 1994, ApJ, in press

Schulz, N. S., Hasinger, G., & Trümper, J. 1989, A&A, 225, 48

van den Heuvel, E. P. J., Bhattacharya, D., Nomoto, K., & Rappaport S. A. 1992, A&A, 262, 97

van der Klis, M., Jansen, F., van Paradijs, J., Lewin, W. H. G., van den Heuvel,

E. P. J., Trümper, J., & Sztajno, M. 1985, Nature, 316, 225
van der Klis, M. 1989, Ann. Rev. Astr. Ap., 27, 517
———. 1991, in *Neutron Stars: Theory and Observation*, ed. J. Ventura and D. Pines (Dordrecht: Kluwer), p. 319
———. 1994, in *Lives of Neutron Stars*, ed. A. Alpar and J. van Paradijs (Dordrecht: Kluwer), in press
White, N. E., Stella, L., & Parmar, A. N. 1988, ApJ, 324, 363

SOLAR AND SUPERNOVA NEUTRINOS

Todor Stanev
Bartol Research Institute
University of Delaware
Newark, DE 19716, U.S.A.

1. INTRODUCTION

Ever since it was realized in the 1920's that nuclear fusion reactions are the only viable source that could power the Sun for a time compatible with the age of the oldest rocks found on Earth, the production and emission of neutrinos is closely related to the stellar evolution. During the time when a star is on the main sequence, i.e. follows the most common evolution pattern, neutrinos are constantly emitted by the same fusion reactions, that provide the energy for its luminosity. For all ordinary stars, similar to the Sun, the emitted neutrino fluxes are almost directly proportional to the stellar luminosity. A very simple estimate gives the magnitude of the neutrino flux from the Sun. If we assume that the first fusion reaction that puts together protons to create a deuteron,

$$p + p \rightarrow {}^2H + e^+ + \nu_e + 1.44\text{MeV} \tag{1}$$

is the only energy source of the Sun, we can calculate the flux of ν_e at Earth. The total amount of energy created by this reaction is $\sim$ 10 MeV. The reaction rate is then

$$\text{Rate} = L_\odot/10\text{MeV} = 2.5 \times 10^{38} s^{-1}, \tag{2}$$

and the neutrino flux F_ν

$$F_\nu \simeq \frac{2.5 \times 10^{38}}{4\pi D^2} \simeq 10^{11}\text{cm}^{-2}\text{s}^{-1}. \tag{3}$$

This number is surprisingly close to the real flux of solar neutrinos.

In the end of the stellar evolution, when the star has run out of nuclear fuel, it may produce one last spectacular display: **a supernova** explosion. The core of the star collapses to a neutron star (or a black hole). For that to happen, stars have to radiate all the binding energy of its constituent nuclei in a short powerful burst. Neutrinos, with their small cross–section for interaction on matter, are the only known particles which can penetrate through the ultra–dense matter in a neutron star and relieve the forming neutron star of that extra energy. The spectacular supernova explosion is accompanied by a much more quiet, but also much more luminous neutrino burst.

M. M. Shapiro et al. (eds.), Currents in High-Energy Astrophysics, 55–67.

In two lectures at this years Course on Cosmic Ray Astrophysics I introduced these important topics. Solar neutrinos and supernova neutrinos have been now observed by various detectors. Here I will briefly review the physics behind the neutrino production in the stellar interior, the observational techniques, and the main features of the observations and their interpretation.

2. SOLAR NEUTRINOS – GENERATION.

The current standard theory of the neutrino production in the Sun is developed mostly by J.N. Bahcall and his collaborators. It is presented in detail in his book "Neutrino Astrophysics", which I shall follow for most of the discussion.

The main energy source of the Sun, at its current evolutionary stage, is the *pp* cycle, which fuses protons into ^{4}He nuclei. The neutrino producing reaction of the *pp* cycle are:

1. $p + p \rightarrow {}^2H + e^+ + \nu_e$
2. $p + e^- + p \rightarrow {}^2H + \nu_e$
3. ${}^7Be + e- \rightarrow {}^7Li + \nu_e$
4. ${}^8B \rightarrow {}^8Be + e^+ + \nu_e$
5. ${}^3He + p \rightarrow {}^4He + e^+ + \nu_e$

The jargon for neutrinos from these processes is pp, pep, Be, B, hep neutrinos. The maximum energy for *pp* neutrinos is 0.42 MeV, for *pep* – 1.44 MeV, for *B* – 15 MeV, and for *hep* – 18.8 MeV. *Be* neutrinos are emitted in two lines of 0.383 and 0.861 MeV. There are 6 other reactions in the *pp* cycle. Although they do not directly generate neutrinos, they are very important for the calculation of the neutrino fluxes since they generate the parent nuclei.

The calculation itself is a complicated iterative procedure which consists of chain of successive stellar evolution calculations. The whole procedure is called Standard Solar Model (SSM). The iterations continue until a satisfactory agreement with all measured basic solar parameters is achieved. These include the solar luminosity ($L_\odot$ = 3.86×10^{33} ergs/s), mass ($M_\odot = 1.99\times10^{33}$ g), age (4.55×10^9 yrs), as well as much less obvious data, such as the strength of the *p* modes of solar oscillations. 0.984 of the solar neutrino luminosity is produced by the *pp* cycle, the rest come from the *CNO* cycle, which fuses He nuclei into medium heavy nuclei. SSM gives detailed information about the fluxes and the energy spectra of neutrinos from different channels and the regions of the Sun where they are generated. The neutrino flux from the *pp* reaction is 6×10^{10} cm^{-2}s^{-1}.

3. DETECTION OF SOLAR NEUTRINOS

There are two basic methods for detection of low energy neutrinos – radiochemical method, which uses inverse β–decay, and one based on neutrino–electron interactions.

The radiochemical method utilizes such nuclei A(N,Z), which will convert to unstable A'(Z+1,N-1) + e^- after interaction with ν_e. The decays of the unstable A' will then be counted to determine the rate of solar neutrino interactions during the exposure. Each such reaction has a characteristic threshold, which determines the sensitivity of the measurement to different solar neutrino species. The first solar neutrino experiment was realized in the Homestake gold mine in South Dakota by R. Davis Jr. and collaborators. It is based on the reaction $\nu_e + {}^{37}Cl \rightarrow {}^{37}Ar + e^-$ and has a threshold of 0.814 MeV. It cannot thus detect *pp* neutrinos. The Homestake chlorine tank contains 615 tons of C_2Cl_4 which are usually exposed for 1–2 halflives of ^{37}Ar, which is 35 days. A longer exposure does not yield higher signal because of the ^{37}Ar decay. The produced radioactive nuclei are then extracted in a complicated chemical process and the number of decays counted for about a year – 10 $\tau_{1/2}$. Such long counting period is required for a precise determination of the background. Low radioactivity counters of special design are used, which have incidental rates lower than 0.1/day. 'Signal' counts have to have energy within the width of the well known ^{37}Ar decay line (2.82 KeV). The number of counts as a function of time has also to be consistent with the halflife of the isotope. Because of the small number of signal counts this last requirement is difficult to follow on a single exposure, but is certainly satisfied for larger statistics. The Homestake detector is working reliably for more than 23 years and has first established that the observed neutrino rate is lower than the SSM predictions by about a factor of 3 (Davis, 1993).

Two currently running experiments, Gallex (Anselmann *et al.* 1994) and SAGE(Gavrin *et al.*, 1994), are based on the reaction $\nu_e + {}^{71}Ga \rightarrow {}^{71}Ge + e^-$. It has several important features, that are different from the chlorine reaction. The threshold energy is 0.232 MeV, which allows the detection of *pp* neutrinos. The cross–section is higher, thus the experiment requires less target material. The halflife of germanium is 11.4 days, which allows for a shorter exposure and counting cycle. Currently Gallex exposes 30 tons of Gallium and SAGE – 57 tons. The basic difference between the two experiments is the extraction procedure, with Gallex' chemistry being more simple.

The electron scattering reaction is $\nu_e + e^- \rightarrow \nu_e + e^-$, where the scattered electron carries a large fraction of the neutrino energy and moves close to the initial neutrino direction. The reaction cross–section is 9.3×10^{-45} $cm^2\times$ $(E_\nu/MeV)^2$. The biggest detector is the water Cherenkov detector in Kamiokande(Hirata *et al.* 1991) which contains 3000 tons of water. The walls of the big water tank are covered with photomultiplier tubes, which detect the Cherenkov radiation of the scattered electron. The timing and the amplitude of the detected signals allow for the reconstruction of the direction and energy of the electron track. The Solar neutrino signal in Kamiokande appears in the form of an excess of events in the anti–Sun direction. The detection threshold was set to 7.5 MeV, i.e. only *B* neutrinos were detected.

4. NEUTRINO FLUXES OBSERVED FROM THE SUN

Table 1 gives the expectations from several different SSM and the fluxes detected by the four experiments. For the radiochemical detector the values given are in SNU (Solar Neutrino Unit, 10^{-36} ν captures per target atom per second). For Kamiokande all predictions and the observations are scaled to the predictions of the SSM of Bahcall and Ulrich (1988).

Obviously all experiments see a smaller number of neutrinos, than are predicted by any of the models. The Ga experiments are detecting enough neutrinos to account for the basic *pp* reaction, which should contribute about 70 SNU. It is most likely then that the deficit is due to the emission of higher energy neutrinos through other channels, mostly *Be, B* and *CNO* neutrinos. The basic theory that nuclear reactions are the source of energy is confirmed. There is, however, a puzzling disagreement between the results of Homestake and Kamiokande. Homestake should observe mostly *B* neutrinos (6.2

Table 1. Predictions and observations of solar neutrino fluxes.

	^{37}Cl	^{71}Ga	Kamiokande
Bahcall&Pinsonneault	8.0 ± 1.0	132.0 ± 7.0	1.00
Turck–Chieze&Lopes	6.4 ± 1.4	123.0 ± 7.0	0.75
Shi&Schramm	$4.5 \pm$??	$114.0 \pm$??	0.54
Dar&Shaviv	$3.9 \pm$??	???	0.49
Experiment	2.3 ± 0.3	79±12(Gallex) 73±17(SAGE)	0.50 ± 0.08

SNU according to Bahcall and Ulrich (1988)), with a small contribution from *Be, pep, N,* and *O* neutrinos. Kamiokande should be only sensitive to *B* neutrinos. The data shows instead that the flux at Homestake is suppressed by a factor of 2.8, while Kamiokande measures 0.5 of the predicted values. An easy way out would, of course be to say that experimentalists are wrong, have lower detection efficiency than advertized, or that the extraction procedures are somehow flawed. Indeed none of the three experiments have been calibrated under conditions similar to the solar neutrino detection. A calibration of Gallex is scheduled in the fall of 1994. The same procedure will eventually be applied to SAGE, but the production of an isotope, suitable for calibration of the ^{37}Cl experiment is almost impossible. Apart from direct calibration, all three radiochemical groups have made all possible checks of the extraction and counting efficiencies.

There are two possible sets of reasons for the small number of solar neutrinos: astrophysical solutions, that require altering of the inputs in the SSM, and particle physics solutions, that assume that SSM are correct, and the difference is due to particle physics processes taking place after the neutrinos are generated in the Sun.

4.1 Astrophysical Solutions.

There are several possible astrophysical solutions and two very recent papers (see entries for SS (Shi and Schramm, 1994) and DS (Dar and Shaviv, 1994) in Table 1) claim that they can decrease the expectations without altering the general picture of the Sun. The most general solution is

Lowering the central temperature of the Sun. Since the production of B, Be neutrinos is concentrated in the core of the Sun, they are mostly affected by small changes in T_c. A small change in T_c would decrease the cross–section for the *B* and *Be* reactions and correspondingly the expected neutrino flux. Bahcall and Ulrich (1988) has shown that the sensitivity to T_c is very high, T_c^{-8} for the *Be* neutrinos and even higher for the *B* ones. The tolerance for changing T_c is, however, extremely small and even changes as small as 1% are difficult to reconcile with other calculated solar parameters. Shi and Schramm (1994), however, decreased T_c by 1.4% together with a change of the

Primordial composition of the Sun. They took the fraction of heavy elements at zero age to be 0.0150 instead of 0.0196 in Bahcall and Ulrich (1988). The change of the composition affects strongly the opacity of the Sun, and thus the rates of different reactions and their neutrino yields.

Fusion cross–sections are amongst the most suspected ingredients of the SSM. The reason is that all laboratory measurements are made at energies exceeding 100 - 200 KeV, while T_c corresponds to less than 15 KeV. All cross-sections are subject of very strong energy dependence, and extrapolations at more than one order of magnitude have to be made. Among all reactions the one that generates ^{8}B is the least known. At this time many heated discussion deal with quality of the experimental data and the extrapolation techniques to low energy. Dar and Shaviv (1994) use 70% of the Bahcall and Ulrich value for the (^{7}Be + p) process and achieve a significant reduction of the B neutrino flux.

The predictions of the Solar neutrino flux, even if they are within a factor of 2 away from some experimental values, are a great theoretical success. In his book J.N. Bahcall has documented the changes in the solar ν predictions as a function of the changing input parameters during the last 30 years. It appears now possible that further experimental improvements could bring predictions in a better agreement with experiment. It does not look like the whole story, however. Even the extreme values of Dar and Shaviv (1994) are in quite of a discrepancy with the Gallex, SAGE and Homestake results. The solution might very well be a part of a different set.

4.2 Particle Physics Solutions.

A large class of particle physics solutions are based on the possibility that neutrinos have small masses, and thus an observed neutrino type is a combination of two basic neutrino eigenstates. The probability that a neutrino of flavor e oscillates into a neutrino of flavor μ, e.g. depends on the strength of the coupling ($\sin^2(2\theta)$ where θ is the mixing angle, and on the oscillation length L which is a function of the mass difference squared (Δm^2).

It is now very difficult, if not at all impossible, to reconcile the experimental results with the classical SSM (Bahcall, 1989; Turck–Chieze *et al.*, 1993) using vacuum neutrino oscillations. Another possibility was inspired by that disagreements and the theory of neutrino oscillations in the presence of matter (MSW effect (Wolfenstein, 1979; Mikheev and Smirnov, 1986) was developed during the last ten years. Electron neutrinos of given energy can undergo a resonant transition to a different flavor, presumably muon neutrinos, in the presence of electrons (matter). The position of the resonance depends on the electron density and neutrino energy, and the changing density of the Sun makes the resonant transition possible for all neutrino energies of interest. Reconciling all observations is still difficult, but possible. There are two allowed regions on the oscillation parameters plot: one at small coupling ($\sin^2(2\theta) \sim 10^{-2}$, $\Delta m^2 \sim 10^{-5}$, and one at large coupling and slightly higher values of Δm^2. Such an oscillation model is favored by most weak interaction theorists as a solution of the solar neutrino problem.

If neutrinos have a magnetic moment, their spin can 'flip' as a result of interaction with the solar magnetic field. In such a case ν_e could turn into neutrinos of different type, that does not couple to matter and thus cannot be detected. This possibility was suggested by Voloshin, Vysotskii and Okun (1986). The flip probability is proportional to the product of the neutrino magnetic moment and the strength of the solar magnetic field, presumably in the convection zone. An explanation of the reduced fluxes of B neutrinos requires $\mu_\nu B_\odot$ of O(10^{-7} μ_BGauss). The problem is that the standard model gives neutrinos a much smaller magnetic moment, of order

$$\mu_{\nu_e} \sim 10^{-19}(m_{\nu_e}/1\text{eV})\mu_B. \tag{4}$$

The laboratory limits constrain the magnetic moment of electron neutrinos to 10^{-10} μ_B and the astrophysical constraints are somewhat stronger – 10^{-12} μ_B. This would then require a solar magnetic field of O(10^5 Gauss) or the construction of a non–standard model that generates high neutrino magnetic model.

A beautiful theory was then developed that combines the two effects – matter enhanced spin flip (Akhmedov and Khlopov, 1988). The assumption is that neutrinos have both masses and magnetic moments that are higher than the standard value. A resonance similar to MSW occurs during propagation in the solar interior and the type of transition depends on the origin of the neutrinos. Dirac neutrinos flip into righthanded neutrinos of all flavors. Majorana neutrinos flip into antineutrinos, and even into $\bar{\nu}_e$ for large mixing angles. $\bar{\nu}_e$ would interact on the protons in the Kamiokande detector and such an effect could bring the SSM and observations in agreement. The matter enhanced spin flip is also energy dependent, i.e. can only work in a restricted energy range, which gives it one more degree of freedom in the choice of parameters. The whole scheme, however, is based on large number of non–standard assumptions.

It is yet impossible to say what the correct explanation of the solar neutrino deficit is. If the suspected correlation of the ^{37}Cl neutrino signal with the solar activity (Davis 1993; Bahcall, Field and Press, 1987; Filippone and Vogel, 1990) and with the magnetic field in the convection zone (Bieber *et al.* 1990) is confirmed in future observations, it would indicate some type of interaction with the solar magnetic field. The solution obviously requires more experimental data, which will be coming in just a couple of years, when the much bigger Superkamiokande (30 Kt of H_2O), and the very sensitive SNO detector (1 Kt of D_2O) will be operational. It is also very important to keep the existing radiochemical detectors in operation, and to make independent measurements of the B neutrino flux. There are no doubts that these experimental developments will be accompanied by corresponding theoretical efforts.

5. SUPERNOVA NEUTRINOS: EXPECTATIONS

Stellar evolution goes smoothly until the nuclear fusion proceeding in the core of the star becomes unable to produce enough energy and support the star from gravitational collapse. The condition for a stable star is the existence of hydrostatic equilibrium. Everywhere inside the star the radiative and thermal pressure is compensated by the gravitational attraction.

$$\frac{dP(r)}{dr} = -\frac{GM(r)\rho(r)}{r^2}, \tag{5}$$

where $M(r)$ is the enclosed mass. When the Hydrogen in the stellar core is exhausted, the energy of the star is replenished by fusion of heavier nuclei. After the *pp* cycle comes the

CNO cycle, then CNO nuclei are fused to Si, and finally Si nuclei are fused to produce Fe. All these processes generate energy and support the hydrostatic equilibrium. The central temperature of the star, T_c, increases. After the Si fuel is exhausted the nuclear fusion can not proceed anymore, because the fusion of Fe nuclei does not release energy, it absorbs energy. The hydrostatic equilibrium does not exist anymore. T_c at that time is ~0.7 MeV and $\rho_c \sim 10^{10}$ g/cm^2.

When the mass of the core M_c exceeds the Chandrasekhar mass M_{Ch}

$$M_{Ch} \;=\; 5.8 \times Y_e^2 M_\odot \;\sim\; 1.4 M_\odot \tag{6}$$

the pressure in the core starts decreasing because the conditions are ripe for two new processes to occur:

Electron capture – $e^- + \mathrm{p} \rightarrow \mathrm{n} + \nu_e$. This is important, because the electron thermal motion is the prevailing form of pressure.

Photodisintegration of Fe, which requires energy.

The core collapses, separates itself from the rest of the star and turns itself into a neutron star or a black hole. The 'deleptonization' ν_e's from the e^-p reaction are the first sign of the collapse. A core of of $1.4 M_\odot$ contains 10^{57} protons. Since the average energy per ν_e is ~10 MeV, the total energy released in neutrinos while all protons are turning into neutrons is $\sim 10^{52}$ ergs. The process feeds on itself, because the central density increases and the electron capture speeds up. The stellar core quickly turns itself into a neutron star with typical radius of 10 km.

When the density approaches 10^{12} g/cm^3 the column density of the proto–neutron star is already so big, than neutrinos cannot escape without scattering. This is estimated to happen $\sim$ 1 s after the start of the collapse. The collapse continues however until nuclear densities of 10^{14} g/cm^3 are reached. The shock bounces back, blows out the stellar envelope and the spectacular stellar explosion starts. At this time the core is still hot and lepton rich – about 40% of the electrons of the original core are still not captured. The star has to also radiate away its binding energy, which is 2–4×10^{53} ergs. The cooling proceeds through the emission of $\nu\bar{\nu}$ pairs of all flavors, created in thermal equilibrium with $T_c \sim 200$ MeV. Such neutrinos have large enough cross-sections, that they can leave the neutron star only after a large number of scatterings during which they cool down. $\nu_e(\bar{\nu}_e)$ undergo $\sim 10^4$ scatterings and leave the neutron star with an average energy of 10–20 MeV. $\nu_\mu, \bar{\nu}_\mu, ...$ have smaller cross–sections, have correspondingly $\sim 10^3$ scatterings and have slightly higher energy. It is easy to show (Schramm, 1987) that the neutrino energy depends very slowly on the central temperature as $T_c^{1/5}$ Most of the binding energy of the star is irradiated in neutrinos in 10 s, although the neutrino emission continues slightly longer. The neutron star is complete when not more than 5% of the original number of electrons remains (Burrows and Latimer, 1986; Woosley, Wilson and Mayle, 1986).

Because of the electron capture, the 'deleptonization' ν_e are emitted in slightly greater numbers. The energy carried by the different neutrino types is distributed as

$$E_{\nu_e} \;:\; E_{\bar{\nu}_e} \;:\; E_{\nu_\mu} = 1.2 \;:\; 1 \;:\; 4, \tag{7}$$

where ν_μ also includes $\bar{\nu}_\mu$, ν_τ and $\bar{\nu}_\tau$.

6. Supernova neutrinos – Observations

Although the letters of the alphabet are often not enough for designation of all detected optical supernovae (the standard is the year + a letter in the order of detection during the year), most of them happen in distant galaxies and the neutrinos cannot be detected by existing detectors. So when SN1987a exploded in the Large Magelanic Clouds, a small neighbour galaxy only 50 Kpc away, it was for the first, and still only time when we had the chance to examine experimentally the theory of stellar collapse.

There are four published accounts for the observation of the neutrino signal from SN1987a:

- The Kamiokande detector (Hirata *et al.* 1987), using 2.1 Kt of water target as a fiducial volume, reported the detection of 12 neutrino events in as many seconds on 07:35:55 UT, 23 February 1987. The chance probability of such an event is one in 10^5 years. The detection occurred $\sim$3 hours before the optical detection of the supernova by Ian Shelton. These events are shown with circles on the energy – time plot on Fig. 1. Nine of twelve events are seen in less than 2 s. A gap of 7 s is followed by another three events. Subsequent analyses challenge one of the events, which is much too low in energy and could be a part of the background.

- The IMB detector (Bionta *et al.* 1987), with a total volume of 10 Kt, but with a significantly higher energy threshold of 20 MeV observed simultaneously seven neutrino events. Part of the reason for the small number of events is that only a fraction of the detector was active because of a power supply failure. Both detectors were built and run for the purpose of a detection of proton decay. For this reason they did not maintain exact synchronized clocks. The time coincidence, shown on Fig. 1 is only precise within 1 minute. Still the chance probability for any of these detections is so low, that there is no doubt in their actual time coincidence.

- The Baksan detector (Alekseev *et al.* 1987), consisting of scintillator tanks (200 Kt available for SN detection) reported 5 events of energy between 12 and 23 MeV. The exact time of the observation is difficult to assess, but in all accounts it is within a minute of the Kamiokande/IMB detection. This observation is sometimes dismissed on the basis that the neutrino rate is much higher than expected and detected by the water Cherenkov detectors. The Baksan scintillator contains 2×10^{31} free protons versus 1.4×10^{32} in Kamiokande. One roughly expects to see a signal, which is smaller by a factor of 7, i.e. less than 2 events. Generally the observations by water Cherenkov detectors are considered more reliable, since they consist of amplitude and timing data from many phototubes, which allow the reconstruction of the energy and direction of the event. In contrast, the scintillator tanks give only one number – the amplitude of the detected signal, which distinguishes a signal from the background.

- The Large Scintillator Detector in the Mont Blanc tunnel reported an outburst of five events in five seconds (Aglietta *et al.* 1987) about 4.7 hours before the Kamiokande/IMB detection. This result has been very contraversial, because it does not fit in with the other observations, or with the standard collapse theory (de Rujula, 1988). The chance probability for such a detection by LSD is very small. No neutrinos were detected at the time of the Kamiokande/IMB burst, but not more than 1 event is expected. There were reports for a coincidental signals in gravity waves.

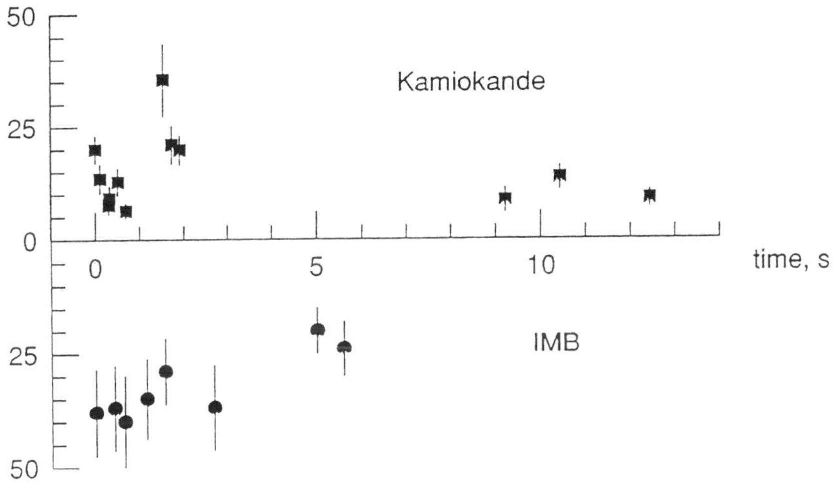

Fig. 1. Energy – time plot of the neutrinos detected by Kamiokande and IMB. Note the higher energy threshold of the IMB detector.

In the further discussion we shall concentrate on the results from IMB and Kamiokande. Both detectors use the same detection technique and give both energy and direction for every event.

In water neutrinos can interact on electrons, free protons and O nuclei. Neutrino – electron interactions (same as used in the detection of solar neutrinos) have cross-sections proportional to the neutrino energy and strengths between 10^{-44} and 10^{-45} cm^2 for 1 MeV neutrinos of different flavor. The electrons are scattered forward, i.e. approximately track the neutrino direction. The highest (and the fastest growing) cross–section process is $\bar{\nu}_e$ + p $\rightarrow$ n e$^+$ – $\sigma = 7.5\times10^{-44}$(E/MeV) cm^2. The angular distribution of the emitted positron is isotropic, does not indicate the direction of the primary neutrino. In ν_e^{16}O scattering an electron is emitted backwards, and the cross–section is a factor of 10 lower. Because of these cross-section differences, the signal in water cherenkov detectors is expected to be dominated by $\bar{\nu}_e$ scattering.

$$S_{\nu_e} \; : \; S_{\bar{\nu}_e} \; : \; S_{\nu_\mu} \; = \; 2 \; : \; 40 \; : \; 1, \tag{8}$$

because charge current interactions of the higher neutrino flavors cannot proceed at that energy range. An inspection of the signal in Kamiokande and IMB shows a distribution consistent with isotropic, with maybe a slight excess of 1$\pm$1 neutrino in backward direction.

An interesting question was asked during the meeting in Erice: since a Galactic supernova is likely to be obscured by intervening matter and not optically seen, is it possible to use the timing of the neutrino burst to locate its position? Not very accurately. The situation is very similar to γ–ray bursts. With three well separated detectors of Kamiokande size, one could triangulate the burst position to within several degrees, if the detectors are equipped with μs GPS synchronized clocks.

6.1 Supernova properties from the neutrino burst observations.

Even with a small number of observed events, one can attempt to analyze the observations in terms of neutrino flux and energy. Fig. 2 shows my own attempt to reconstruct the neutrino energy spectrum from the Kamiokande and IMB observations. All events were assumed to be $\bar{\nu}_e$p scatterings. Every event was scaled with the cross-section and the corresponding detector efficiency. The supernova neutrinos, reconstructed in this way, were then grouped in energy bins and compared to Plank spectra of different temperatures. Fig. 2 shows that this is not easy to do. I was only able to fit the observations with a two temperature spectrum containing T_1 = 1.5 MeV and T_2 = 5 MeV.

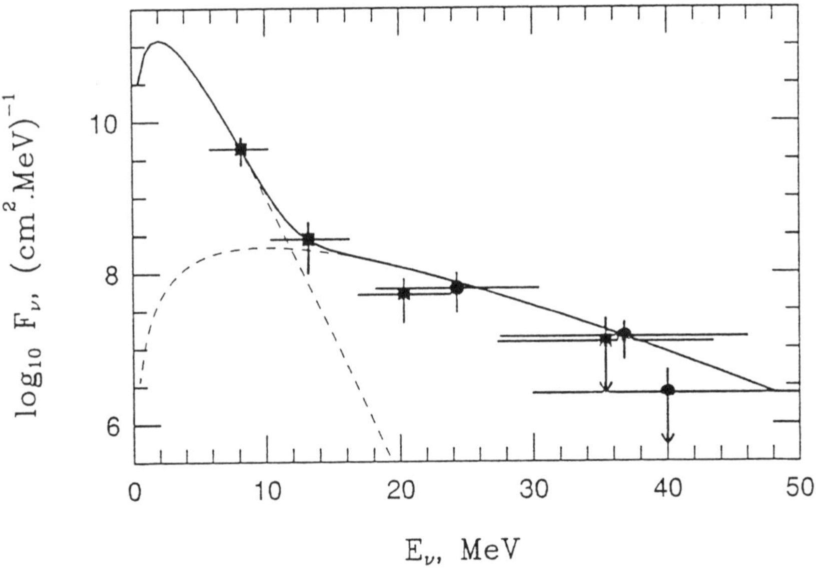

Fig. 2. Energy spectrum of the supernova neutrinos. The Kamiokande data are plotted with squares, those of IMB – with dots. The two dashed lines represent blackbody spectra with temperatures 1 and 5 MeV, and the solid line is their sum.

Much more detailed analysis was done by Bahcall *et al.* (1987), who used Monte Carlo simulations of the detection of neutrinos emitted with different spectra to determine the region in a T_ν – E_ν plot, which is allowed by each experimental observation. E_ν here is the total amount of energy emitted by SN1987a in $\bar{\nu}_e$'s. Kamiokande and IMB allow different bands on this plot, which intersect for T_ν = 4.1±0.1 MeV. This corresponds to E_ν = 4.5±1.5×10^{52} ergs. For all six neutrino species the total amount of energy is 2.7±0.9×10^{53} ergs, in very good agreement with the expectations.

The very detection of the neutrino burst from SN1987a, the neutrino flux and luminosity were a strong confirmation of the theory of stellar collapse. Combined with observations in the optical, X–rays and γ–ray lines, these observations gave a strong boost to the theoretical development of much more exact models of stellar collapse and supernova explosions.

6.2 The supernova as a neutrino properties laboratory

The detection of the SN1987a neutrino burst also produced invaluable information on the basic neutrino properties. The propagation and survival of neutrinos in a relatively narrow energy window from a distance of 50 Kpc allowed the setting of better limits on the neutrino mass, lifetime, electric charge, magnetic moment, etc.

The suggestion that the detection of a supernova neutrino burst will give an estimate of the neutrino mass is due to G.T. Zatsepin (1968). The time delay of a neutrino emitted at 50 Kpc is

$$\Delta t = 2.57\mathrm{s} \left(\frac{D}{50\mathrm{Kpc}}\right) \left(\frac{10\mathrm{MeV}}{E_\nu}\right)^2 \left(\frac{m_\nu}{10\mathrm{eV}}\right)^2 \tag{9}$$

One would then have a time delay in a reverse proportion to E_ν^2, which is a measured quantity. The actual determination of the mass is much more complicated because of its dependence on the stellar collapse theory and the small experimental statistics. The duration of the neutrino burst is supposed to be in excess of 10 s. The exact calculations also show a time dependence of the average neutrino energy during the cooling of the neutron star. There are many publications discussing the neutrino mass limits, some of which use only a part of the available statistics. An excellent review (Kolb, Stebbins and Turner, 1988) discusses the most interesting results, which give a range of 6 to 30 eV as a limit for the $\bar{\nu}_e$ mass. The more conservative estimates give a limit of $\sim$15 eV, comparable with the best laboratory limits. In a very interesting and important paper Cowsik (1988) showed that these limits are not very sensitive to neutrino oscillations. If $\bar{\nu}_e$ have converted into a different neutrino type, with a much lower interaction cross-section, that would require much higher neutrino emission during the collapse and violate limits on the binding energy.

Neutrino lifetime can be similarly estimated to $> 5 \times 10^5\mathrm{s} \times m_\nu(\mathrm{eV})$. The electric charge, which can delay the neutrinos because of interactions on the galactic magnetic field, can also be related to the neutrino mass. Magnetic moment limits come from the fact that electron antineutrinos have not spin flipped in the vicinity of the neutron star and during propagation. Many other neutrino and particle physics limits were set, including one on the coupling of weak interacting particles to matter (Raffelt and Seckel, 1988) , which I shall use as an example.

If weakly interacting particles (such as axions) exist, they could have contributed to the cooling of the supernova. The amount of energy carried away by such x particles is however limited by the total amount of available energy, i.e. the binding energy of the stellar core. Raffelt and Seckel reason in the following way – see Fig. 3. If the coupling g_x is small, x particles will be emitted by the total volume of the neutron star and their luminosity $L_x \sim g_x^2$. For large couplings the neutron star will be optically thick to x particles (as it is for neutrinos) and the emission will only come out of a surface layer of thickness r_x corresponding to $\tau_x = 1$. The luminosity in this case is $\sim r_x^2 T^4(r_x)$ and decrease with g_x. The region where $L_x > L_\nu$ is forbidden from energy considerations. Large couplings are limited by the non–observation of x particles in laboratory experiments. Using the proper numerical values Raffelt and Seckel set strong limits on the cross-sections of light exotic particles.

Acknowledgements I wish to thank the organizers of this course, Drs. M.M. Shapiro and J.P Wefel, and the Center for Scientific Culture 'E. Majorana', for their hospitality and for the creative atmosphere in Erice.

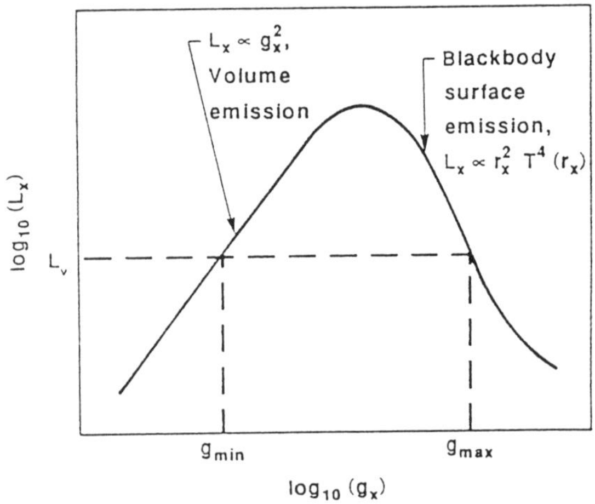

Fig. 3. Schematic dependence of L_x on the coupling strength g_x (from Raffelt&Seckel). The horizontal line reflects the nominal neutrino luminosity L_ν, which limits the luminosity of X particles.

References

Aglietta, M. *et al.* (1987) *Europhys. Lett.* **3**, 1315.
Akhmedov, E.K. and M. Yu. Khlopov, (1988) *Mod. Phys. Lett.* **A3**, 451.
Alekseev, E.N. *et al.* (1987) *JETP Lett.* **45**, 589.
Anselmann, P. *et al.* (1994) *Phys. Lett.* **B327**, 377.
Bahcall, J.N. *Neutrino Astrophysics* (Cambridge University Press, 1989).
Bahcall, J.N., G.B. Field and W.H. Press (1987) *Ap. J.* **320**, L69.
Bahcall, J.N. *et al.* (1987) *Nature*, **327**, 682.
Bahcall, J.N. and M.H. Pinsonneault (1992) *Rev. Mod. Phys.* **64**, 885.
Bahcall, J.N. and R.K. Ulrich (1988) *Rev. Mod. Phys.*, **60**, 297.
Bionta, R.M. *et al.* (1987) *Phys. Rev. Lett.* **58**, 1494.
Bieber, J.W. *et al.* (1990) *Nature* **348**, 407.
Burrows, A. and J.M. Latimer (1986) *Ap. J.* **307**, 178.
Cowsik, R. (1988) *Phys. Rev.* **D37**, 1685.
Dar, A. and G. Shaviv (1994) Astro-ph/940435.
Davis, R. (1993) in Proc. Int. Symposium on Neutrino Astrophysics, eds. Suzuki, Y. and N. Nakamura (Universal Academic Press, Tokyo, 1993).

De Rujula, A. (1987) *Phys. Lett.*, **B193**, 514.
Filippone, B. and P. Vogel (1990) *Phys. Lett.* **B246**, 546.
Gavrin, V.N. *et al* (1994) *Nucl. Phys. B (Proc. Suppl.)* **35**, 412. Hirata, K.S. *et al.* (1987) *Phys. Rev. Lett.* **58**, 1490.
Hirata, K.S. *et al.* (1991) *Phys. Rev.* **D44**, 2241.
Kolb, E.W., A. Stebbins and M.S. Turner (1987) *Phys. Rev.* **D35**, 3598.
Mikheev, S.P and A.Yu. Smirnov (1986) *Nuovo Cim.* bf 9C, 17.
Raffelt, G. and D. Seckel (1988) *Phys. Rev. Lett.* **60**, 1973.
Schramm, D.N. (1987) *Comm. Part. Nucl. Phys.* **17**, 239.
Shi, X. and D.N. Schramm (1994) *Part. World* **3**, 151.
Turck–Chieze, S. *et al.* (1993) *Phys. Rep.* **230**, 57.
Turck–Chieze, S. and I. Lopez 1993 *Ap. J.* **408**, 347.
Voloshin, M.B., M.I. Visotskii and L.B. Okun (1986) *Sov. J. Nucl. Phys* **44**, 440.
Wolfenstein, L (1978) *Phys. Rev.* **D17**, 2369.
Woosley, S.E., J.R. Wilson and R. Mayle (1986) *Ap. J.* **309**, 19.
Zatsepin, G.T. (1968) *ZETP Lett.* **8**, 205.

NEUTRINO MASSES IN ASTROPHYSICS

Raj Gandhi
Center for Theoretical Physics, Texas A&M University,
College Station, TX 77843-4242
and
The Mehta Institute of Mathematics and Mathematical Physics,
10, Kasturba Gandhi Marg, Allahabad, 211002 INDIA

Introduction

The standard model of particle physics (SM) [1] as it stands today can best be described as a highly impressive but crucially incomplete edifice. Many particle physicists would perhaps agree that the most important deficiency is the lack of understanding of the mechanism that provides masses for the quarks and leptons. Although the well-known Higgs mechanism provides a consistent and theoretically acceptable way of parametizing the generation of masses, it raises more questions than it answers. A measure of this is the fact that of the seventeen free parameters that are associated with the SM, the Higgs sector of the theory accounts for fully fifteen. An intriguing aspect of the mass problem is that neutrinos seem to have masses which maybe zero, or perhaps, orders of magnitude smaller than those of quarks or charged leptons. Progress on understanding fermion masses will thus certainly require an understanding of why neutrino masses are so small compared to the other energy scales in the theory.

A second important aspect that sets neutrinos apart from other fermions is the fact that they interact only weakly (neglecting their gravitational interactions) [See Table I.]. A neutrino can pass through almost a light-year length of lead before it will have one interaction.

Both its small (or zero?) mass, and its feeble interaction are responsible for making the neutrino a highly elusive particle in terrrestial experiments. However, these very properties enable it to be an ideal coolant for astrophysical objects and to reach the earth almost unaltered from remote parts of the universe, carrying the fingerprint of high energy particle processes that maybe occurring there. It is thus not surprising that the role of neutrinos in astrophysics has been a subject of active study by particle physicists, in the hope that information currently unobtainable from accelerators can be gleaned, and clues to physics beyond the SM may perhaps be obtained.

M. M. Shapiro et al. (eds.), Currents in High-Energy Astrophysics, 69–86.

Table I

Particle	Weak	EM	Strong	Gravity
Quarks	√	√	√	√
Charged leptons, *e,u,t*	√	√		√
Neutrinos	√			√
Relative Strength	10^{-5}	10^{-2}	1	10^{-40}
Associated Gauge Bosons	$W^{\pm}, Z^{o}$	Photon	Gluons	Gravitons
Associated Gauge Group	SU(2)	U(1)	SU(3)	---

Table I: The interaction of the elementary particles and their associated gauge groups.

In this short and informal lecture, we attempt to provide a flavor for this approach by discussing the information on neutrino masses which has been obtained from (a) SN1987A and (b) the sun. After a qualitative discussion of the particle content and important general features of the SM, we focus on (a) and (b) above, and end with a brief description of what may be expected in the future from this active area of theoretical and experimental research. Our aim is to give a feel for the important phsyics aspects of this area, without being bogged down by detailed aspects of problems still under investigation.

The physics of the SM is based on local invariance under the direct product of three gauge groups, *SU(3)* for the strong interactions and $SU(2)_L \otimes U(1)_Y$ for the electro-weak interactions. As shown in Figure 1 and Table I, the known elementary particles fall into two main classes - fermions and gauge bosons. The former are spin-1/2 dirac particles, and are grouped into leptons and quarks. The leptons (6 in total) are put into 3 doublets, each consisting of a neutrino (υ_i) and its charged lepton partner (l_i) $(i \equiv e, \mu$ or $\tau)$. The six quarks are similarly grouped in three doublets, with the upper members (u,ct,) having a charge of +2/3, while the lower ones (d,s,b) have a charge of -1/3. Note that both for quarks and leptons, the charge difference between

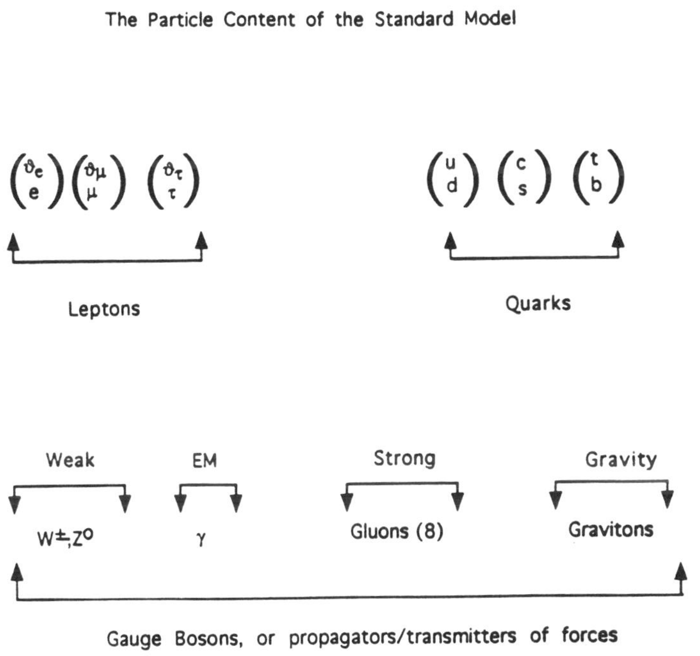

Fig. 1: The particle content (gauge bosons and fermions) of the Standard Model.

partners in a doublet is one unit. The grouping into doublets is a consequence of the transformation properties under the weak interaction gauge group $SU(2)_L$, and the charged weak gauge bosons ($W^{\pm}$), which transmit the weak force between members of the doublet thus carry one unit of electric charge. A third weak gauge boson, the Z^0, is responsible for propagating neutral weak interactions. The quarks, in addition to their weak charges, which determine their transformation under $SU(2)_L$, also carry color (red, blue or green), which is the strong $SU(3)$ charge. As indicated in the table; this force is propagated by 8 gluons, which can also carry the color charge.

The two underlying principles which dictate the dynamics for all the above particles are (a) Gauge Invariance under the product group $SU(3) \otimes SU(2)_L \otimes U(1)_Y$, and (b) Renormalizability, or the finiteness requirement for the theory. When these are wedded to the particle content above and the theoretical framework provided by quantum field theory, one obtains the impressive structure known as the standard model. It provides powerful agreement of theoretical predictions with all known accelerator data, and impressive cross-checks across widely separated sectors of the theory. Albeit, there are seventeen free parameters and serious shortcomings remain, the three most vexing being (a) the lack of a finite theory that incorporates quantum gravitational interactions; (b) the mass problem discussed above; and (c) the lack of a proper understanding of the mechanism for CP violation, which was observed experimentally many years ago, and has crucially important implications not only for particle physics but also for cosmology. The essence of the problem facing particle physicists has remained unchanged for more than a decade: the standard model works very well, and is capable of *parametrizing* all known strong, weak and electromagnetic physics at the elementary level, but we do not have a viable theory (despite great efforts) that takes us beyond it and explains why the seventeen free parameters in the theory have the values they do, and how they connect to a finite theory of quantum gravity. For a more complete and detailed description of the physics of the standard model, see Ref. [2]

Neutrino Mass and SN1987A

After this brief and somewhat sketchy review of the status of the standard model and the role of neutrinos in it, we narrow our focus to what has been learned about neutrino masses from SN1987A.

The general features of gravitational star collapse and the subsequent neutrino emission which carries away ~99% of the gravitational binding energy of the collapse (~2-4 x 10^{53} ergs) are well established and were confirmed by the observation of the neutrino burst from SN1987A [3].

The first few milliseconds of collapse are associated with a deleptonization burst of electron neutrinos (υ_e) associated with electron capture on protons

$$p+e \rightarrow n+\upsilon_e \tag{1}$$

The protons are formed by the dissociation of nuclei by the outwardly moving hot shock that is created when collapse is halted as the inner core stiffens to supra-nuclear densities of ~10^{15} gms/cm^3. These electron neutrinos carry away roughly 1-3 x 10^{51} ergs within ~10ms. The bulk of the energy is released soon after by the production and diffusion of neutrino pairs of all three species, via various pair production mechanisms like $e^+e^- \rightarrow \upsilon_i\overline{\upsilon}_i$, $n+p \rightarrow n+p+\upsilon_i\overline{\upsilon}_i$ etc. $(i \equiv e,u,t)$. This release of energy by pair production and the consequent neutrino diffusion through the core occurs over a period of several seconds, since the ~10 km (superdense) core is opaque to neutrinos due primarily to neutral current scattering interactions. This is the time it takes neutrinos to arrive at a core density which is transparent to them, at the emitting (blackbody) surface often called the *neutrinosphere*, beyond which their travel is at light time scales.

Two detectors, Kamiokande II [4] and IMB [5] reliably detected the anti-electron neutrinos, $(\overline{\upsilon}_e)$,from the burst. The energy and event characteristics of these fit the general expected profile very well. Eleven events were detected by Kamiokande II and eight by IMB over a period of ~10 secs. The distribution of the scattering events in the detector is isotropic and in conformity with the reaction

$$\overline{\upsilon}_e + p \rightarrow e^+ + n \tag{2}$$

which has the largest cross-section among the reactions that neutrinos can have with the detection medium, which is H_2O (for instance, the $\upsilon_e e$ reaction is forward peaked has a much lower value of the cross-section). The event energies, when extrapolated to the source at 50 kpc, compare well with the gravitational binding energy expected to be released on collapse.

A bound on the mass of the electron neutrino (mass m_υ, energy E_υ)can be obtained by a very simple time of flight calculation. For a relativistic neutrino, the velocity v is

$$v = 1 - \frac{1}{2}\frac{m_\upsilon{}^2}{E_\upsilon{}^2} \tag{3}$$

Hence the time of travel t is given by

$$t = d\left(1 + 1/2\frac{m_\upsilon{}^2}{E_\upsilon{}^2}\right) \tag{4}$$

where d is the distance to SN1987A (~50 kpc). For two neutrinos with energies E_1 and E_2, the difference in their times of travel, $\Delta t = |t_1 - t_2|$ is then

$$\Delta t = \tfrac{1}{2} d m_\upsilon{}^2 \left(\frac{|E_1{}^2 - E_2{}^2|}{E_1{}^2 E_2^2}\right) \tag{5}$$

Using a time spread of $\Delta t \le 10s$, one obtains a bound of $m_{\upsilon_e} \lesssim 15$ eV. It is interesting to note that this back-of-the-envelope calculation yields a bound that was as stringent as the existing lab bound at the time of observation.

A little more effort is required to obtain reliable bounds on the masses of mu and tau neutrinos from the supernova observations. To do this, we first need to clarify the difference between the helicity and chirality of neutrinos. The *chirality* operators $P_R = \frac{1}{2}\left(1+\gamma^5\right)$ and $P_L = \frac{1}{2}\left(1-\gamma^5\right)$ when acting upon a neutrino field ψ^υ project out the *right-handed* $\left(\psi_R^\upsilon\right)$ and *left-handed* $\left(\psi_L^\upsilon\right)$ components of the field respectively. The weak interactions couple to only the left-handed components of the field, and as far as the standard model is concerned, the ψ_R^υ, if they exist at all, are "sterile", or non-interacting. The *helicity* operator, $\vec{\Sigma}\cdot\vec{p}/|\vec{p}|$, on the other hand, projects out the component of the neutrino spin in the direction of its momentum $(\vec{p})$. Here $\vec{\Sigma} = \frac{1}{|\vec{p}|}\begin{pmatrix} \vec{\sigma}\cdot\vec{p} & o \\ o & \vec{\sigma}\cdot\vec{p} \end{pmatrix}$ (where the $\vec{\sigma}$ are the usual Pauli matrices) and its eigenstates are ψ_+^υ (right-helicity) and ψ_-^υ, (left-helicity). For the present discussion, the relevant point is that for massless neutrinos, $\psi_+^\upsilon \equiv \psi_R^\upsilon$ and $\psi_-^\upsilon \equiv \psi_L^\upsilon$. If neutrinos do have a mass, then the states differ by corrections of order $\sim m_\upsilon/2E_\upsilon$

$$\psi_L^\upsilon \approx \psi_-^\upsilon + \frac{m_\upsilon}{2E_\upsilon}\psi_+^\upsilon + ---$$

$$\psi_R^\upsilon \approx \psi_+^\upsilon - \frac{m_\upsilon}{2E_\upsilon}\psi_-^\upsilon + --- \tag{6}$$

Essentially, the difference between these states arises because the (free) massive Dirac hamiltonian does not commute with γ^5 (and hence with P_L and P_R), while it does commute with $\vec{\Sigma}$. Hence the free particle eigenstates that one normally deals with in scattering processes (the "in" and "out" states) are eigenstates of the helicity operator but not the chirality operators, if neutrinos have a mass.

This difference, of no consequence in terrestrial experiments due to the highly relativistic nature of neutrinos, can be exploited with profit in the highly dense core of the collapsed star, where the neutrino has so many interactions that it is trapped for a period of several seconds in a region of a few (~10) km. Due to this unusual (for the feebly interacting neutrino) situation, an incoming ψ_-^υ interacting with an electron, proton or neutron in the core via Z° exchange can every once in a while flip its spin and emerge as an outgoing ψ_+^υ state. From. Eqn. 6 above, the cross-section for this to occur, i.e. σ_Z^{flip}, should roughly be $\frac{m_\upsilon^2}{4E_\upsilon^2}\sigma_Z^{weak}$ where σ_Z^{weak} is the usual weak scattering probability for the neutrino, which does not involve any change in its spin direction. Also from Eqn. 6, we can deduce that once the scattered state is transformed to a flipped state, it is non-interacting up to corrections of order $m^2/4E_\upsilon^2$ (i.e. the cross-section is zero up to this order) due to its being predominantly a ψ_R^υ. Such a neutrino would not be trapped in the core over a timescale of several seconds but would escape on light travel timescales. If σ_{flip}^{Z} is large enough (or, in other words, if m_υ is large enough) then this process could drain the star of energy on much shorter timescales, in contradiction with the observation that the cooling timescale represented by the neutrino burst was on the order of several seconds, thus leading to a mass bound. We now make this qualitative argument slightly more precise by putting in some of the relevant numbers [6,7].

We compare the expected luminosity of flipped neutrinos to the maximum neutrino luminosity from a cooling neutron star in the standard, non-helicity-flip case, about 10^{53} ergs/sec. As argued above, a larger luminosity would result in cooling of the star that is too rapid to explain the signal observed from SN 1987A which, in the standard case, is due to thermal $\overline{\upsilon}_e$'s emitted form the neutron star's surface and was observed to last 5-10 sec.

The expected luminosity of flipped neutrinos can be expressed as an integral over the volume of the star,

$$L \lesssim 2N_\nu \int \langle \epsilon_\nu \rangle \left(n\sigma_Z^{flip} c \right) n_\nu dV, \tag{7}$$

where N_ν is the number of neutrino flavors that flip (the factor of 2 includes the antineutrinos). n_ν and $\langle \varepsilon_\nu \rangle$ are the neutrino number density and the average energy, and are about 1.8 $(2\pi)^{-1}(kT/\hbar c)^3$ and 3.15 kT, respectively for μ and τ neutrinos, which have a thermal distribution. The quantity in parentheses is the rate of neutrino flipping per neutrino. For the present situation we find that the flip rate from nucleons gives an effective cross section approximately 20 times larger than that from electrons, so here n refers to the nucleon density and $\sigma_Z^{flip} \geq 0.75 \times 10^{-47} (m/50keV)^2 cm^2$ (This cross-section is calculated in Ref. [7].). The integral in (7) is proportional therefore to the mass average T^4 in the neutron star:

$$L \lesssim 0.35 \times 10^{53} N_\nu (m/50keV)^2 \times \langle T/30MeV \rangle^4 ergs/\sec, \tag{8}$$

where a 1.4 M_{sun} neutron star is assumed. Setting $L<10^{53}$ ergs/sec, we find

$$m \leq 10.5 N_\nu^{-1/2} \langle T/30MeV \rangle^{-2} keV. \tag{9}$$

For the reference value of T = 30 MeV this translates to a limit of 60 keV, which is roughly three orders of magnitude better than the current laboratory bound on tau neutrinos.

We note that this limit applies to relativistic mu and tau neutrinos with masses below several MeV. The calculation above [7] glosses over complications, but is given here to bring out the essential physical arguments involved. A firmer and rigorous bound can be obtained by a proper numerical calculation of the cooling of a hot neutron star incorporating the standard cooling mechanism (thermal neutrino emission) and the additional cooling provided by wrong-helicity emission. It also would incorporate other flipping processes, the effects of neutrino mixing and degeneracy and pion condensation in the core. When this is done [6,8] the bound conservatively improves by roughly a factor of 2, to 30 keV. Fig. 2, from Ref. [6], demonstrates how the

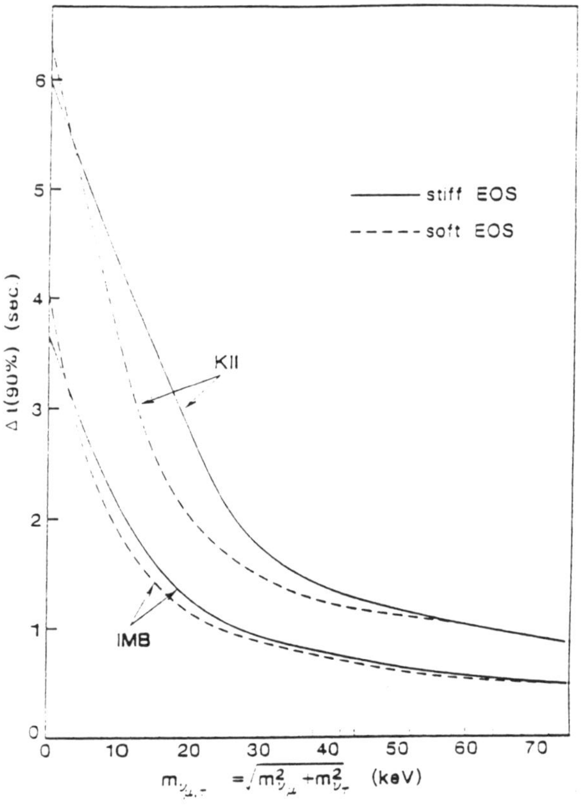

Fig. 2: The time taken for the accumulated number of events to reach 90% of the final number of events as a function of $m_{\upsilon_{\mu,\tau}}$, for both Kamiokande (KII) and IMB, (from Ref. 6). The calculations utilize the proto-neutron-star evolution codes of Burrows and Lattimer [20] and Burrows [21]. EOS refers to equation of state.

burst duration drops in both detectors as the flip reaction causes energy drain on smaller time-scales. The criterion used to obtain the mass bound is that the burst duration falls to ~ 1 sec in both detectors for a mass of ~30 keV, clearly in contradiction with the observed duration in K II and IMB.

Solar Neutrinos

As a second illustration of information obtained on neutrino masses from extra-terrestrial sources, we discuss the well known solar-neutrino problem. Again, in view of the fact that our purpose here is to give a flavor of how astrophysics and particle physics may be merged fruitfully rather than to be rigorous, we gloss over several important, but technical, details in what follows. For a more detailed review, see [9,10].

Table II

Experiment	Process	Observed/Prediction
Homestake	$\upsilon_e + {}^{37}Cl \rightarrow e + {}^{37}Ar$	0.27±0.04
Kamiokande II	$\upsilon_e + e \rightarrow \upsilon_e + e$	0.46±0.05±0.06
SAGE and GALLEX	$\upsilon_e + {}^{71}Ga \rightarrow e + {}^{71}Ge$	0.59±0.10

Table II: The results of the solar neutrino experiments, along with the ratio of observed fluxes to those predicted by Bahcall et al [15].

It is well-known that at present four detectors currently observe the electron neutrinos produced in the sun. Table II summarizes the results from the Homestake [11], Kamiokande [12], SAGE [13] and GALLEX [14] experiments by comparing the ratio of observed flux of each to that predicted by the solar model calculations of Bahcall and collaborators [15]. In addition, the relevant detection reaction is also listed. For our purpose, we note that all detectors show a non-trivial deficit compared to the theoretical predictions of Bahcall et al [15]. Whether this is due to new particle physics or simply to astrophysics or related nuclear physics that has not been fully incorporated into the theoretical predictions is an open question at present. It is fair to say, however, that the theoretical predictions are as thorough as present-day knowledge allows, and have been checked for consistency and refined over a long time period. Differences between predictions for the expected flux do exist [16], however, and it is not possible to say with certainty where the origin of the deficit may lie at present. The consensus has gradually been tilting, however, towards the view that the answer lies within the realm of particle physics and massive neutrinos, and it is this viewpoint that we shall adopt for the following discussion.

If neutrinos have mass, then it is natural to assume that the free particle states (or in other words, propagating or mass-eigenstates) are not the same as the eigenstates of the weak interaction hamiltonian. Hence these two sets of basis states are expressible in terms of each other via mixing matrices. This enables, in principle, an electron neturino produced in the sun via a weak interaction to oscillate to a muon neutrino as it propagates through the region between the sun's core and the detector on earth, and thus go undetected, since at present the experiments are unable to detect mu (or tau) neutrinos to an appreciable extent due to the relatively small cross-sections for these species. The oscillation, or conversion to a different flavor can occur either in vacuum (after the neutrino leaves the sun) or in the sun's core itself, due to the difference in the interactions of mu and electron neutrinos as they propagate through the sun. This difference in the forward, coherent scattering potentials 'seen' by the two different flavors of neutrino leads to the exciting possibility of resonant conversion as the electron neutrino propagates through the varying density of the sun from the core to its surface. It is this resonance, discovered by Mikheyev, Smirnov and Wolfenstein [17] and called the MSW effect, which today is thought to be the most likely candidate for a solution to the solar deficit. Below we first discuss vacuum oscillations and then the incorporation of the matter effects, leading to the above mentioned resonance.

Let propagating (or mass/energy) eigenstates be $\upsilon^{(p)} \equiv \binom{\upsilon_1}{\upsilon_2}$ and flavor (or interacting) eigenstates be $\upsilon^{(f)} \equiv \begin{pmatrix} \upsilon_e \\ \upsilon_\mu \end{pmatrix}$. Let U be a unitary matrix such that $\upsilon^{(f)} = U\upsilon^{(p)}$. E_1, E_2, m_1, m_2 and p_1, p_2 are the energies, masses and three-momenta of the states υ_1 and υ_2 respectively.

Assuming relativistic neutrinos,

$$E_{1,2} = \sqrt{|p|^2 + m_{1,2}^2} \cong p + \frac{m_{1,2}^2}{2p} \quad (p = |p|). \tag{10}$$

Here we have made the approximation that $p_1 = p_2 = p$, i.e. the energies differ only in the masses.

$$\therefore H = \begin{pmatrix} E_1 & \\ & E_2 \end{pmatrix} = p + \frac{1}{2p}\begin{pmatrix} m_1^2 & \\ & m_2^2 \end{pmatrix}$$

$$= \left(p + \frac{m_1^2 + m_2^2}{4p}\right) - \frac{\Delta\sigma^3}{4p} \tag{11}$$

where $\sigma^3 \equiv \begin{pmatrix} 1 & \\ & -1 \end{pmatrix}$ and $\Delta = m_2^2 - m_1^2$.

In the flavor basis, with $\upsilon^{(p)} = U^+\upsilon^{(f)}$ and $t \approx x$ for relativistic neutrinos, we write

$$i\frac{d}{dx}\left(U^+\upsilon^{(f)}\right) = HU^+\upsilon^f. \tag{12}$$

For 2 flavors,

$$U = \begin{pmatrix} \cos\Theta & \sin\Theta \\ -\sin\Theta & \cos\Theta \end{pmatrix} \text{ and } H' = UHU^+ = p + \frac{m_1^2 + m_2^2}{4p} - \frac{\Delta}{4p}U\sigma^3U^+$$

$$= p + \frac{m_1^2 + m_2^2}{4p} + \frac{\Delta}{4p}\begin{pmatrix} -\cos 2\Theta & \sin 2\Theta \\ \sin 2\Theta & \cos 2\Theta \end{pmatrix} \tag{13}$$

hence $\tan 2\Theta = 2H'_{12}/\left(H'_{22} - H'_{11}\right)$

The solution to the equation of motion is

$$\upsilon^F(x) = e^{-iH'x}\upsilon^F(0)$$

$$= \exp\left[-i\frac{\Delta}{4p} \times (\sigma_1 \sin 2\Theta - \sigma_3 \cos 2\Theta)\right]\upsilon^f(0) \tag{14}$$

Let

$$\Sigma = \sigma_1 \sin 2\Theta - \sigma_3 \cos 2\Theta = \begin{pmatrix} -\cos 2\Theta & \sin 2\Theta \\ \sin 2\Theta & \cos 2\Theta \end{pmatrix} \tag{15}$$

Using $\Sigma^2 = 1$, it is easy to see that

$$\upsilon^f(x) = \left[\cos\frac{\Delta x}{4p} - i\Sigma \sin\frac{\Delta x}{4p}\right]\upsilon^f(o)$$

and hence

$$\upsilon^e(x) = \left[\cos\frac{\Delta x}{4p} + i\cos 2\Theta \sin\frac{\Delta x}{4p}\right]\upsilon^e(0) \tag{16}$$

$$-\left[i \sin 2\Theta \sin\frac{\Delta x}{4p}\right]\upsilon^\mu(0)$$

leading to

$$P_{\upsilon_e \to \upsilon_\mu} = \left| \langle \upsilon_\mu(0) | \upsilon_e(x) \rangle \right|^2 = \sin^2 2\Theta \sin^2 \frac{\Delta x}{4p}$$

where $P_{\upsilon_e \to \upsilon_\mu}$ is the probability for an electron neutrino produced in the sun to convert to a muon neutrino by the time it arrives on earth.

The next step is to add in matter effects. Specifically, the coherent interaction of matter at rest contributes to additional potential terms to H'. For instance, the charged current contribution of neutrino scattering off electrons in the sun's core to the Lagrangian becomes:

$$L_{eff} = \frac{4G_F}{\sqrt{2}} \left[\bar{e}_L(p_1) \gamma^\mu \bar{\upsilon}_{eL}(p_2) \right] \left[\bar{\upsilon}_{eL}(p_3) \gamma_\mu e_L(p_4) \right] \tag{17}$$

For coherent scattering, one averages over the electron background, and assumes that the neutrino momentum is unchanged. In addition, the electrons are at rest and spin effects are unimportant (i.e. average to zero). Hence $Leff$ obtains contributions from the $\mu = 0$ part only, and we have

$$Leff = \sqrt{2} G_F n_e \left(\bar{\upsilon}_{eL} \gamma^0 \upsilon_{eL} \right)$$

$$n_e \equiv \text{ electron number density} \tag{18}$$

Similarly, one can evaluate the contribution due to neutral current scattering of neutrinos on electrons (e), protons (p), and neutrons (n). The contributions due to e and p neutral currents cancel exactly and the potential for neutral current scattering on neutrons is

$$V_{NC,n} = \frac{-1}{\sqrt{2}} G_F n_n \qquad n_n \equiv \text{neutron number density} \tag{19}$$

Hence the new effective hamiltonian is

$$\tilde{H} = H' + \begin{pmatrix} \sqrt{2} G_F \left(n_e - \frac{1}{2} n_n \right) & 0 \\ 0 & \frac{-1}{\sqrt{2}} G_F n_n \end{pmatrix}$$

$$= p + \frac{m_1{}^2 + m_2{}^2}{4p} + \frac{1}{4p} \begin{pmatrix} \Delta \cos 2\Theta + 4A & \Delta \sin 2\Theta \\ -\Delta \sin 2\Theta & \Delta \cos 2\Theta + 4A - 4B \end{pmatrix} \tag{20}$$

where $\qquad A = \sqrt{2} G_F \left(n_e - \frac{1}{2} n_n \right) p$

and $\qquad B = \left(\sqrt{2} G_F n_e \right) p$

Hence one obtains an "effective"mixing angle in matter, $\tilde{\Theta}$

$$\tan 2\tilde{\Theta} = \frac{2\tilde{H}_{12}}{\tilde{H}_{22} - \tilde{H}_{11}}$$

$$= \frac{\Delta \sin 2\Theta}{\left(\Delta \cos 2\Theta - 2\sqrt{2} G_F n_e p\right)} \quad (21)$$

Hence, $$P_{\upsilon_e \to \upsilon_u}^{matter} = \sin^2 2\tilde{\Theta} \sin^2 \frac{\Delta x}{4p}$$

$\equiv \frac{1}{2}\sin^2 2\tilde{\Theta}$, if one averages over the x dependent term.

Also, from (20) above, $$\sin^2 2\tilde{\Theta} = \frac{\Delta^2 \sin^2 2\Theta}{\Delta^2 \sin 2\Theta + \left(\Delta \cos 2\Theta - 2\sqrt{2} G_F n_e p\right)^2} \quad (22)$$

and $$\sin^2 2\tilde{\Theta} \to 1 \text{ as } \left(\Delta \cos 2\Theta - 2\sqrt{2} G_F n_e p\right)^2 \to 0$$

The last equation expresses how the probability that a υ_e will oscillate to a υ_μ can be maximized due to the matter (MSW) effect due to the mixing angle, neutrino masses and electron number density in the sun. The vacuum oscillations, discussed previously, are less attractive since only a very small parameter space (i.e., Δm^2 vs $\sin^2 2\Theta$) has not yet been ruled out by the existing experiments. On the other hand, the allowed MSW parameter space spans a large range of mixing angles and masses, as Figure 3 demonstrates [18]. (Figure 3 represents the results of detailed calculations and comparisons with experimental data. Our presentation above merely scratches the surface by presenting some of the important physical concepts. See the review articles mentioned at the beginning of this section for more details).

In conclusion, when one assumes neutrinos are massive and combines this with the physics of the MSW effect and our knowledge of the nuclear astrophysics of the sun, a fascinating solution to the solar deficit emerges. The solution predicts neutrino mass differences (and hence, reasonably, neutrino masses themselves) in a range of $10^{-2} \sim 10^{-4}$ eV. Clearly, laboratory experiments are currently nowhere near this range. Whether the MSW effect is what is causing the deficit only time and more experimental results (especially from the currently under construction SNO detector in Sudbury, Canada, [19]) can tell. In the meantime, the physics that has emerged from it is certainly nothing short of beautiful.

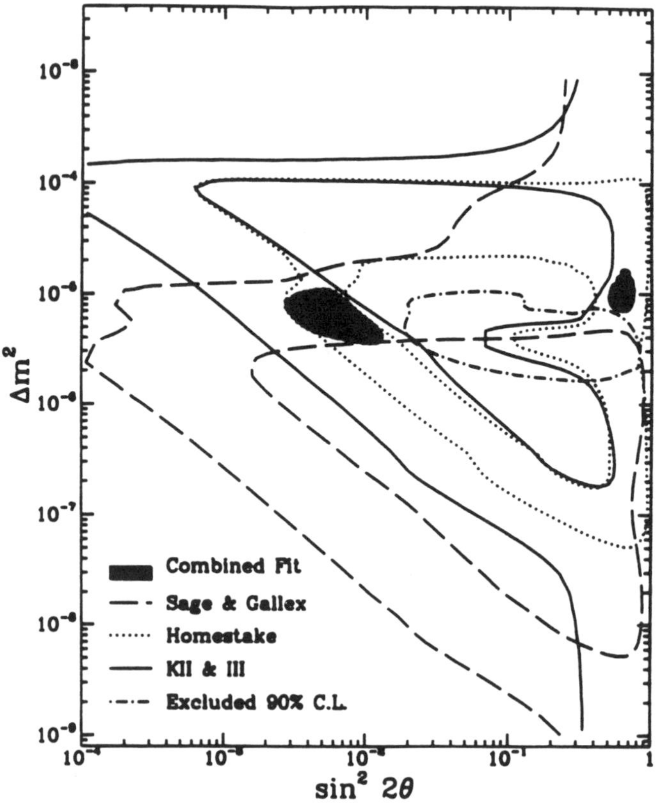

Fig. 3: The MSW solution space for the solar neutrino deficit, from Ref. [18]. The ranges of the vacuum mixing angle and the mass squared difference values which are compatible with the various detectors are shown.

The Future

That the marriage between particle physics and astrophysics is taken seriously is demonstrated by the large number of extant or planned detectors looking for either solar, supernova or neutrinos from other extra-terrestrial sources.* Deserving special mention among these is SNO [19], and we briefly discuss its principal features due to its importance to both particle physics and astrophysics.

SNO will consist of one kton of heavy water (D_20) surrounded by 4 m of light water (with a fiducial H_2O mass for supernova detection of approximately 1.6 kton).

*For more on *high energy neutrino detectors*, which we have not discussed, see the lecture by M. Shapiro in this volume.

Designed primarily to detect solar neutrinos, it is uniquely suited to detect $\upsilon_\mu(\overline{\upsilon}_\mu)$'s and $\upsilon_\tau(\overline{\upsilon}_\tau)$'s via the neutral-current breakup reactions of deuterium:

$$\upsilon_i + d \to n + p + \upsilon_i \quad (\in_{th} = 2.22 MeV)$$
$$\overline{\upsilon}_i + d \to n + p + \upsilon_i \quad (\in_{th} = 2.22 MeV) \qquad (23)$$

The secondary neutrons will be detected through the Cerenkov light from the pair cascades of the γ rays $(\Sigma E_\gamma \sim 8.6 MeV)$ from neutron capture on the ^{35}Cl slated throughout the D_2O in the form of NaCl (2.5 ton) or with ^{3}He counters suspended in the D_2O tank. Whichever neutron detection scheme is employed, SNO cannot measure energy or angles for reactions of Equation 23, but can measure event times.

The charged-current reactions

$$\upsilon_e + d \to p + p + e^- \quad (\in_{th} = 1.44 MeV) \qquad (24)$$

and

$$\overline{\upsilon}_e + d \to n + n + e^+ \quad (\in_{th} = 4.03 MeV) \qquad (25)$$

are at least as important in SNO as neutral-current reactions. The coincidence between the e^+ and neutron signals from process (25) can, in principle, be used to identify $\overline{\upsilon}_e$ events in D_2O, while the absence of coincident neutron signals should point to the υ_e process (24).

Its sensitivity to υ_μ's, υ_τ's and υ_e's, its low anticipated detection threshold of 5 MeV, its great depth of 6000 mwe, and it H_2O/D_2O design will make SNO a premier supernova and solar sentinel when it goes on line in 1995 or 1996.

Finally, Table 3 (from Ref. 22) provides a summary of detectors (present and future) which will be looking out for a *galactic* supernova burst in the near future and their principal features. One need only glance at the column that lists the number of expected events from a burst at 10 kpc to get a feel for their observational potential, and consequently, for the amount that we may expect to learn from this particular interface of two immensely exciting fields of scientific endeavor.

Table III

Detector	Total mass (ton) [Fiducial mass (ton)	Composition	Depth (mwe)	Threshold (MeV)	No. events at 10 kpc	Resolution (e,t,a)
Cerenkov						
IMB3	8000 (6800)	H_2O	1570	10	940	*e,t,a*
KII	3000 (2140)	H_2O	2700	7^a	370	*e,t,a*
Super Kamiokande	40000 (32000)	H_2O	2700	7(?)	5510	*e,t,a*
SNO	1600/1000	H_2O/D_2O	6000	5	781	*e,t,a* (*t* for NC)
Scintillation						
LVD	1800 (1200)	Kerosene	3100-4000	5-7	375	*e,t,(a?)*
MACRO	1000	"CH_2"	3100-4000	10	241	*e,t*
Baksan	330 (200)	"White Spirits"	850	10	70	*e,t*
LSND	200	"CH_2"		5	70	*e,t*
Borex(ino)	1760(300)	$(BO)_3(OCH_3)_3$	3100-4000	4.7	~325	*e,t*
CalTech	1000		?	2.8	290	*e,t*
Drift chamber						
ICARUS	3600	^{40}Ar	3100-4000	5	120	*e,t,a*
Radiochemical						
Homestake ^{37}Cl	610	C_2Cl_4	4200	0.814	4	
Homestake ^{127}I	1000	NaI	4200	0.789	~25	
Baksan ^{37}Cl	3000	C_2Cl_4	5000	0.814	22	
Extragalactic						
SNBO	100000	$CaCO_3$	?		10000	*t*
JULIA	40000	H_2O	?		10000	*e,t,a*

Table III: Supernova neutrino telescope characteristics: *e,t,* and *a* mean the detector has or will have energy, time, or angle resolution: NC means neutral current [22].

References

[1] S. Weinberg, Phys. Rev. Lett. **19**, 1264 (1967); S. Glashow, Nucl. Phys. **22**, 579 (1961); A. Salam in *Elementary Particle Physics: Relativistic Groups and Analyticity (Nobel Symposium 8)*, ed. by N. Svartholm (Almqvist and Wiksell, Stockholm, 1968).

[2] F. Halzen and A. Martin, *Quarks and Leptons*, Wiley (1984).

[3] The SN1987A neutrino detection by KII and IMB and its agreement with basic supernova theory is discussed in A. Burrows and J. Lattimer, Astrophys.. J. **318** (1987) L63; J. Bahcall, T. Piran, W. Press and D.N. Spergel, Nature (London) **327** (1987) 682; L.L. Krauss, Nature (London) **329** (1987) 689; S. Bludman and P. Schinder, Astrophys. J. **326** (1988) 265; D.N. Schramm, Comm. Nucl. Part. Phys. **17** (1987) 239; D.Q. Lamb, F. Melia and T. Loredo, in : Supernova 1987A in the Large Magellanic Cloud, eds. M. Kafatos and A. Michalisianos, (Cambridge U.P., Cambridge, 1988), p. 204; S. Bruenn, Phys. Rev. Lett. **59** (1987) 938; S. Kahana, J. Cooperstein and E. Baron, Phys. Lett. B **196** (1987) 259; R. Mayle and J.R. Wilson, Astrophys. J. 334 (1988) **909**; K. Sato and H. Suzuki, Phys. Rev. Lett. **58** (1987) 2722.

[4] Kamiokande II Collab., K. Hirata *et al.*, Phys. Rev. Lett. **58** (1987) 1490.

[5] IMB Collab., R.M. Bionta *et al.*, Phys. Rev. Lett. **58** (1987) 1494.

[6] R. Gandhi and A. Burrows, Phys. Lett. B **246**, 149 (1990); *ibid*, **261**, 519 (E) (1991).

[7] K. Gaemers, R. Gandhi, and J. Lattimer, Phys. Rev. D **40**, 309 (1989).

[8] A. Burrows, R. Gandhi and M. Turner, Phys. Rev. Lett. **68**, 3834 (1992).

[9] T.K. Kuo and J. Pantaleone, Rev. Mod. Phys. **61**, 937 (1989).

[10] R. Mohapatra and P. Pal, *Massive Neutrinos in Physics and Astrophysics* (World Scientific, Singapore, 1991).

[11] R. Davis Jr. at the *International Symposium on Neutrino Astrophysics*, Takayama, Japan, 1992 (unpublished); R. Davis Jr. *et al.*, in *Proceedings of the 21st International Cosmic Ray Conference*, Vol. 12, ed. R.J. Protheroe (University of Adelaide Press Adelaide, 1990) p. 143.

[12] K. Hirata *et al.*, Phys. Rev. Lett. **65**, 1297 (1990); **65**, 1301 (1990); **66**, 9 (1991); Y. Suzuki at the *International Symposium on Neutrino Astrophysics*, Takayama, Japan, 1992 (unpublished).

[13] A.I. Abazov *et al.*, Phys. Rev. Lett. **67** 3332 (1991); V.N. Gavrin in *XXVI International Conference on High Energy Physics*, Dallas, 1992, ed. J. Sanford (AIP, New York, 1993).

[14] GALLEX Collaboration; P. Anselmann *et al.*, Phys. Lett. **B285**, 376 (1992); **B285**, 290 (1992).

[15] J.H. Bahcall and M.H. Pinsonneault, Rev. Mod. Phys. **64**, 885 (1992)

[16] S. Turck-Chieze and I. Lopes, Astrophys. J. **408**, 347 (1993).

[17] L. Wolfenstein, Phys. Rev. **D17**, 2369 (1978); **D20** 2634, (1979); S.P. Mikheyev and A. Yu. Smirnov, Yad. Fiz. **42**, 1441 (1985); Nuovo Cim. **9C**, 17 (1986).

[18] N. Hata and P.G. Langacker, University of Pennsylvania preprint UPR-0592T, submitted to Phys. Rev. **D**.

[19] G.T. Ewan, in Proceedings of the Supernova Watch Workshop, Santa Monica, California, 1990 (unpublished).

[20] A. Burrows and J. Lattimer, Astrophys.. J. **309**, 178 (1986).

[21] A. Burrows, Astrophys.. J. **334**, 891 (1988).

[22] A. Burrows, D. Klein and R. Gandhi, Phys. Rev. D **45**, 3361 (1992).

SOURCE COMPOSITION OF COSMIC-RAYS AND MODELS OF ORIGIN

R. SILBERBERG
Universities Space Research Association
Washington, DC 20024

C.H. TSAO
E.O. Hulbert Center for Space Research
Naval Research Lab., Washington, DC 20375-5000

M.M. SHAPIRO
University of Maryland
College Park, MD 20472

ABSTRACT. The observed composition of cosmic-ray nuclei above the earth's atmosphere has been modified by spallation reactions in the interstellar gas. About half of the cosmic-ray nuclei heavier than carbon have suffered spallation reactions. Propagation (or nuclear transport) calculations that apply numerous nuclear cross sections are applied to relate the observed and source composition of cosmic rays, and to deduce the latter. The source composition differs from the composition of normal stars and the interstellar medium (i.e., the general abundance of elements). When normalized relative to Mg, Si and Fe, the cosmic-ray sources are underabundant in H, He and N by a factor of 20 to 30 while the ratio $^{22}Ne/^{20}Ne$ is enhanced by a factor of 4. We favor (a) the origin of cosmic-ray material in the flares of flare stars. These flare particles are boosted up in energy by an average of three orders of magnitude by shock waves of supernova remnants. The latter accelerate some Wolf-Rayet star stellar wind particles. We also explore alternative proposals: (b) a mix of type I and II supernova material with the interstellar gas, (c) acceleration by supernova shock waves of interstellar gas enriched in nonvolatile dust particles that are evaporated by the shock waves, and (d) preferential acceleration of heavy ions. We favor model (a), because the calculated source composition agrees with that deduced from the experimental abundances to within 20 per cent, with a standard deviation of about 10 per cent.

1. Introduction

Cosmic rays (including the anomalous cosmic-ray component of low-energy singly charged He, N, O, Ne and Ar) constitute the only sample of material from outside the solar system. The energy spectrum and the elemental and isotopic composition of cosmic

M. M. Shapiro et al. (eds.), Currents in High-Energy Astrophysics, 87–101.

rays contains important clues (1) on the injection of cosmic rays into the acceleration region, (2) on the acceleration process, and (3) models of nucleosynthesis in stars and supernovae, evolution of stars and the interstellar medium. The "general abundance" of elements is largely deduced from the solar and meteoritic composition, which is representative of the interstellar medium at the time the sun condensed from it, about 4.5X 10^9 years ago. The early solar nebula may also have contained a small admixture of material from supernova(e) that preceded by a relatively short time the condensation of the solar system. This material from supernovae also contained some radioactive newly nucleosynthesized nuclei. Cosmic rays (although with poor statistics) indicate an enhancement of heavy ($Z > 60$) r-process nuclei, by a factor of 2. Nuclei with $Z > 90$ are enhanced by an additional factor of 3, Thompson et al. (1993).

In Section 2, the arriving and source composition of cosmic rays are compared. In Section 3, the flare star model (with First Ionization Potential effects, and rigidity dependent suppression of the light, fully ionized ions) is explored. In Section 4, the origin and acceleration of very high energy cosmic rays (E >1 TeV/u) is discussed. In Section 5, the model of cosmic ray origin in newly synthesized nuclei in supernovae is explored. In Section 6, the model of preferential acceleration of heavy ions is discussed. In Section 7, the model of enhancement of many cosmic-ray nuclei from evaporation of non-volatile grains is discussed. In Section 8, the conclusions are presented.

2. The Arriving and Source Composition of Cosmic Rays

The arriving composition of cosmic rays differs greatly from the composition that left the cosmic ray source regions. The difference is largely due to nuclear breakup when the original cosmic rays break into lighter nuclei upon colliding with the interstellar gas. While the density of the gas is low, about 0.3 atoms/cm^3 averaged over the Galactic disk and the inner halo, the degree of breakup is appreciable during the long confinement time of about 10^7 years or $3X10^{14}$ sec for cosmic rays in the Galaxy. (This time is calculated from the degree of survival of ^{10}Be in cosmic rays; this isotope has a half life of about $1.5X10^6$ years). With diffusion in the outer halo, from where few particles return, the confinement time is about 10^8 years.

The path length traversed by cosmic rays, X, in units of atoms/cm^2, is $X = \rho\,\beta\, c\, T$ where $\rho = 0.3$ atoms/cm^3, c is the velocity of light, 3×10^{10} cm/s, β is the velocity of cosmic rays in units of c, i.e., $\beta \approx 1$, T = confinement time of cosmic rays = 3×10^{14} sec. Thus, $X \approx 3 \times 10^{24}$ atoms/cm^2 is the path length traversed by cosmic rays.

The number of interstellar hydrogen nuclei encountered by, e.g., a cosmic-ray carbon nucleus of cross section σ (C,p) = 200 mb or 2×10^{-25} cm^2 is $\sigma X \approx 2 \times 10^{-25} \times 3 \times 10^{24} \approx 0.6$. Hence, about half of cosmic-ray nuclei heavier than helium have suffered nuclear collisions in the interstellar medium, breaking into lighter nuclei. Fig. 1 (from Simpson, 1983) illustrates the large effects of nuclear breakup: the light nuclei Li, Be, B are built up by a factor of about 10^5, F, Sc, V by about 10^3, Ti by 10^2, and P, Cl, K, Cr, Mn by about 10.

The source composition is derived from the arriving composition by use of measured and semiempirical partial cross sections, as constructed by Silberberg and Tsao

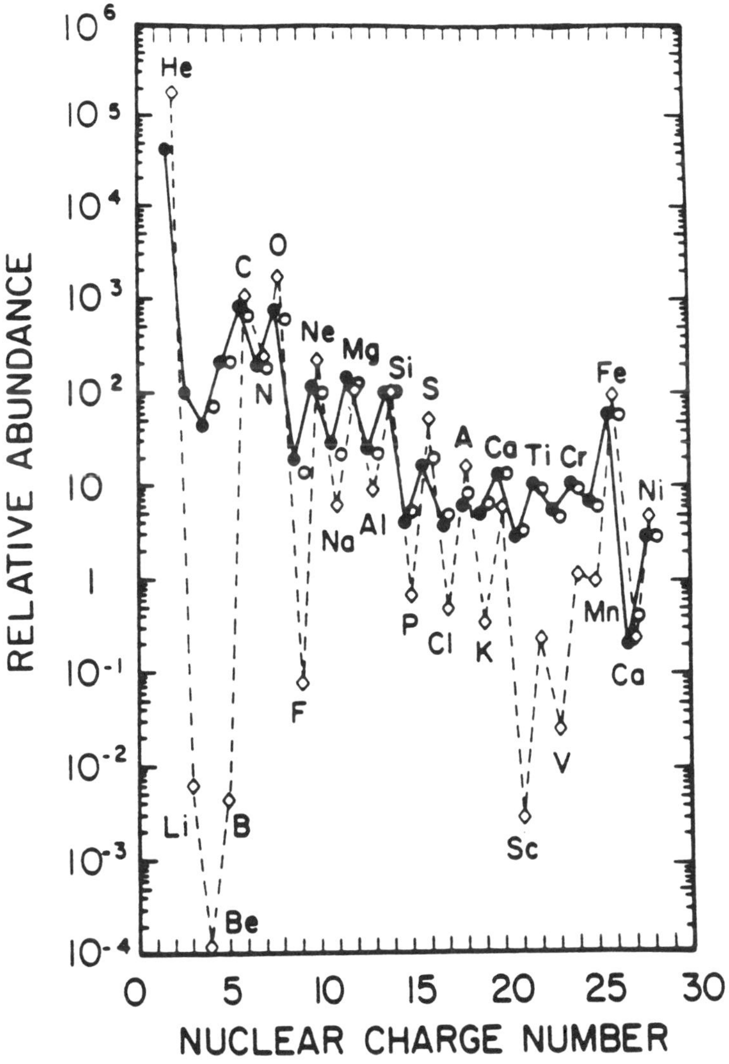

Fig.1. The cosmic-ray elemental abundances (He to Ni) measured at earth, compared to solar system abundances, normalized at Si. Solid circles show show the data at 70-280 MeV/u, open circles at 1000-2000 MeV/u, and diamonds show the solar system abundances. (From Simpson, 1983).

(1973), and updated by Silberberg and Tsao (1987, 1990a). Many cross sections have recently been measured, e.g. by Webber, Kish, Schrier (1990) who also constructed parametric empirical equations. Precise calculations include also the effects of (a) radioactive decay, (b) ionization losses, (c) rigidity-dependent leakage from the galaxy, and (d) solar modulation.

The solar-system (or general) abundances of Suess and Urey (1956) have been updated by Cameron (1982) and Grevesse and Anders (1989). The general abundances for the more abundant elements are now known to 10%. The general abundances reflect nucleosynthetic processes in stars and supernovae, i.e. nuclear fusion processes, while the arriving cosmic rays reflect nuclear breakup processes.

The source composition of cosmic rays and the general galactic abundance of elements is more striking in their similarity than in their differences, as shown by Silberberg, Tsao and Letaw (1989) in their Figure 1, reproduced here as Figure 2. While the general abundances of Pt and Pb relative to H is about 10^{-10}, the relative abundances of cosmic-ray source to the general abundances of elements (normalized at silicon) agree to within a factor of 5, except for the elements H, He, and N which are underabundant in cosmic rays (relative to the solar H/Si, He/Si, N/Si ratios) by a factor of 20 to 30.

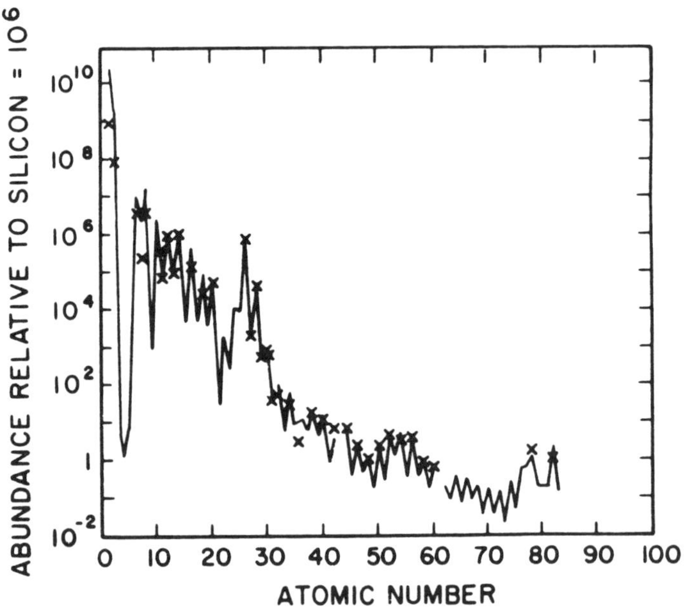

Fig.2. The general abundance of elements in nature compared with the cosmic-ray source abundances (shown by the crosses), normalized at 10^6 for silicon.

The differences between the cosmic-ray source and the general abundances are displayed in Figure 3, which shows the ratios of cosmic-ray source abundances to the general abundances. These ratios are shown as a function of the first ionization potential (FIP). The ratios are normalized at Si. Clearly, the FIP selection process provides a remarkable ordering of most of the abundances, but it leaves some notable differences to be explained.

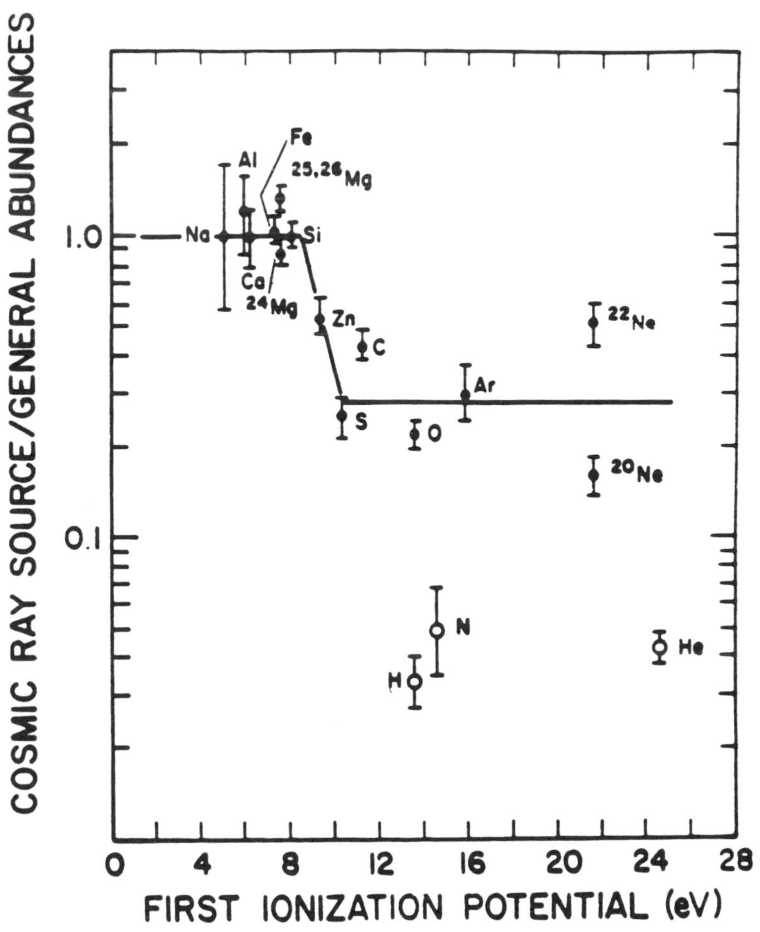

Fig.3. The ratio of cosmic-ray source abundances from spallation - corrected cosmic-ray observations to the general abundances.

3. Our Preffered Model for Processes at the Sites of Origin of Cosmic Rays

3.1 FIRST IONIZATION POTENTIAL

Elements whose first ionization potential exceeds 10 eV are about 4 times less abundant in cosmic rays. The process which brings particles from stellar photospheres to the corona is subject to FIP selection. The solar coronal composition and both the solar wind and solar flare particles display a suppression of elements with FIP > 10 eV, similar to that of the line shown in Figure 3. The physical meaning of the first ionization potential is: if the energy input to an atom equals or exceeds the first ionization potential, the atom can be stripped of an electron from its outermost filled energy level, producing a singly ionized atom. The temperature at which such ionization occurs is like that of the photosphere of a normal star like the sun. Most elements, including Na, Mg, Al, Si, Ca, Fe and Ni have FIP < 10 eV. Elements with FIP > 10 eV are: the noble gases He, Ne, Ar, Kr and Xe, the relatively light elements H, C, N, and O, the halogens F, Cl, Br and I, the non-metallic elements P and S, and in addition Hg. The FIP-dependent suppression was proposed by Havnes (1971), and a detailed investigation was carried out by Casse, Goret and Cesarsky (1975). The solar flare particles (derived from the corona) have a composition that is more similar to the cosmic-ray source particles than to the solar photosphere; this was noted by Webber (1975) and emphasized by Meyer (1985). Meyer (1985) points out that a further stage of injection (stellar flare particles near shock waves of supernova remnants) at energies near 1 MeV is possible, as relative ionization loss effects on particle range and composition cancel due to the effective charges of atoms near energies of 1 MeV/u. Shapiro (1990) presents several strong arguments that the very numerous flare stars are likely to be injectors of cosmic rays.

We suggest a model of three stages of acceleration and two "filters". The first stage of acceleration brings the particles from the stellar photospheres (energies near 1 eV) to the corona (energies near 10^3 eV), with the first FIP-dependent "filter" acting on elemental composition. (Possibly the singly ionized ions leak out more easily along the magnetic field lines than the neutral high-FIP atoms). The second stage of acceleration energizes some coronal particles (10^3 eV/u) to flare particles (~ 10^6 eV/u or 1 MeV/u). The second "filter", proposed by Silberberg and Tsao (1990b), suppresses the lighter nuclei (Z < 12) prior to acceleration to cosmic-ray energies. We shall next explore the second "filter" and the physical processes giving rise to it.

3.2. RIGIDITY-DEPENDENT SUPPRESSION OF LIGHT NUCLEI

A major difference between the cosmic-ray and general abundances, only partially explained by FIP, is the underabundance of H, He and N in cosmic rays, as seen in Figure 3. The explanation proposed here suppresses C and O, but this is compensated by enhancement of C and O, derived from Wolf-Rayet stars. (The latter produce also much of ^{22}Ne, some $^{25,\,26}$Mg, and probably ^{18}O.) A rigidity-dependent suppression of light nuclei was proposed by Silberberg and Tsao (1990). Particles with a high rigidity have a larger radius of curvature in a magnetic field, thus can escape more readily. (Rigidity is momentum/Z_{eff} where Z_{eff} is the effective charge of an incompletely ionized ion, i.e., the nuclear charge minus the number of electrons attached to the nucleus). At flare-particle energies near 1 MeV/u, multiple electron pickup becomes appreciable in the interval C to Mg. From Northcliffe and Schilling (1970) we can deduce that $A/Z_{eff} = 2.0$ for He and 2.95 for Mg. Thus, the escape of He across a rigidity cutoff at the boundary shocks of astrospheres and the interstellar space is suppressed relative to Mg. If the exponent of the integral momentum spectrum of stellar flare particles is 5, i.e. $J(P > P_o) \propto P_o^{-5}$ then He is suppressed relative to Mg by a factor $(2.95/2)^5 = 7$. Additional details of this rigidity-dependent suppression of nuclei with $Z < 12$ are given by Silberberg and Tsao (1990b). The determination of the suppression factor from first principles would require detailed knowledge of the energy spectrum of stellar flares and of processes at the astrosphere boundary shocks. In the absence of such knowledge, we shall adopt an empirical factor S(Z), where $0.12 \leq [S(Z) = 0.15\,Z - 0.93] \leq 1.0$, i.e., $S(Z) = 0.12$ for $Z \leq 7$ and 1 for $Z \geq 13$. This factor explains the low values of H, He, N, and ^{20}Ne in Figure 3. The ratio of cosmic-ray to general abundances is given by I S(Z), except for the nuclides with a major component from Wolf-Rayet stars. The term $I = 0.27$ is the factor that represents the correction for FIP, if FIP ≥ 10.4 eV. $I = 1$ for FIP ≤ 8.4, and there is a transition region in between, shown by the line in Figure 3, and the element Zn.

3.3 CONTRIBUTION OF WOLF-RAYET STARS TO COSMIC RAYS

In Fig. 3 we note several more discrepancies from the solid-line fit: the enhancement of ^{22}Ne and slightly of $^{25,26}Mg$, and the deviation of C and O from the line-fit. All of these are explainable in terms of nucleosynthesis in Wolf-Rayet stars, with about 2% contribution of Wolf-Rayet energetic wind particles to the total cosmic-ray pool. The Wolf-Rayet contribution was proposed by Meyer (1981). Prantzos et al. (1985) showed that the abundance ratio C/O in cosmic rays (about twice the solar one) can be explained in terms of Wolf-Rayet star contribution to carbon. (The winds of Wolf-Rayet stars are energetic, close to 0.1 MeV per nucleon, relatively close to energies of flare particles. The acceleration of these wind particles thus is plausible).

The Wolf-Rayet stars go through two phases, the WN stars in which (or in whose precursors) the hydrogen-burning CNO cycle operates, with much of H burning into He, and C into N. Thereafter, during the WC phase, He burns into C and O, e.g. $3(^4He) \rightarrow {}^{12}C$ and ^{14}N into ^{18}F ($^{14}N + {}^4He \rightarrow {}^{18}F$) which yields ^{18}O, by β decay. (The presence of enhanced ^{18}O in cosmic rays was recently reported by Gibner et al. (1992). However, the exact degree of enhancement should still be questioned until the secondary contribution $p + {}^{22}Ne \rightarrow p + {}^{18}O + He$ is measured rather than semiempiracally calculated. The cross sections of reactions in which ^4He or 2p + 2n are produced have large values. In Wolf-Rayet stars $^4He + {}^{18}O \rightarrow {}^{22}Ne$ is an important reaction, which, to some extent is followed by burning into ^{26}Mg and ^{25}Mg. These reactions explain the remaining apparent anomalies in Fig.3.

3.4. THE CALCULATION OF SOURCE ABUNDANCES AND COMPARISON WITH THE EXPERIMENTAL SOURCE ABUNDANCES

Based on the procedures discussed above, the calculated cosmic-ray source abundances (CR) are given by the equation (CR) = 10^{-4} (G) I S + 2.1 (10^4 A^{-1}) [1.0 or 0.63] (WC + aWN).

Here (G) is the general abundance of Grevesse and Anders (1989), normalized to 10^6 for Si. Hence multiplication by 10^{-4} changes it to Lund's normalization of 100 for Si. The first ionization potential factor I and light-ion suppression factor S have been defined above. WC and WN are the compositions of ejecta from the WC and WN stars, tabulated in Table 2 of Prantzos et al. (1986). The factor "a", also from Table 2, is the fraction of mass loss during the WN phase. Integrating over the initial mass function given by Prantzos et al. (1986), a = 0.5. Since the table is in mass fractions while cosmic-ray abundances are in numbers of nuclei, the mass number A is introduced for conversion. The factor 0.63 is the correction for metallicity, discussed by Silberberg and Tsao (1990); it applies, e.g., to nitrogen and for nuclei produced via nitrogen-burning in helium and nuclei heavier than ^{16}O in the Wolf-Rayet stars. For helium and for nuclei produced via He-burning (e.g., ^{12}C, ^{16}O, and ^{4}He), the factor 1.0 is used. The factor 2.1 X 10^4 is for normalization.

With the three processes described above, the source composition of cosmic rays can be calculated from the general abundances following the procedures of Silberberg and Tsao (1990 b).

Fig. 4 shows the ratio of the source abundances (from the spallation-corrected

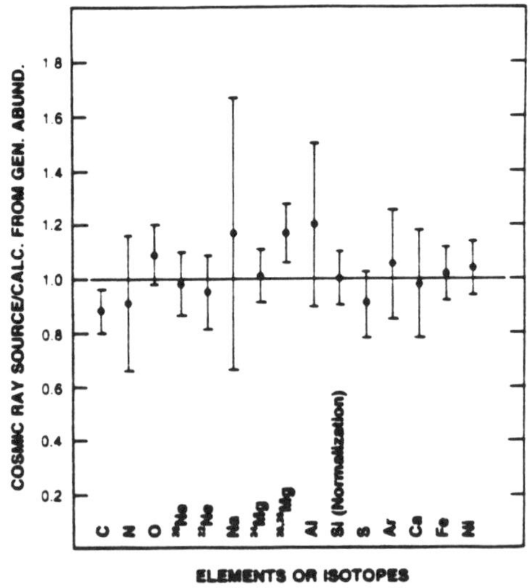

Fig.4. The ratio of cosmic-ray source abundances to the general abundances modified by FIP, suppression of (light) nuclei that have $Z_{eff} \approx Z$, and W-R nucleosynthesis.

cosmic-ray observations) to the calculated cosmic-ray source abundances (deduced from the general abundances, with the three processes described above). We note that the calculated cosmic-ray source abundances agree within a standard deviation of 10% with the observed ones. This is better than the degree of agreement obtained with that of other proposed procedures. This is a strong argument in favor of the model advanced here.

3.5 TESTS FOR THE FLARE STAR AND ESCAPE FROM ASTROSPHERE MODEL

Are the fully ionized light ions subject to a lesser degree of escape from the astrosphere as proposed in Section 3.2.? An investigation of the composition of solar flare particles at, and just beyond the boundary of the heliosphere may be analogous to the conditions at the astrosphere boundaries of flare stars. Within a couple of decades spacecraft will be able to explore the boundary of the heliosphere, and flare particles there.

The relative abundance of Ge and Pb in cosmic rays is somewhat lower relative to the meteoritic abundance as discussed in Section 7. But for these elements the solar and meteoritic relative abundances are in disagreement, and the solar relative abundances agree with those of cosmic rays. It could be argued that the solar spectral measurements are uncertain. A test would be to measure the relative abundance of Ge in solar flare particles, with a large detector that overcomes the small number of Ge nuclei.

4.Composition and Acceleration of Cosmic Rays at $1 \leq E$ (TeV/u) $\leq 10^3$

The observational data of Burnett et al. (1990) imply an enhanced He/H ratio and even greater enhancement of (CNO)/H. Silberberg et al. (1990) proposed that the acceleration of the pre-supernova stellar wind material, with the acceleration mechanism of Völk and Biermann (1988) at very young supernova remnants provides an explanation.

Our calculation weighted (a) the relative abundances of red supergiants, blue supergiants and Wolf-Rayet stars, and (b) their respective mass loss rates. The surfaces of these stars display (1) enhancement of hydrogen-burning products of the CNO cycle (He and N), and (2) for W-R stars, of the He-burning products ^{12}C and ^{16}O as well as those of ^{14}N and ^{4}He burning that yields ^{18}F, with $^{18}O + {}^{4}He \rightarrow {}^{22}Ne$. The enhancements (relative to H) are: by a factor or 3 for ^{4}He; 5 to 6 for ^{12}C, ^{16}O, ^{22}Ne; 8 for ^{14}N. Thus a ratio $^{14}N/^{16}O$ of 4 per cent at lower energies, becomes 6 or 7 per cent. Most Ne is predicted to be ^{22}Ne at these very high energies.

The acceleration model of Völk and Biermann (1988), involving pre-supernova wind material, acts in a very compact region, i.e., in a very young supernova remnant. Freshly nucleosynthesized matter, as in the model of Yanagita (1990) may become admixed with the wind material and contribute to the composition of cosmic rays at very high energies ($E \geq 1$ TeV/u). If acceleration occurs early-on in type I supernovae, (which are not likely to have abundant wind material), the elements $16 \leq Z \leq 28$ may be enhanced in cosmic rays at very high energies. A test would be whether the flux of Ar would be suppressed. This should not occur, as the FIP should not operate at the high temperatures of supernova ejecta. As the shock waves of the supernova remnant pass beyond the stellar wind domain, they become somewhat weaker and accelerate mainly the particles that end up in the energy interval 10^8 to 10^{12} eV/nucleon.

Figure 5 illustrates the respective acceleration regions, the stellar wind domain, and the outer acceleration region in which probably the flare particles of neighboring stars and higher-energy wind particles, such as those of Wolf-Rayet stars get accelerated. Figure 6, based on a review by Linsley et al. (1989), shows the energy spectrum of cosmic rays. In the present model, the strong shock waves in the immediate vicinity of the supernova generate the flatter region of energy spectrum near 10^5 and 10^6 GeV, and its dashed

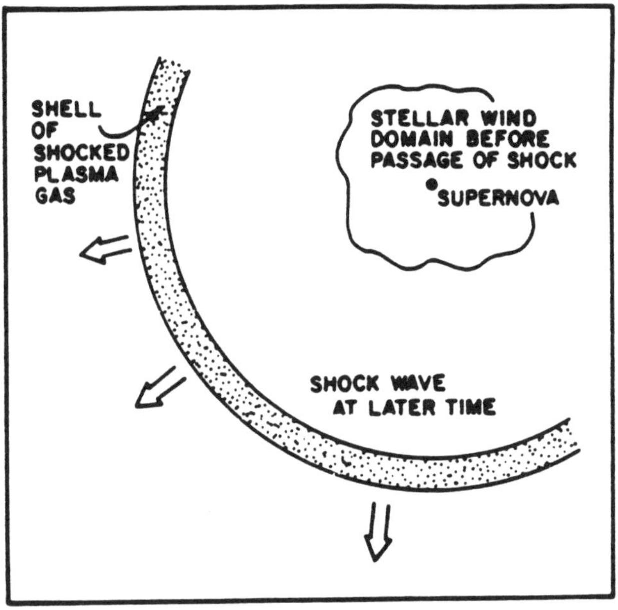

Fig.5 Illustration of the regions of acceleration of cosmic rays. First the very strong shock waves of the young supernova remnant accelerate the presupernova stellar wind particles. Later, energetic (flare) particles, possibly near 1 MeV/nucleon, are accelerated in interstellar space.

extension toward lower energies, while the steeper spectrum below about 10^4 GeV is attributed to the outer acceleration region shown in Figure 5. Correcting the spectra for rigidity-dependent leakage from the Galaxy yields flatter source spectra, with the respective exponents α of $dJ/dE = KE^{-\alpha}$ reduced from 2.75 to about 2.4, and from 2.50 to about 2.15 (without weak reacceleration, the reduction of α would be larger, about 0.5).

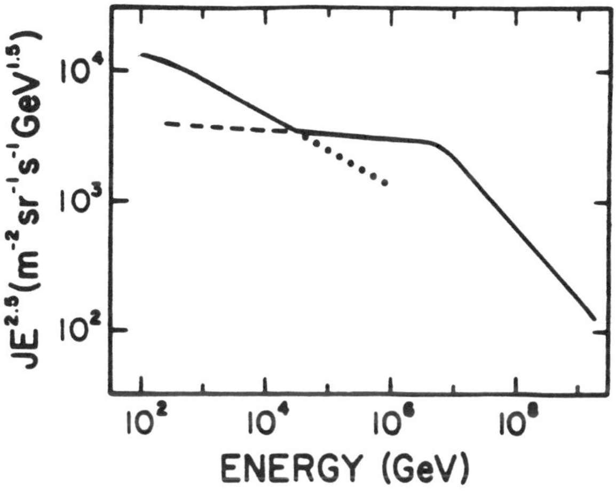

Fig.6 The energy spectrum of cosmic rays, multiplied by $E^{2.5}$ at energies 10^2 to 10^9 GeV, based on Linsley (1983) and Burnett et al. (1989). The significance of the slopes at 10^2 to 10^4 GeV and 10^4 to 10^6 GeV is discussed in the text.

5. The Model of Origin at Supernovae, Compared with Origin in Flare Star Material

The model of origin in supernovae, with admixture of matter of the interstellar medium has been explored by Hainebach et al. (1976), Shapiro (1985) and recently by Yanagita et al. (1990). The latter applied the nucleosynthesis calculations of Nomoto et al. (1984) for type I supernovae and of Hashimoto et al. (1989) for type II. While the composition of the material in our model described in Section 3 and that of Yanagita et al. differ, both assume that the principal acceleration to cosmic-ray energies is accomplished by the shock waves of supernova remnants.

We shall now explore the model of Yanagita et al. (1990). The mixing ratios are 1.0 for material of SN I and 6.2 to 9.3 of SN II, mixed with about 10 to 20 M_0 of local galactic material per supernova. The elements $16 \leq Z \leq 28$ as well as half of Si are attributed to SN I; the elements $8 \leq Z \leq 12$ and half of Si, to SN II; and H, He, N to the interstellar medium. All three components contribute to carbon.

Fig. 7 compares our model (Silberberg and Tsao, 1990b normalized at Si), with that of Yanagita et al. (1990). While our model has a standard deviation of 10% for elements $1 \leq Z \leq 28$, that of Yanagita et al. (1990) has about 80%. We note that the ratio of calculated ratios to those inferred from observations, ^{16}O/(Mg, Al, Si) = 2 and that of ^{20}Ne/(Mg, Al, Si) = 4. Furthermore, the isotopic ratio ^{22}Ne/^{20}Ne relative to the observed value is 1/3. But if one introduced some W-R contribution to make the latter ratio 1, then the ratio of calculated to experimental ratios ^{22}Ne/(Mg, Al, Si) = 4 just as for ^{20}Ne above.

It is possible that the actual errors of the supernova model are smaller, and that the relatively large errors are due to errors in nucleosynthesis calculations.

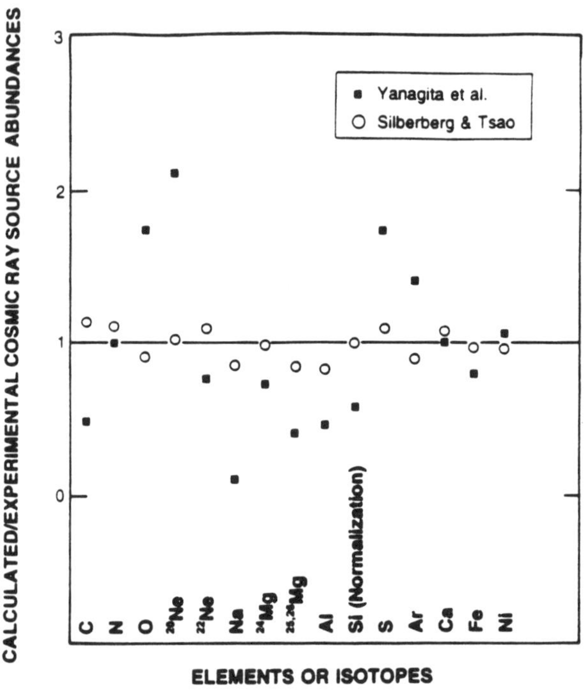

Fig.7. The ratio of calculated to observational abundances. The circles represent the model of Silberberg and Tsao (1990) and the squares the nucleosynthesis-in-supernovae model of Yanagita et al. (1990). The ratio is the inverse of Figure 6, to make it directly comparable with the figures of Yanagita et al.

Yet the model uses several ad-hoc parameters: (a) the relative contribution of type I to type II supernovae, (b) the admixture of interstellar gas that supplies H, He and N, and (c) may need a contribution of Wolf-Rayet stars, to adjust the $^{22}Ne/^{20}Ne$ ratio and the abundance of C.

There are some new experimental data that are not consistent with the model of origin in supernovae: (1) the ratio of ^{57}Co to ^{59}Co is about 1 rather than 8, (Leske, Milliken, and Wiedenbeck, 1992) which implies the decay of ^{59}Ni into ^{59}Co, hence acceleration > 10^5 years after synthesis in supernovae. (2) The early analysis of LDEF data (Thompson et al., 1993) indicates Th/U $\approx$ 2 and Th/Pu $\approx$ 5/0.5, i.e., an age of > 10^9 years for the accelerated material, based on Fig. 8 taken from Blake and Schramm, (1974). For $\leq 10^7$ years, the expected Th/Pu ratio $\approx$ 0.5, to be compared with the observed value of ~ 10. Within 2 years, the number of actinides analyzed should be larger by an order of magnitude, and permit a statistically firm conclusion.

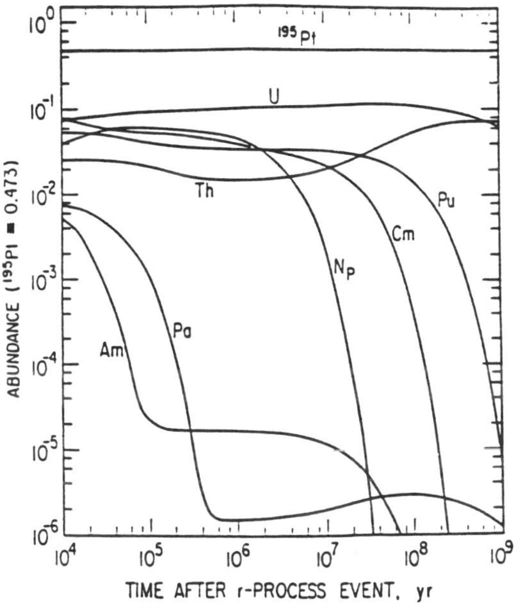

Fig.8. The relative abundances of the actinide elements, as a function of time after nucleosynthesis. (From Blake and Schramm, 1974).

6. Preferential Acceleration of Heavy (High-Z) Nuclei

This model explains the dearth of H and He, and the more limited dearth of C, O and Ne. But it does not explain: (a) the lack of relative enhancement of He/H, predicted by Fig. 1, of Eichler and Hainebach (1991), (b) the low $^{14}N/^{12}C$ ratio (see Fig. 1, Shapiro et al. 1993), (c) the low $^{36}Ar/^{40}Ca$ ratio (see Fig.1 Shapiro et al. 1993) (d) the $^{22}Ne/^{20}Ne$ enhancement, and (e) C/O in cosmic rays being 2x that of the general abundances.

7. Enhanced Contribution of Non-Volatile Elements to Cosmic Rays

This model (recently explored by Sakurai, 1990) predicts a suppression of volatile elements, with results rather similar to the predictions of FIP. It also predicts suppression of Ge whose abundance is controversial. Using the meteoritic abundances as reference, Ge in cosmic rays is under-abundant. Using the solar spectra, no suppression of Ge is required. This model does not explain (a) the low abundance of N (relative to C or O or Ar), (b) the greater suppression of H, or He relative to Ar, and (c) the relative enhancement of $^{22}Ne/^{20}Ne$.

Also this model needs several ad-hoc assumptions: (a) The moderately volatile elements Na, Ge, Pb are reduced by a factor of 0.6. (b) Most of the volatile elements have an abundance of ~ 0.3 relative to the "grain component" e.g. S, Ar, Kr, Xe. (c) Need Z-dependent acceleration to explain the large underabundance of H, He and N relative to Ne, Ar. (d) Need Wolf-Rayet star contribution to enhance the ^{22}Ne/^{20}Ne ratio, and probably also to enhance ^{12}C.

8. Conclusions

The model of Silberberg and Tsao (1990b) and Silberberg et al. (1990 and 1991) of acceleration of flare star particles (after a rigidity dependent escape from the flare star astrosphere) by supernova shock waves, with an admixture of 2% of energetic Wolf-Rayet star wind particles appears to yield the best fit, to about 10%. The differences in composition above 1 TeV/u favor a contribution of acceleration of pre-supernova wind particles, with a slightly flatter energy spectrum, as illustrated in Figure 5.

References

Blake, J.B. and Schramm, D.N. 1974, Ap. Space Sci. 30, 275.
Burnett, T.H. et al. 1990 Ap. J. (Letters) , 349 L25, and 21st Internat. Cosmic Ray Conf. (Adelaide) 3, 101.
Cameron, A.G.W. 1982, in "Essays of Nuclear Astrophysics," ed. C. Barnes, R.N. Clayton and D.N. Schramm, Cambridge Univ. Press, p.23.
Casse, M., Goret, P., and Cesarsky, C.J.: 1975, 14th Intl. Cosmic-Ray Conf. (Munich) 2, 646.
Eichler, D. and Hainebach, K. 1981, Phys. Rev. Letter, 47, 1560.
Gibner, P.S. et al. 1992, Ap. J. Letters 391, L89.
Grevesse, N. and Anders, E.1989, in "Cosmic Abundances of Matter," ed. C.J. Waddington, (New York): AIP Conf. Proc. 183, p.1.
Hainebach, K.L., Norman, E.B. and Schramm, D.N. 1976, Ap.J. 203, 245.
Hashimoto, M., Nomoto, K., and Shigeyama 1989, Astron. Ap. 210, L5.
Havnes, O. 1971, Nature 229, 548.
Leske, R.A, Milliken, B., and Wiedenbeck, M.E. 1992, Ap. J. Letters, 390, L99.
Linsley, J. 1983, Rapporteur Paper, 18th Intl. Cosmic Ray Conf. (Bangalore) 12, 135.
Meyer, J.P. 1981, 17th Intl. Cosmic-Ray Conf. (Paris) 2, 265.
Meyer, J.P. 1985, Ap.J. Suppl. 57, 173.
Nomoto, K. Thielemann, F.K., and Yokoi, K. 1984, Ap.J. 286, 644.
Northcliffe, L.C. and Schilling, R.F. 1970, Nuclear Data Tables A7, 233.
Prantzos, N. et al. 1985, 19th Intl. Cosmic-Ray Conf. (LaJolla) 3, 167.
Shapiro, M.M. 1985, 19th Intl. Cosmic-Ray Conf. (LaJolla) 3, 163.
Shapiro, M.M. 1990, 21st Intl. Cosmic-Ray Conf. (Adelaide)4, 8.
Shapiro, M.M., Silberberg, R., and Tsao, C.H. 1993, 23rd Intl. Cosmic Ray Conf. (Calgary), 2, 386.
Silberberg, R., and Tsao, C.H. 1973, Ap.J. Suppl. 25, 315.
Silberberg, R., and Tsao, C.H. 1987, 20th Intl. Cosmic Ray Conf. (Moscow) 2, 133.
Silberberg, R. and Tsao, C.H. 1990a, Physics Reports , 191, 351.

Silberberg, R. and Tsao, C.H. 1990b, Ap.J. (Letters), 352, L49.
Silberberg, R. Tsao, C.H. and Letaw, J.R. 1989, in "Cosmic Gamma Rays Neutrinos and Related Astrophysics" ed. M.M. Shapiro and J.P. Wefel, Publ. by Kluwer Acad. Publ. p. 491.
Silberberg, R. Tsao, C.H., Shapiro, M.M. and Biermann, P.L. 1990, Ap.J. 363, 265.
Silberberg, R. Tsao, C.H., Shapiro, M.M. and Biermann, P.L. 1991, in "Cosmic-Rays, Supernovae and the Interstellar Medium", ed. M.M. Shapiro, R. Silberberg, and J.P. Wefel, Kluwer Acad. Publ., p.97.
Simpson, J.A. 1983, Ann. Rev. Nucl. Particle Sci., 33, 323.
Suess, H.E., and Urey, H.S. (1956), Rev. Mod. Phys. 28, 53.
Thompson, A. et al. 1993 Internat. Cosmic Ray Conf. (Calgary) 1, 601.
Völk, H.J., and Biermann, P.L. 1988, Ap.J. (Letters), 333, L65.
Webber, W.R. 1975, 14th Intl. Cosmic-Ray Conf. (Munich) 5, 1597.
Webber, W.R., Kish, J.C. and Schrier, D.A. Phys. Rev. 41, 520, 533, 547 and 566.
Yanagita, S., Nomoto, K., and Hayakawa, S., 1990, 21st Intl. Cosmic Ray Conf. (Adelaide), 4, 44.

HIGHLIGHTS OF THE 1993 ICRC ON COMPOSITION AND PROPAGATION OF COSMIC RAYS; AND ON NEUTRINOS

REIN SILBERBERG
Universities Space Research Association
Washington, DC 20024

CHEN HSIANG TSAO
E.O. Hulbert Center for Space Research
Naval Research Laboratory
Washington, DC 20375-5352

ABSTRACT. The isotopic abundances of elements Li to Ca, and Fe are now reasonably well measured. The new data imply: (1) no overabundance of ^{29}Si and ^{30}Si, (2) small or no overabundance (by about 1.15 ± 0.15) of ^{25}Mg and ^{26}Mg. Possible overabundance of ^{18}O, but the conclusion is highly dependent on the subtraction of the large secondary component from the total ^{18}O. The calculated secondary component depends on whether one uses the cross sections estimated by Webber et al. or Silberberg and Tsao. The escape rate of cosmic rays from the Galaxy (previously estimated α $R^{0.6}$, can be reduced to $R^{1/3}$) as expected from Kolmogorov theory, if one corrects for (a) the energy dependence of cross sections, newly measured by Webber et al. and (b) weak distributed reacceleration of cosmic rays. Progress on cosmic-ray origin, deduced from the source composition is presented by us and Prof. Shapiro in a separate paper. The neutrino detectors DUMAND, AMANDA, NESTOR and at lake Baikal show significant progress, hence experimental high-energy neutrino astronomy is only a few years in the future.

1. INTRODUCTION

Cosmic rays were discovered by Hess in 1912. Progress in the field has been steady, with many discoveries still to come. It took 36 years, until 1948 that heavy nuclei (heavier than helium) were discovered in cosmic rays by Freier et al (1948). It took another 25 years before isotopic resolution was achieved (except for ^{3}He/^{4}He, which was measured by O'Dell et al. (1965). Every bi-annual International Cosmic Ray Conference, starting in the 1940-ies, has up-dated discoveries in the field. The present paper deals with the 23rd Cosmic-Ray Conference (ICRC) in Calgary, Canada. Only some topics will be explored, namely those that the authors are best familiar with: elemental and

M. M. Shapiro et al. (eds.), Currents in High-Energy Astrophysics, 103–121.

isotopic composition, nuclear transformations, cross sections, origin and propagation, including reacceleration, and the newly starting field of neutrino astrophysics.

Section 2 presents the Calgary Conference data on cosmic-ray isotopes. Section 3 presents the new data on elemental abundances. Section 4 deals with distributed reacceleration of cosmic rays. Section 5 is on recent cross section measurements that are of importance to cosmic-ray propagation or source composition calculations. Section 6 is on neutrino astronomy and neutrino detectors.

2. COSMIC RAY ISOTOPES

The isotopes in cosmic rays are (1) primary, from cosmic-ray sources that tell us about the source composition and (2) secondary (spallation products) that tell us about cosmic-ray propagation and interactions in the interstellar gas. Both classes have long-lived radioactive isotopes, e.g. ^{244}Pu for the primaries (not yet observed) that set limits on the time of nucleosynthesis of nuclei that later were accelerated to cosmic-ray energies, and secondaries, e.g. ^{10}Be, whose abundance permits us to deduce the confinement time of cosmic rays in the Galaxy.

Among the light primaries, ^{1}H and ^{4}He and the secondaries ^{2}H and ^{3}He were studied on Voyager 2 for energies <100 MeV/u, at a distance of 23 AU, and presented by Seo et al. (1993) at Calgary. They concluded that the secondary/primary ratios of these very light cosmic rays are consistent with the mean path length of heavier cosmic rays like B/C.

Seo et al. (1993) used the cross sections of Webber (1990) in their analysis. The production of deuterium has been calculated by Papini, Grimani and Stephens (1993), for (1) cosmic-ray ^{4}He breakup in interstellar gas, (2) $^{1}H + ^{1}H \rightarrow ^{2}H + \pi^{+}$, and (c) proton interactions with nuclei, including atmospheric nuclei.

The $^{3}He/^{4}He$ ratio was measured at higher energies, 100 to 1600 MeV/u by Ficenec et al. (1993), using a balloon flight of SMILI (Superconducting Magnet Instrument for Light Isotopes). Fig. 1 shows the ratio $^{3}He/^{4}He$ as a function of energy. We note that their value agrees well with that of O'Dell et al. (1965), measured near 300 MeV/u. The path length agrees with that deduced from the B/C ratio. It is 9.2 g/cm^{2} near 1 GeV/n. They rule out the simple "closed Galaxy" model, which would generate 2-3 times too much ^{3}He.

The confinement of cosmic rays in the Galaxy was explored by Lukasiak et al. (1993), using ^{10}Be at low energies, 35 to 113 MeV/u. They conclude an average density 0.275 ± 0.12 nucleons/cm^{3} for the propagation medium. With the diffusion-in-halo model of Ptuskin (1974), they estimate a time of 1.5×10^{8} years before escape from the halo. Of course, the particles far out in the halo have a minute probability to return to the central plane of the Galactic disk where the sun is located.

Hindi et al. (1993) suggest another isotope, ^{91}Nb for measuring the confinement time of cosmic rays. Normally ^{91}Nb decays by electron capture with a half life of 680 years. When it is stripped of electrons, it decays by e^{+} emission. Hindi et al. have measured the half-life of stripped ^{91}Nb. It is $(8.8 \pm 1.9) \times 10^{6}$ years. However, also ^{92}Nb should be produced, and they plan to measure the respective production rates, so

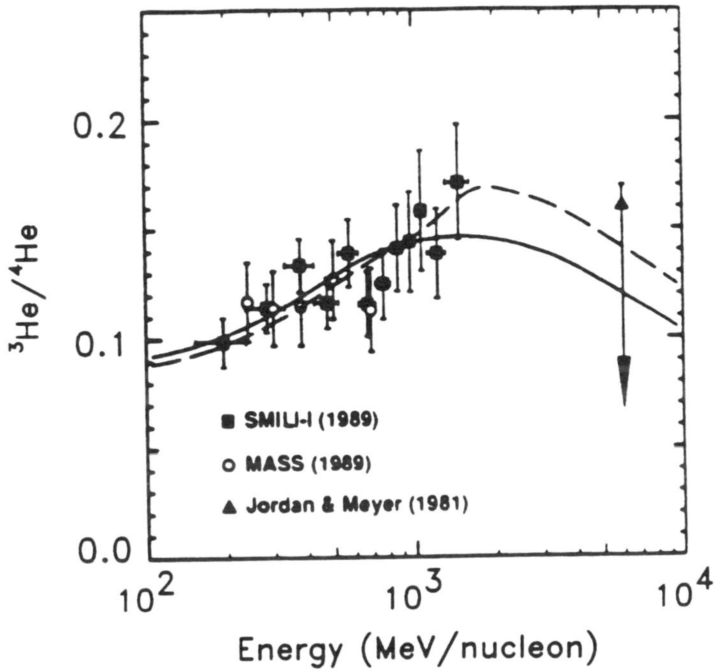

Figure 1. The ratio $^3He/^4He$ as a function of energy from Ficenec et al.

that the ratio Nb/Mo could be used to deduce the confinement time.

The isotopic composition of C, N, and O, i.e. ^{12}C, ^{13}C, ^{14}N, ^{15}N, ^{16}O, ^{17}O, and ^{18}O has been measured by 3 groups: (a) Lukasiak et al. (1993), with the Voyager 1 and 2 spacecrafts at 10 to 30 AU, b) Garcia-Munoz et al. (1993) with the CRRES satellite, and Connell and Simpson (1993) with the Ulysses spacecraft. The mass resolution is excellent, 0.1 to 0.2 amu. Fig. 2 from Connell and Simpson (1993) shows the measured isotopic abundance ratios $^{13}C/^{12}C$, $^{15}N/N$, $^{17}O/^{16}O$ and $^{18}O/^{16}O$, measured by 5 groups. The propagated solar system composition is also shown. The experimental ratios agree, but the values calculated-at-source disagree, as shown in table 1 below.

Table 1. Isotopic Composition

	Measured		Calculated at Source	
Ratio	Lukasiak	Connell	Lukasiak	Connell
$^{13}C/^{12}C$	0.060±0.005	0.071±0.005	0.003±0.005	0.006±0.005
$^{18}O/^{16}O$	0.015±0.002	0.020±0.003	0.004±0.002	-0.003±0.002

The data of Table 1 imply (a) a lower ^{13}C abundance at the source, possibly a suppression similar to ^{14}N. Correspondingly, the $^{12}C/^{13}C$ ratio is enhanced relatively to

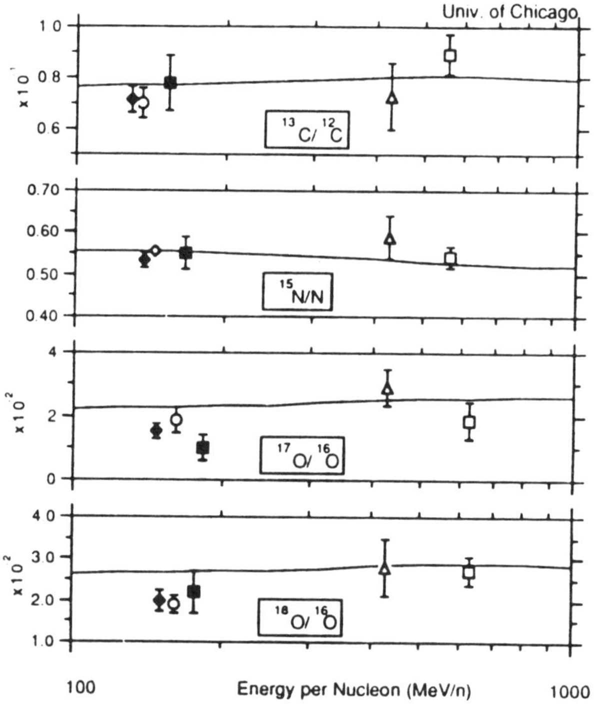

Figure 2. The isotopic abundance ratios $^{13}C/^{12}C$, $^{15}N/N$, $^{17}O/^{16}O$ and $^{18}O/^{16}O$. The propagated solar system composition is also shown (Connell and Simpson).

solar abundances. This is consistent with much of ^{12}C coming from Wolf-Rayet stars. The suppression of ^{13}C, similar to that of ^{14}N is consistent with our model of suppression of fully ionized light flare particles at the shock boundaries of the astrospheres.

The source abundance of $^{18}O/^{16}O$, and the large difference between Lukasiak et al. (1993) and Garcia-Munoz et al., or Connell and Simpson is due to the subtraction of secondaries, using (a) the cross sections of Webber et al. (1990) vs. those of Silberberg

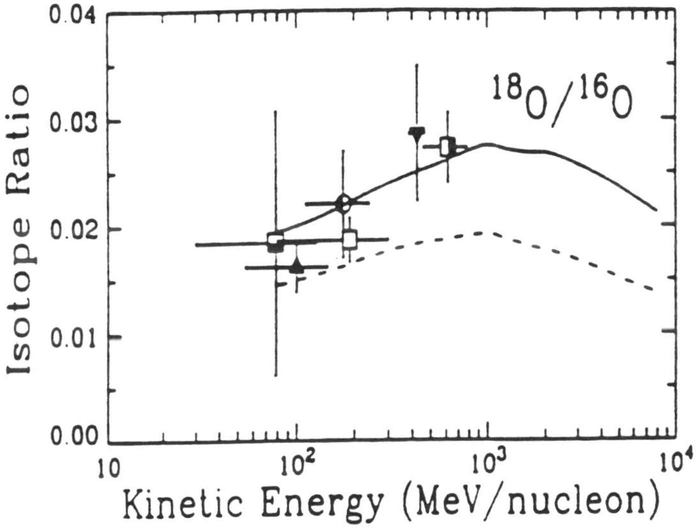

Figure 3. The calculated and measured $^{18}O/^{16}O$ ratio. The calculated curves are based on the cross sections of Silberberg and Tsao vs. Webber et al. (Guzik et al.).

and Tsao (1990). Note that the crucial ^{22}Ne into ^{18}O cross section has not yet been measured, that of Silberberg and Tsao is 4 x larger.

The effect of cross sections is discussed by Guzik et al. (1993). Fig. 3 shows their calculation of the $^{18}O/^{16}O$ ratio, using the S&T vs. Webber cross sections, and solar-like source composition. We see that extra ^{18}O is needed when the cross sections of Webber are used.

Blake and Dearborn (1993) suggest that the helium burning of ^{14}N in Wolf-Rayet stars into ^{22}Ne, via ^{18}O could result in an ^{18}O overabundance.

Also the Ne, Mg and Si isotopic abundances were measured by Connell and Simpson (1993) and DuVernois et al. (1993), with Ulysses and CRRES spacecrafts, respectively. Figure 4 shows the ratios $^{21}Ne/^{20}Ne$, $^{22}Ne/^{20}Ne$, $^{25}Mg/^{24}Mg$, $^{26}Mg/^{24}Mg$, $^{29}Si/^{28}Si$ and $^{30}Si/^{28}Si$, of Connell and Simpson, including earlier measurements. They confirm the $^{22}Ne/^{20}Ne$ overabundance by a factor of 3 or 4 in the cosmic-ray sources, consistent with a significant contribution from Wolf-Rayet stars. However, they do not find the overabundances of ^{25}Mg, ^{26}Mg, ^{29}Si and ^{30}Si reported earlier. The results thus argue against the super-metallicity model. The Wolf-Rayet model, with a slight enhancement of heavy Mg, and negligible enhancement of heavy Si is still compatible with the average of all groups: enhancement of heavy Mg by 1.15±

0.15. The isotopic measurement of Si by Hesse et al. (1993) is consistent with that reported above.

Leske and Wiedenbeck (1993) have developed improved techniques to analyze the 10-year old ISEE-3 data. They report on the elemental abundances of 9≤Z≤20 at 220 MeV/u, and on isotopic composition of S, Cl, Ar, K and Ca. Figure 5 shows the

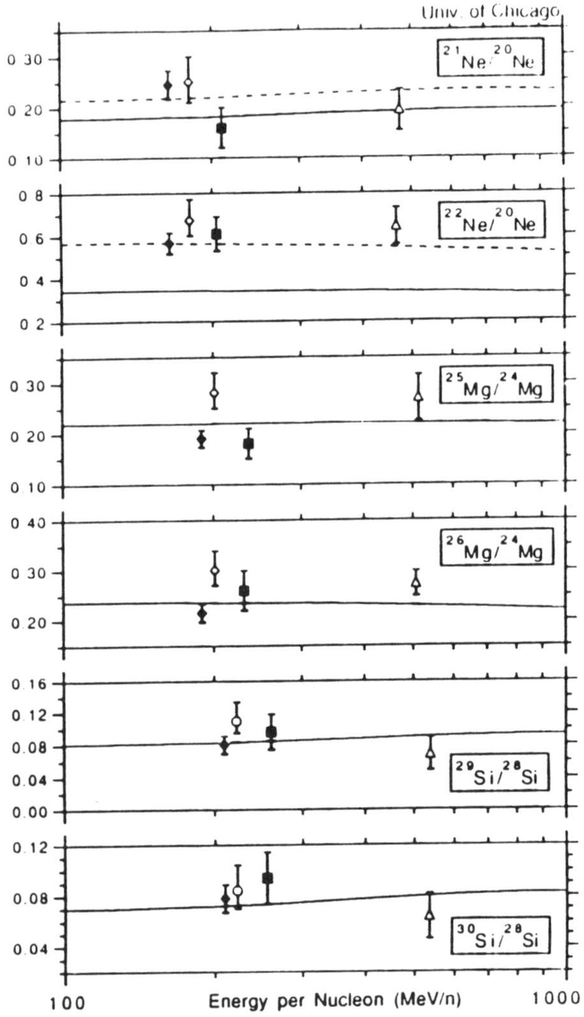

Figure 4. The ratios $^{21}Ne/^{20}Ne$, $^{22}Ne/^{20}Ne$, $^{25}Mg/^{24}Mg$, $^{26}Mg/^{24}Mg$, $^{29}Si/^{28}Si$ and $^{30}Si/^{28}Si$. (Connell and Simpson).

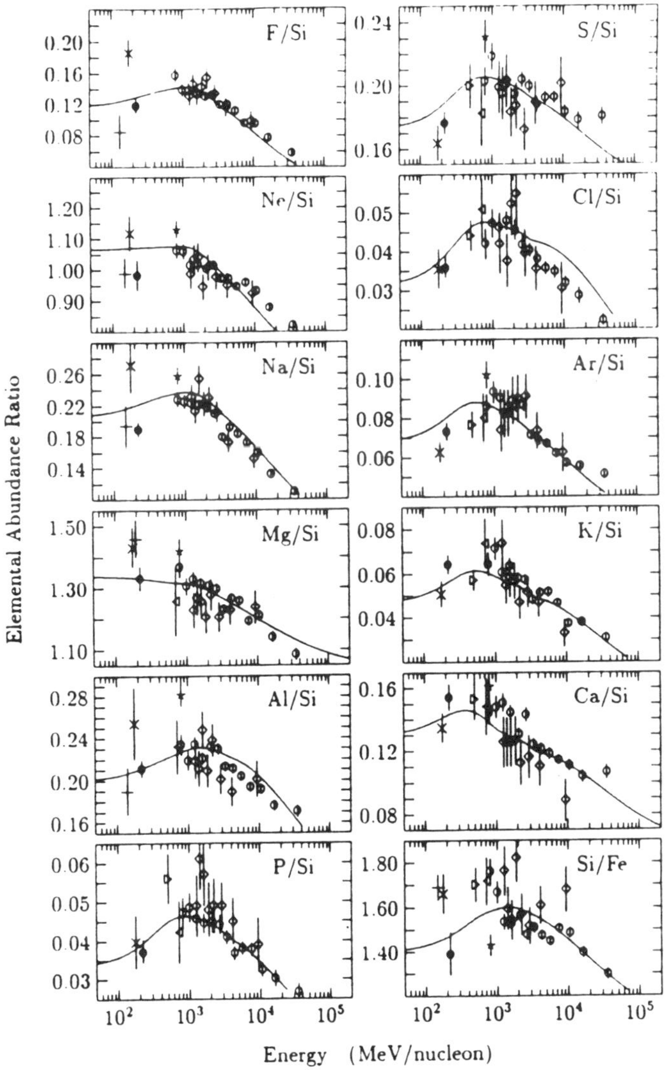

Figure 5. The elemental abundances of 9≤Z≤20 relative to Si. (Leske and Wiedenbeck).

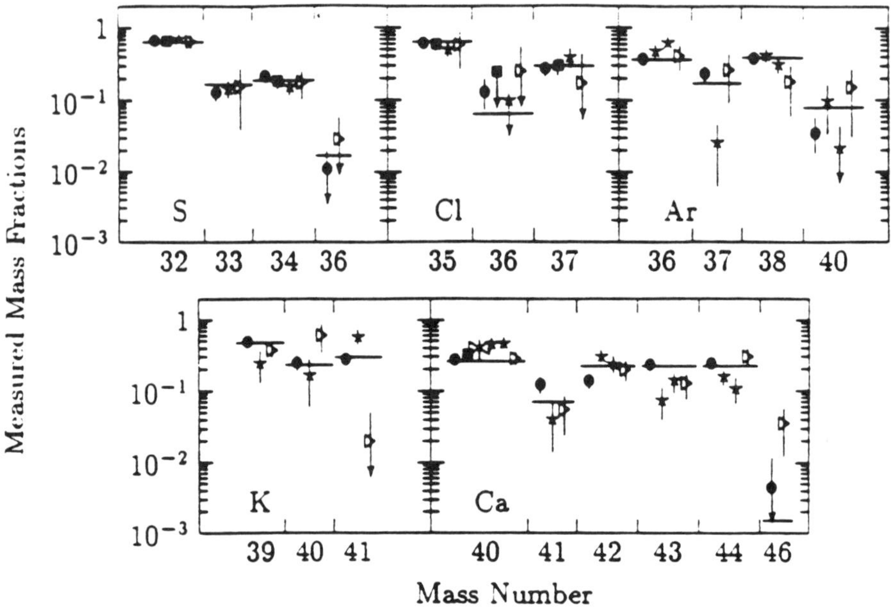

Figure 6. The isotopic abundances of S, Cl, Ar, K and Ca. (Leske and Wiedenbeck).

elemental abundances relative to Si, as a function of energy. We note that the data agree well with the curves based on propagation calculations. Only the data for Cl imply that the calculated cross sections for the production of Cl should be reduced 10%. Figure 6 shows the isotopic abundances of S to Ca. We note that the propagation model reproduces the data rather well.

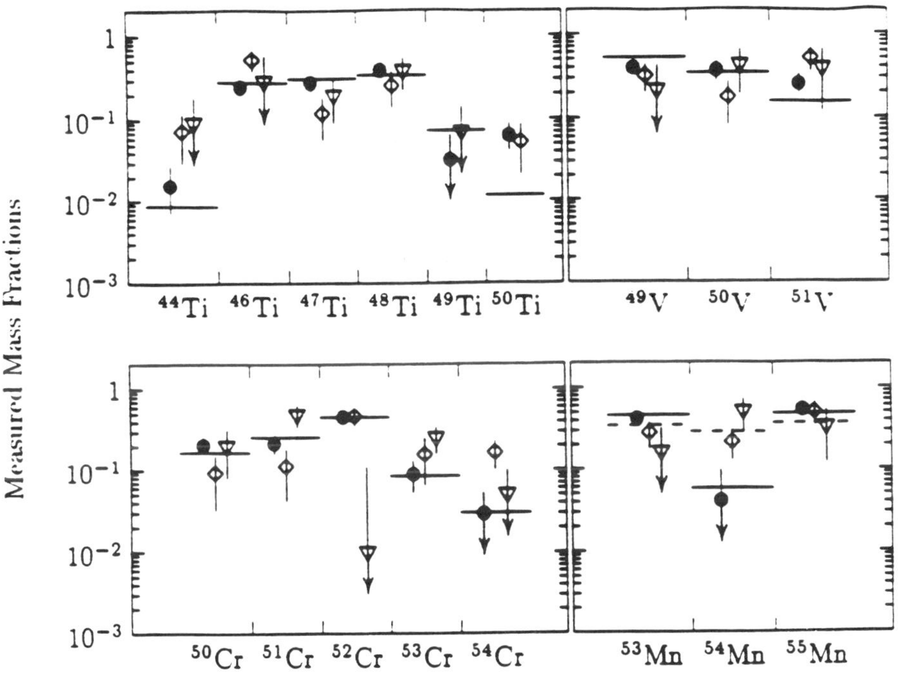

Figure 7. The isotopic abundances of Ti, V, Cr and M. (Wiedenbeck and Leske).

Wiedenbeck and Leske, (1992) COSPAR Conference at Washington, have extended their above-described work to isotopes Ti through Ni. Figure 7 shows the abundances Ti through Mn. The agreement with propagation calculations is good. One interesting feature is that ^{54}Mn has probably decayed, and that its half life is of the order of 10^6 years. Figure 8 shows the isotopic abundances of Fe and Ni, ^{60}Ni may be overabundant by 1.5 or 2. This could be due to some overshooting of the standard

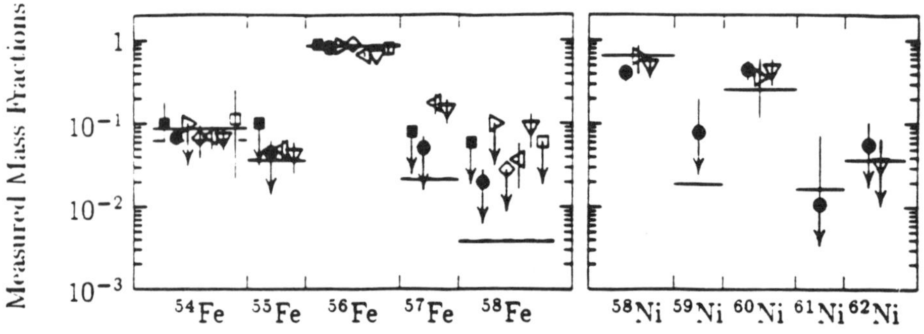

Figure 8. The isotopic abundance of Fe and Ni. (Wiedenbeck and Leske).

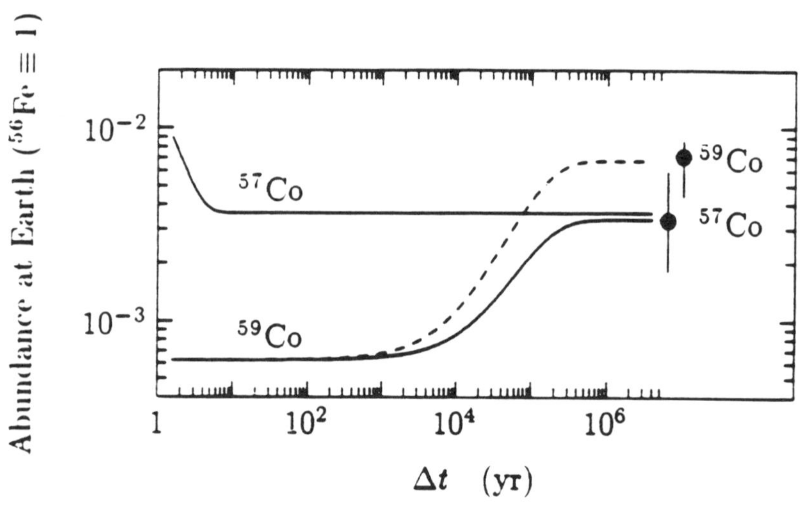

Figure 9. The abundance of isotopes of Co, with the effect of decay of ^{59}Ni shown. (Wiedenbeck and Leske).

nucleosynthesis pathway ^{44}Ti-^{48}Cr-^{52}Fe-^{56}Ni-^{60}Zn, with a decay into ^{60}Ni. Figure 9 shows that ^{59}Ni has decayed into ^{59}Co, and that cosmic rays are likely to be accelerated more than 10^5 years after nucleosynthesis.

To sum up the isotopic anomalies of cosmic-ray sources: ^{13}C is underabundant, ^{22}Ne is overabundant (both consistent with some contribution of Wolf-Rayet star material to cosmic-ray sources, and a nucleosynthesis environment that leads to extra ^{60}Zn(^{60}Ni).

3. ELEMENTAL ABUNDANCES

Lithium in cosmic rays is difficult to measure, because the hundred times more abundant helium nuclei swamp it with background. Aizawa, Evenson, Meyer and Yanagita (1993) have measured it, adding significantly to the few earlier measurements. Figure 10 shows the Li/C ratio from 0.1 to 100 GeV/u, with a comparison of the curve based on propagation calculations. The agreement, while not perfect, is satisfactory.

Buckley et al. (1993) have measured He, C and O at 40 to 200 GeV/u, and present graphs from 1 to 10^4 GeV/u including data from earlier experiments.

Thompson et al. (1993), using the plastic stacks on LDEF spacecraft, with only 10% of the data analyzed, report on about 250 ultra-heavy nuclei, $Z \geq 70$. The abundance of actinides, about 10 tracks, yields $Z \geq 88/(74 \leq Z \leq 87) = 0.038 \pm 0.012$. This implies an enhancement of the heaviest r-process nuclei at the source. The very preliminary Th/U ratio is about 2 ± 1.

The elemental composition is fairly well known to energies of 1 TeV/u (i.e. 10^{12} eV/u). But some reliable data extend to 10^{15}eV/nucleus. Ichimura et al. (1993) have measured the energy spectra of p, He, Si, S, $17 \leq Z \leq 25$ and Fe up to about 10^5 GeV-per-nucleus, using X-ray film and emulsions as detector. They find the energy spectrum of iron at 30 to 1000 GeV/u to be relatively flat, $\sim E^{-2.6}$ while that of Si is appreciably steeper, $\sim E^{-3}$ above 100 GeV/u. They thus confirm the observation of Muller et al. (1991) regarding the spectrum of Si.

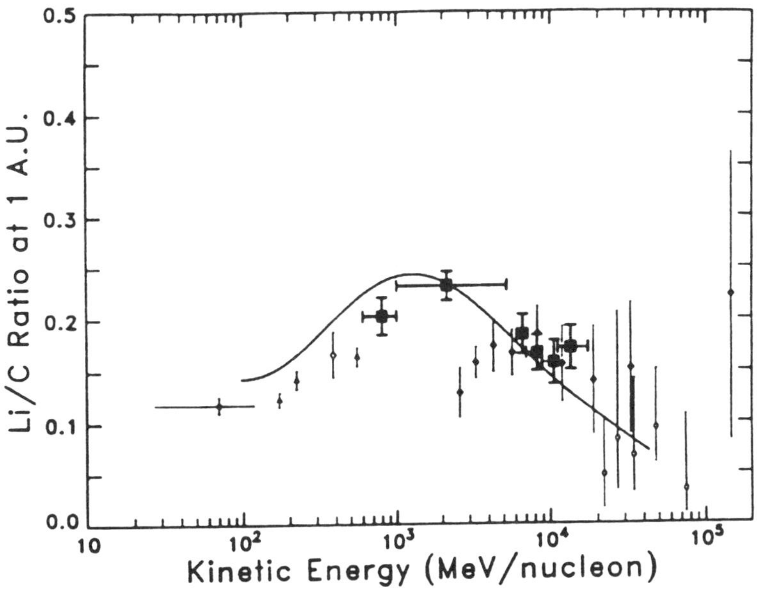

Figure 10. The ratio Li/C as a function of energy. (Aizawa et al.)

Asakimori et al. (1993), of the JACEE collaboration find that the energy spectrum of protons steepens above 20 TeV, changing from $E^{-2.6}$ to $E^{-3.2\pm0.3}$. They also find that the C-O spectrum is particularly flat between 1 to 100 TeV/n, about $E^{-2.3}$. This implies an enhancement of helium-burning products, as predicted by Silberberg et al. (1990) in the model with the pre-supernova giant star stellar wind composition. These wind particles are accelerated to a relatively flat spectrum that extends to high energies. This is briefly described also in our second paper at this conference.

At still higher energies 3×10^6 to 3×10^8 GeV per nucleus, the composition estimates are quite uncertain. Bird et al. (1993) believe that heavy nuclei (C to Fe) are significant here, while somewhat older estimates by Acharya et al. (1983) imply a light composition.

4. DISTRIBUTED REACCELERATION

Two papers, Silberberg et al. (1993) and Heinbach and Simon (1993) were presented at the Calgary conference. Both find that with reacceleration the pathlength for rigidity-dependent escape from the Galaxy is $\lambda_{esc} = \lambda_{esc(o)} R^{\varepsilon}$ (ε = -0.4 to -0.5 for Silberberg et al. and -1/3 by Heinbach and Simon). The standard leaky box model requires a discontinuity in the mean free path, as shown by the dashed line of Figure 11. A second advantage of the distributed acceleration model is a smaller change in leakage rate with rigidity than with the value ε = -0.6 for the standard leaky box model -- i.e. more consistent with the small anisotropy of cosmic rays.

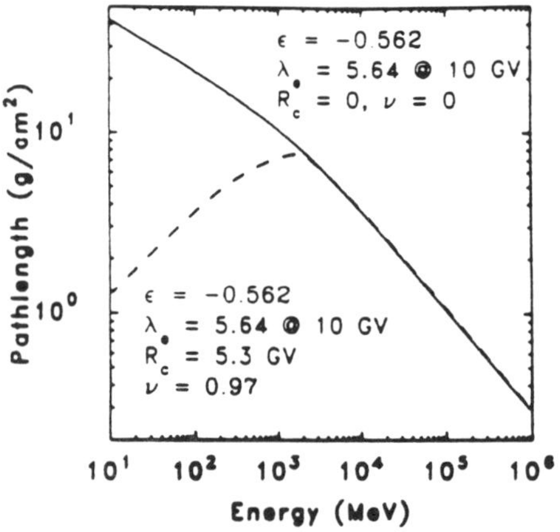

Figure 11. The mean path length as a function of energy for the reacceleration model (solid curve) and the standard leaky box (dashed). (Silberberg et al.)

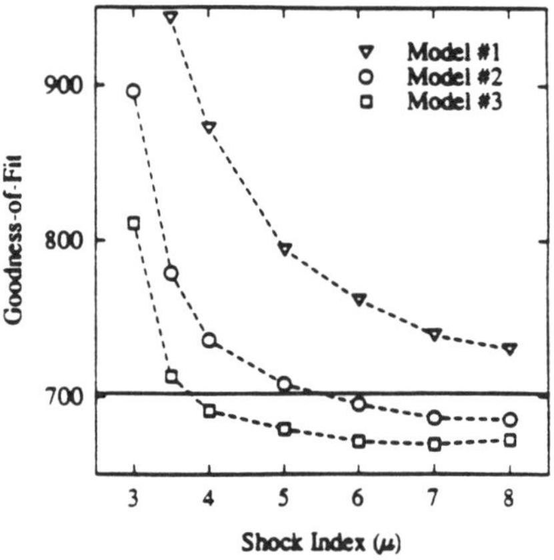

Figure 12. The goodness of fit of 3 reacceleration models, and of the standard model, the latter is shown by the solid line, (Silberberg et al.).

Figure 12 shows that the goodness -of-fit for the distributed reacceleration model is equally good or slightly better than for the standard model. Figure 13 shows the experimental ratios of B/C as a function of energy, for the standard and for the reacceleration models. Precise measurements at 1 TeV/u can also distinguish between the standard and reacceleration models.

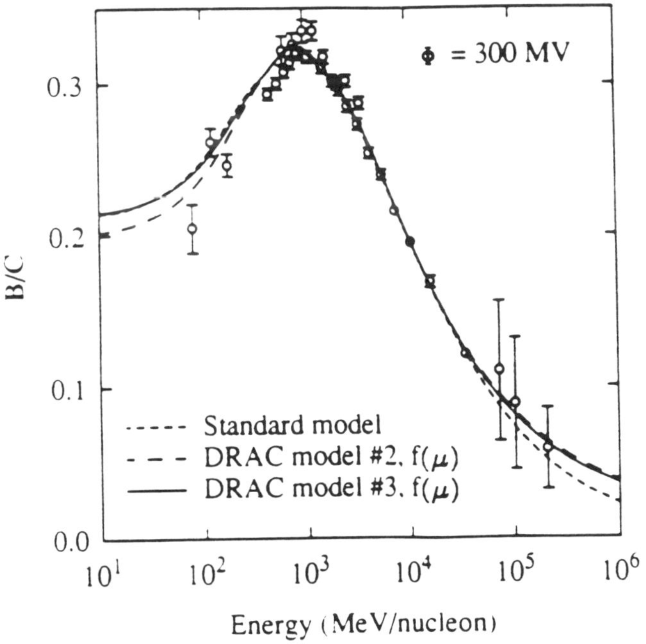

Figure 13. The experimental B/C ratio and the predicted ones for the reacceleration models and standard model. (Silberberg et al.).

Osborne and Ptuskin (1987) deduced that λ_{accel} (R) λ_{escape} (R) = const. We have introduced a similar rigidity dependence for $\lambda_{accel.} \sim R^{\alpha}$ while $\lambda_{escape} \sim R^{\varepsilon}$ with $\varepsilon \approx -0.5$ and $\alpha \approx 0.5$. This implies less reacceleration at high energies, since the high-rigidity particles escape more readily from the reacceleration region.

Currently we are working on electron reacceleration. The preliminary calculations reproduce the electron spectrum with appropriate weighting factors for the shock strength distribution parameters.

5. CROSS SECTIONS

The energy dependence of some partial cross sections extends to higher energies than believed earlier. Webber et al. (1993) and Hirzebruch et al. (1991) found that the cross section of Si into Al and Mg, of Ne into F and O, and of O into N are smaller at 15 GeV/u than at 1.5 GeV/u by about 30%. This is displayed in Figure 14. Webber et al. suggest that the crucial C into B cross section is similarly affected. This has a major effect on propagation calculations, reducing the exponent -0.6 of the path length α $R^{-0.6}$ relationship to $R^{-0.3}$ according to Webber. In our opinion it is reduced to $R^{-0.4}$, and when combined with distributed reacceleration, will be reduced to $R^{-0.3}$, in agreement with Kolmogorov scattering. Thereby the difficulty of the small anisotropy up to at least 10^{15} eV is also resolved.

Isbert et al. (1993) have measured the spallation cross section of ^{28}Si at 245 MeV/u. Figure 15 shows the production cross sections of ^{26}Al, ^{18}F, ^{22}Na and ^{24}Na. The solid curve is based on the calculations of Silberberg and Tsao (1987) and the dashed curve on Webber, Kish and Schrier (1990). For ^{18}F and ^{24}Na, the predictions of Silberberg and Tsao appear superior.

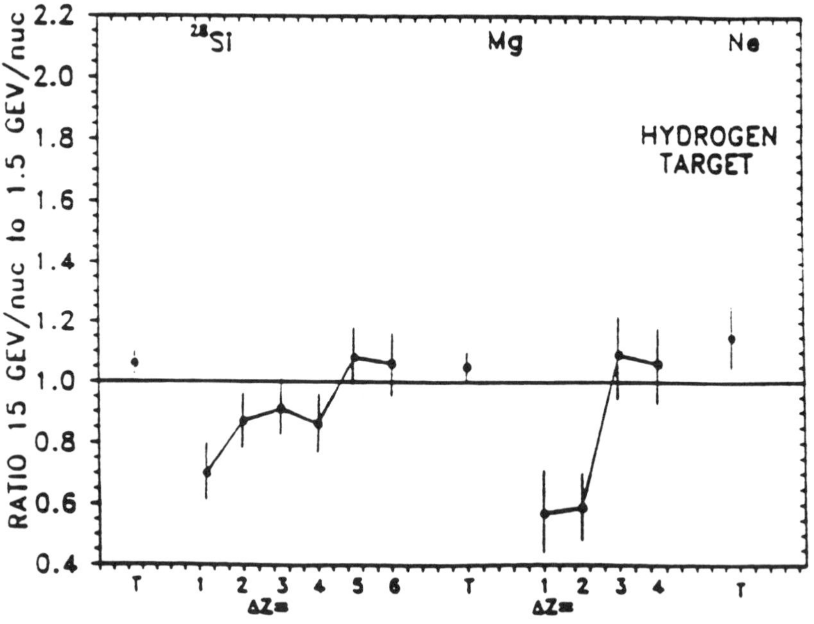

Figure 14. The ratio of total and partial cross sections at 15 GeV/u to 1.3 GeV/u (Webber et al.).

Tull et al. (1993) of the Transport Collaboration have measured the projectile fragmentation of ^{32}S at 400 and 600 MeV/u. The cross sections for the production of the n-rich isotopes ^{26}Mg, ^{25}Na, ^{22}Ne and ^{18}O are about 40% less at 400 MeV/u than at 600 MeV/u. Also ^{25}Mg, ^{24}Na, ^{21}Ne and ^{17}O are less at 400 MeV/u, but ~20%.

Knott et al. (1993) of the Transport Collaboration point out significant systematics related to isospin (T=Z-A/2) of the beam. For T=0, e.g. ^{32}S, ^{36}Ar, ^{40}Ca products with ΔZ=2, 4, or 6 are enhanced. For T=-2, e.g. ^{40}Ar, ^{52}Cr, ^{56}Fe the cross sections decrease exponentially with increasing ΔZ. Figures 16 and 17 display these relationships.

Hirzebruch, Winkel and Heinrich (1993) explored the spallation of Fe in H, C and Pb targets. For fragments with masses $M_F < M_{Fe}/2$ the fragmentation cross sections with C and Pb (relative to H) increase with decreasing fragment masses. The spallation cross sections of Fe on H were found to agree better with the Webber et al. (1990) formula for $\Delta Z \leq 14$, and better with that of Silberberg and Tsao for $\Delta Z \geq 15$.

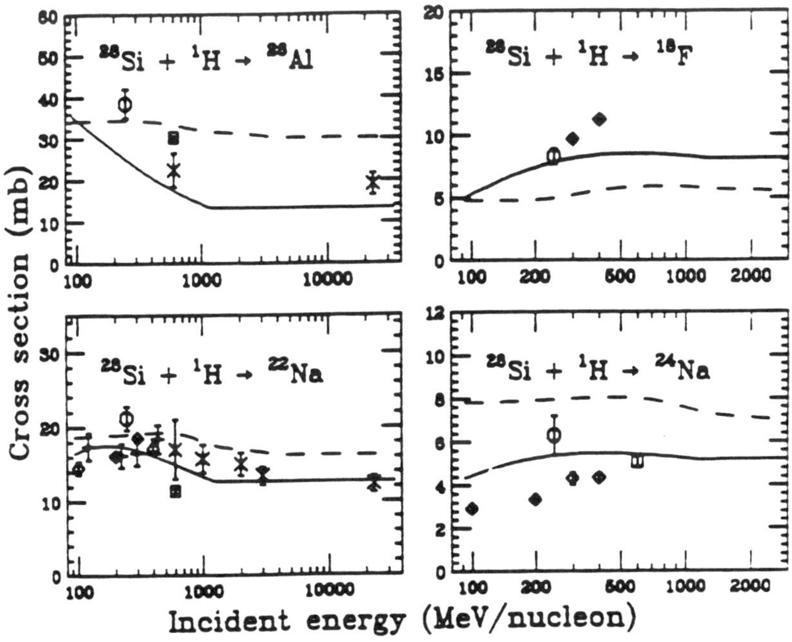

Figure 15. The cross sections of ^{28}Si into ^{26}Al, ^{18}F, ^{22}Na and ^{24}Na. These are compared to the calculation of Silberberg and Tsao (1987) and Webber et al. (1990). (From Isbert et al.).

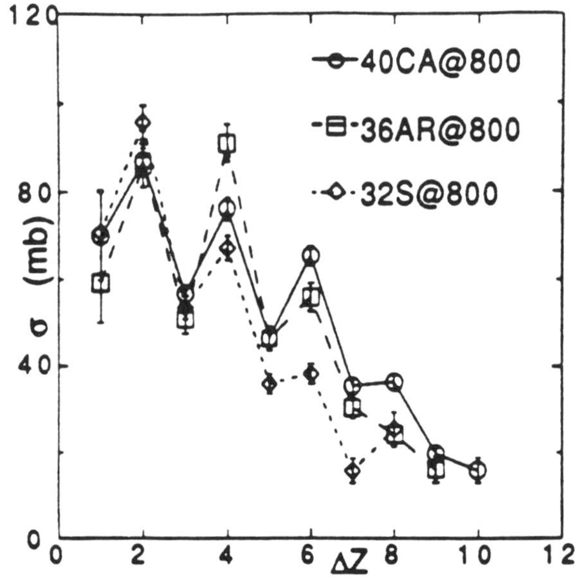

Figure 16. The partial cross sections of ^{32}S, ^{36}Ar and ^{40}Ca. (Isospin T=0 nuclides). (Knott et al.).

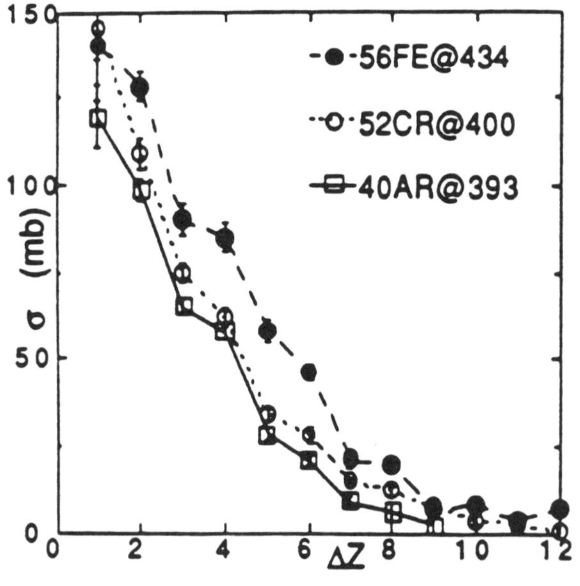

Figure 17. The partial cross sections of ^{56}Fe, ^{52}Cr and ^{40}Ar (Isospin T=-2 nuclides). (Knott et al.).

Barghouty, Tsao and Silberberg (1993) have developed a procedure for calculating energy and momentum loss in nucleus-nucleus collisions, based on the participant-spectator model. Energy and momentum loss distributions are calculated for Fe-Al and Fe-C collisions, for products of Mn, Ca and Si.

The spallation of heavier nuclei, like Kr and Ag was explored by Nilsen et al. (1993). They developed an equation for the total inelastic cross section, based on an optical model. For partial cross sections, neither the parametric equations of Cummings et al. (1990), nor the equations of Silberberg and Tsao (1990) provided a good fit. The authors had to construct a new fit for their spallation cross sections of Kr and Ag, in the energy interval 0.5 to 1.4 GeV/u.

The interactions of Au at 11.4 GeV/u were studied by Price and He (1993). They explored the electromagnetic contribution to the charge changing ($\Delta Z \approx 1$ or 2) cross section and found that it increases appreciably with target charge $Z_{t.}$: $\sigma_{e.m.} \sim 4 \times Z_t^{1.4}$ mb. The nuclear charge pickup (probably associated with π^- production) increases as $\sim A_t^{0.37}$.

6. NEUTRINO ASTROPHYSICS AND NEUTRINO DETECTORS

Szabo and Protheroe (1993) have developed improvements in their model of neutrino production (10^{12} to ~ 10^{17} eV) at active galactic nuclei by ($p\gamma$) and (pp) reactions with pion production and decay.

The MACRO collaboration, DiCredico et al. (1993) have recorded 4.2×10^6 muons deep underground (4000 meters water equivalent) between June 1991 and January 1993. They did not find any point sources and set an upper limit $< 2 \times 10^{-13}$ muons per cm^2 sec for Cyg X-3.

The LVD (Large Volume Detector, 368 tons) at Gran Sasso, Aglietta et al. (1993) has been operational since June 1991, to search for neutrinos from Galactic supernovae, at energies 3 to 10 MeV. For Galactic supernovae (from a distance like the Galactic Center) about 100 neutrino interactions would be expected. Also the MACRO detector (Grassi et al. 1993) has a massive liquid scintillator that is being used to search for Galactic supernovae. The IMB detector of Becker-Szendy et al.(1993) has not observed supernovae between May 1986 and March 1990, except SN1987a.

An update on DUMAND II was given by Alexander et al. (1993). A muon array of 216 optical modules, that will record neutrinos in a detector of area 20,000 m^2 at a depth of 4760m ocean water is in the final construction phase. Three strings with photo tubes (out of a total of nine) will be deployed late in 1994.

Learned and Roberts (1993) point out the feasibility of further expansion of DUMAND to 1 km^3, with 100 vertical strings and with a total of 7848 optical modules. The angular resolution with 10^3 m long strings is 0.25°, and energy resolution, integrating over energy-loss interactions, is 50%. The cost is estimated at 80 million dollars. About 500 neutrinos per year should be observable from the Seyfert galaxy NGC 4151.

Stenger (1993) finds that the early DUMAND detector with 3 strings has a marginal probability to observe the AGN neutrinos.

Learned and Wilkes (1993) explore the acoustical detection of neutrino-initiated cascades in DUMAND. There will be 53 hydrophones to determine the positioning (to 10 cm) of the optical array. However, with new realtime Fourier methods, the hydrophones (with detection sensitivity that extends to a distance of a km) can detect neutrinos of 10^{16} eV energy. Karaevsky et al. (1993) estimate that the acoustic threshold is about 10^{15} eV.

Several other detectors of neutrinos are under construction:

A Russian-Greek collaboration, NESTOR (Neutrinos From Supernovae and Teraelectron volt sources Ocean Range) is in the early stage of deployment in the Mediterranean, at a depth near 4000m. A report was given by Anassonzis et al. (1993).

A detector in Antarctic ice, AMANDA, (Antarctic Muon and Neutrino Detector Array) is designed to be placed into bore holes, about 1000 m deep in the ice. Three strings were planned to be deployed early in 1994. This detector is discussed by Miller et al. (1993). They find that the attenuation length in ice is over 18m, maybe even 25m. Tilav et al. (1993) discuss a first phase of 9 strings, with 20 downward-looking modules on each string. The array is designed to look for upward coming muons, since the downward going atmospheric muons swamp the neutrino signal by a factor of 10^6 to 1. A detector in ice, like AMANDA does not suffer from ^{40}K decay background like ocean water does.

In Lake Baikal, the large NT-200 array at a depth of 1 km under water is under design. The attenuation length is 20 m. The array consists of 196 optical modules along 8 vertical strings, three to be completed in 1993. This array is described by Belolaptikov et al. (1993). The effective area for 1 TeV muons is 2000 m^2. The phototubes look alternatively up and down, so that the downward atmospheric neutrino flux can also be measured. About 4 x 10^8 atmospheric neutrinos will be recorded per year, of these 2000 at energies E > 100 TeV.

References

The notation 23rd ICRC stands for the 23rd International Cosmic Ray Conference (Calgary)

Acharya, B. et al. 1983 Proc. ICRC (Bangalore) 9, 191.
Aglietta, M. for LVD at Gran Sasso, 1993 23rd ICRC 4, 468.
Aizawa, I. et al. 1993, 23rd ICRC 1, 595.
Alexander, C.M. et al., for DUMAND, 1993 23rd ICRC 4, 515.
Anassontzis, E.G., et al. for NESTOR, 1993 23rd ICRC 4, 554.
Asakimori, K. et al. 1993 23rd ICRC 2, 21 and 25.
Barghouty, A.F., Tsao, C.H. and Silberberg, R. 1993 23rd ICRC 2, 171.
Becker-Szendy, R., et al. for IMB, 1993 23rd ICRC 4, 476.
Belolaptikov, I.A., for NT-200, Lake Baikal, 1993 23rd ICRC 4, 573.
Bird, D.J. et al. 1993 23rd ICRC 2, 38.
Blake, J.B. and Dearborn, D.S. B. 1993 23rd ICRC 2, 394.
Buckley, J.H. Et al. 1993 23rd ICRC 1, 599.
Connell, J.J. and Simpson, J.A. 1993a 23rd ICRC 1, 547.
Connell, J.J. and Simpson, J.A. 1993b 23rd ICRC 1, 559.

Cummings, J.R. et al. 1990 Phys. Rev C, 42, 2530.
Di Credico, A. for MACRO, 1993 23rd ICRC 4, 450.
DuVernois M.A. et al. 1993 23rd ICRC 1, 563.
Ficenec, D.J. et al. 1993 23rd ICRC 1, 515.
Freier, P. et al. 1948 Phys. Rev. 74, 213.
Garcia-Munoz, M. et al. 1993 23rd ICRC 1, 543.
Grassi, M. for MACRO 1993 23rd ICRC 4, 472.
Guzik, T.G. et al. 1993 23rd ICRC 1, 555.
Heinbach, U. and Simon, M. 1993 23rd ICRC 2, 263.
Hesce, A. et al. 1993 23rd ICRC 1, 567.
Hindi, M.M. et al. 1993 23rd ICRC 1, 531.
Hirzebruch, S.E. et al. 1991 Proc. 22nd ICRC (Dublin) 2, 280.
Hirzebruch, S.E.,Winkel, E. and Heinrich, W. 1993 23rd ICRC 2, 175.
Ichimura, M. et al. 1993 23rd ICRC 2, 5 and 9.
Isbert, J. et al. 1992 23rd ICRC 2, 167.
Karaevsky, S. for SADCO, 1993 23rd ICRC 4, 550.
Knott, C.N. et al. 1993 23rd ICRC 2, 187.
Learned, J.G. and Roberts, A. 1993 23rd ICRC 4, 519.
Learned, J.G. and Wilkes, R.J. 1993, 23rd ICRC 4, 538.
Leske, R.A. and Wiedenbeck, M.E. 1993 23rd ICRC 1, 571.
Lukasiak, H. et al. 1993 23rd ICRC 1, 523.
Miller, T. et al. for AMANDA, 1993 23rd ICRC 4, 557.
Nilsen, B.S. et al. 1993 23rd ICRC 2, 195.
O'Dell, F.W. et al. 1965 Proc. 9th ICRC (London), 1, 412.
Osborne, J.L. and Ptuskin, V.S. 1987 Proc 20th ICRC (Moscow) 2, 218.
Papini, P. Grimani, C. and Stephens, S.A. 1993, 23rd ICRC 1, 503.
Price, P.B. and He, Y.D. 1993 23rd ICRC 2, 199.
Ptuskin, V.S. 1974 Ap. Space Sci. 28, 17.
Seo, E.S. et al. 1993 23rd ICRC 1, 511.
Silberberg, R. and Tsao, C.H. 1987 20th ICRC (Moscow) 2, 133.
Silberberg, R. and Tsao, C.H. 1990 Phys. Reports 191 (6), 351.
Silberberg, R. et al. 1990 Ap.J. 363, 265.
Silberberg, R. et al. 1993 23rd ICRC 2, 251.
Stenger, V.J. 1993 23rd ICRC 4, 527.
Szabo, A.P. and Protheroe, R.J. 1993 23rd ICRC 4, 438.
Thompson, A. et al. 1993 23rd ICRC 1, 603.
Tilav, S. et al. for AMANDA, 1993 23rd ICRC 4, 561, 565.
Tull, C.E. et al. 1993 23rd ICRC 2, 163.
Webber, W.R., Kish, J.C., and Schrier, D.A 1990 Phys. Rev C41, 520, 533, 547.
Webber, W.R. et al. 1993 23rd ICRC 2, 183.
Wiedenbeck, M.E. and Leske, R.A. 1992 COSPAR Conf. Washington, Publ. in Adv. Space Res. 13. Also Leske, R.A. 1993 Ap.J. 405, 567.

RECENT RESULTS FROM THE FLY'S EYE EXPERIMENT

D.J. BIRD,[1] S.C. CORBATÓ,[2,6] H.Y. DAI,[2] B.R. DAWSON,[3] J.W. ELBERT,[2] B.L. EMERSON,[2,7] T.K. GAISSER,[4] K.D. GREEN,[2,5] M.A. HUANG,[2] D.B. KIEDA,[2] S. KO,[2] C.G. LARSEN,[2] E.C. LOH,[2] M. LUO,[2,8] M.H. SALAMON,[2] J.D. SMITH,[2] P. SOKOLSKY,[2] P. SOMMERS,[2] T. STANEV,[4] J.K.K. TANG, [2] S.B. THOMAS, [2] And S. TILAV[4]

1 Department of Physics, University of Illinois at Urbana-Champaign, Urbana,IL 61801 USA
2 High Energy Astrophysics Institute, Department of Physics, University of Utah, Salt Lake City UT 84112 USA
3 Department of Physics and Mathematical Physics, University of Adelaide, Adelaide, South Australia 5001 Australia
4 Bartol Research Institute, University of Delaware, Newark, Delaware 19716 USA
5 Department of Physics, university of Michigan, Ann Arbor, MI 4809 USA
6 Present address : NorthWestNet, Bellevue, Washington 98007 USA
7 Present address : Science Department, Central Oregon Community College, Bend OR 97701, USA
8 Present address : Biology Division, Dugway Proving Ground, Dugway, UT 84002, USA

ABSTRACT. We report on recent results from the Fly's Eye experiment up to July 1992. The detector consists of FE1, running from 1981 to 1992, and FE2 running from 1986 till now. The monocular data comes from a single eye detector, and the stereo data are detected by both detectors and reconstructed by stereo geometry. This report includes results on the spectrum, composition, anisotropy, and a 320 EeV event. The stereo spectrum shows a dip around $10^{18.5}$eV. The composition changes from predominately heavy (below 10^{18}eV) to predominately light (above 10^{19}eV). The anisotropy shows no significant deviation from random fluctuation in both a small scale and large scale anisotropy search. A $320^{+35}_{-40} \times 10^{18}$eV event was detected by FE1 on Oct. 15, 1991.

1 Introduction

The puzzling origin of extremely high energy cosmic rays (EHECR) can be studied through three experimental topics: the energy spectrum, chemical composition and anisotropy in arrival direction. Every well determined feature in these three topics will constrain other topics and theories of the nature of EHECRs.

Due to the extremely low flux of EHECRs, the EHECRs are only detectable by giant ground arrays or fluorescent light detectors by converting cosmic ray primaries into numerous various secondary particles through the extensive air shower. The nitrogen molecules are excited by the secondary particles and emit fluorescent light mainly in the UV region. The Fly's Eye detector collects this fluorescent light along the track of the air shower. The detector provides a direct measurement of shower longitudinal development on an event by event basis.

M. M. Shapiro et al. (eds.), Currents in High-Energy Astrophysics, 123–134.

2 The Fly's Eye Experiment

The details of Fly's Eye experiment have been covered in earlier papers [1] [2]. A brief description is given here. The Fly's Eye detector is located at Dugway, Utah (40° N, 113^oW, atmospheric depth $860g/cm^2$). Fly's Eye 1 (FE I) consists of 67 spherical mirrors, each mirror contains 12 to 14 photomultiplier tubes of 5.5 degree in diameter. The mirrors and PMTs are arranged so the whole night sky is imaged. FE I began full operation in November 1981. Fly's Eye II (FE II), which is located 3.4 km away from FE I, consists of 36 spherical mirrors and covers half of the night sky in the direction of FE I. It began full operation in November 1986. The data used in this paper are the total data set.

The shower-detector plane is reconstructed by knowing the pointing direction cosines of triggered tubes. The monocular data are events recorded by FE I and reconstructed through signal timing. The stereo data are events detected by both FE I and FE II and reconstructed by the intersection of two shower-detector plane. The stereo data generally have better resolution then monocular data except when the opening angle between the shower axis and both FE I and FE II are small, typically $\leq 20°$. The arrival direction can be traced back along the shower axis.

The detected signals are converted to numbers of photoelectrons. The direct Cherenkov light, air molecule (Rayleigh), and aerosol (Mie) scattered Cherenkov light are subtracted. The longitudinal shower profiles are obtained by binning the number of photoelectrons due to fluorescent light (shower size) as function of atmospheric depth. The photoelectron number is then proportional to the number of charged particles at any given depth once atmospheric effects are accounted for.

The atmospheric depth at shower maximum (X_{max}), which is sensitive to the nature of primary nuclei, can be found by fitting the shower profile to a Gaisser–Hillas formula

$$N(X) = N_{max}\left(\frac{X - X_o}{X_{max} - X_0}\right)^{(X_{max}-X_0)/\lambda} e^{(X_{max}-X)/\lambda} \tag{1}$$

where X_o is the depth of first interaction, $\lambda = 70g/cm^2$.

The energy of the primary cosmic rays can be found from the integral of either the Gaisser & Hillas formula.

$$E = \frac{\epsilon_0}{\chi_0}\int N(x)dx \tag{2}$$

where ϵ_0/χ_0 is the ratio of critical energy of an electron to the radiation length in air and is taken as $2.18\ MeV\ g^{-1}\ cm^2$. [3]

The detector performance was carefully modeled through Monte-Carlo simulation. There are two levels of Monte-Carlos. The first is a shower Monte-Carlo which uses high energy interaction models to generate shower profiles. The second is the detector Monte-Carlo which simulates the detector triggering conditions and responses. The direct Cherenkov light, Rayleigh and Mie scattered Cherenkov light and fluctuation from background sky noise are included in tubes signals. The Monte-Carlo writes out simulated data files and these are processed through the same software reconstruction stream as the real data.

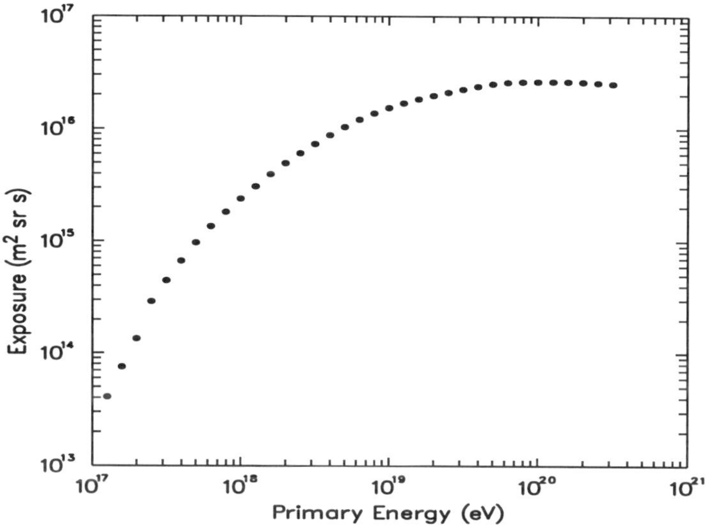

Figure 1: The aperture of monocular and stereo aperture.

3 Spectrum

The aperture energy dependence is calculate from the Monte-Carlo. Figure 1 shows the aperture of the monocular data.

The systematic uncertainty of energy comes from the fluorescent efficiency, the atmospheric scattering and transmission of light, Cherenkov subtraction, and uncertainty in geometric reconstruction. The systematic errors for the first three are $\sim 20\%$, $\sim 10\%$, and $\leq 4\%$ respectively. For the stereo data, the systematic uncertainty in energy due to geometric reconstruction is $\leq 5\%$ [6]. The monocular data has an extra 20% offset in energy when compared to the stereo data and has been corrected in our analysis. In conclusion, we estimate the maximum systematic uncertainty in energy is about 40% in the worst case. The random error is about 20% to 24% for stereo data and 27% to 36% for monocular data.

The stereo data was chosen to study the features of spectrum because of it's smaller energy uncertainty. A loose cut of 100% estimated fractional energy uncertainty has been imposed. The result is shown on figure 2. [4] [5] [6]

The stereo spectrum can be divided into three regions and a dip is clearly seen. Using the weighted least χ^2 fit, the best fit spectral indexes are listed in table 1.

The expected number of events based on the overall fit (renormalized at $10^{17.6}eV$) is 5936.3, while the observed number is 5477. The significance of this deficit is 5.96σ. The expected number of events above $10^{18.5}eV$ based on the overall fit (renormalized at $10^{18.5}eV$) is 230.0, and the observed number is 281. The significance of this excess is 3.4σ. If we use the best fit in $10^{17.6} - 10^{18.5}$, the expected number is 205.9 which is 5.2σ.

This dip can be seen in the event energy distribution. Since this has no aperture folded

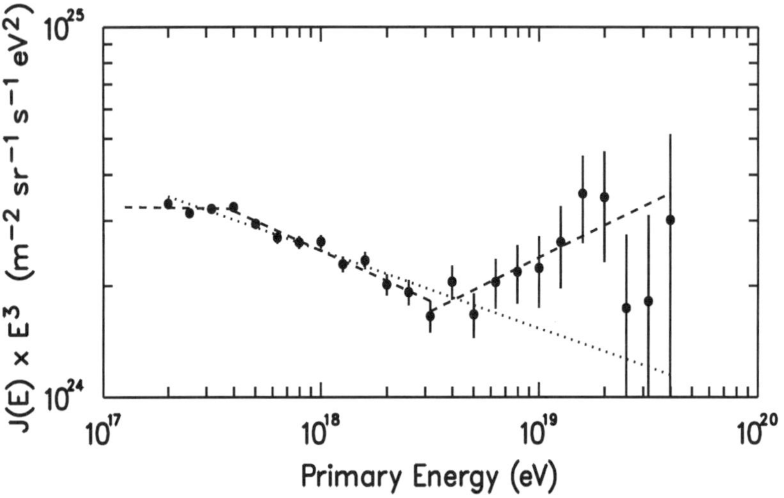

Figure 2: Fly's eye Stereo energy spectrum. Points : data. Dashed line : Best fit in each region. Dotted line : Best fit up to $10^{18.5}eV$.

in, it is not a spurious effect of Monte-Carlo simulation.

The monocular spectrum is shown in figure 3. Due to the worse energy resolution of monocular data, the dip feature is washed out in the monocular spectrum. If we impose tight cuts to insure good geometric reconstruction, the dip is found but the statistics are low. The best fit spectral index is shown in table 1. The expected number of events between $10^{19.0}eV$ and $10^{19.9}eV$ based on the best fit below $10^{19.0}eV$ is 242.35, and 297 events was detected. This is a 3.5σ excess. But the spectrum steepens again after $10^{19.7}eV$. The expected number of events above $10^{19.7}eV$, based on the best fit between $10^{19.0}eV$ and $10^{19.7}eV$, is 20.63, while only 10 events are detected. However, we still need more statistics to determined whether there is a cut off or not.

4 Composition

The stereo data are used in this part of the study. A cut of fractional uncertainty of depth at shower maximum $\delta X_{max}/X_{max} \leq 12\%$ is imposed to insure good X_{max} reconstruction. The opening angle between shower axis and viewing tubes must $\geq 20°$ to avoid the strong Cherenkov contamination. The number of observed events is 8790 for the energy range from $0.1EeV$ to $30EeV$ and 5129 for $0.3EeV$ to $30EeV$. [7]

Since there are no accelerator data in the Fly's Eye energy range, we use three models to extrapolate the accelerator data. Those models are characterized by different predictions Of the energy dependence of *inelasticity*, the fraction in energy that is not carried out by fragment of primary particle. These are the low inelasticity statistical model, large and

Table 1: Spectral slopes and normalization of $J(E)(m^{-2}sr^{-1}s^{-1}eV^{-1})$

Type	Energy range (eV)	Power index	$\log_{10}$(normalization)	Normalized at (eV)
stereo	$10^{17.3} - 10^{19.6}$	-3.18 ± 0.01	-29.593	10^{18}
	$10^{17.3} - 10^{17.6}$	-3.01 ± 0.06	-29.495	10^{18}
	$10^{17.6} - 10^{18.5}$	-3.27 ± 0.02	-29.605	10^{18}
	$10^{18.5} - 10^{19.6}$	-2.71 ± 0.10	-32.623	10^{19}
mono	$10^{17.3} - 10^{19.9}$	-3.07 ± 0.01	-29.55	10^{18}
	$10^{17.3} - 10^{19.0}$	-3.08 ± 0.01	-29.56	10^{18}
	$10^{19.0} - 10^{19.7}$	-2.89 ± 0.15	-29.62	10^{18}

Table 2: Mean X_{max} for three models and primary nuclei.

	H	CNO	Fe
QCD-Pomeron	750±1.9	711±0.9	653±0.8
QCD-Minijet	761±1.9	719±1.0	663±0.8
Statistical	789±2.4	744±1.1	681±0.8

increasing high inelasticity QCD Pomeron model, and slowly increasing high inelasticity QCD minijet model [8], [9]. The predicted mean X_{max}s are listed in table 2. Although different absolute values are predicted, the difference between proton and iron is always about $100g/cm^2$. The statistical model predicts a X_{max} too deep to fit the Fly's Eye data for any composition. The QCD-Pomeron and QCD-minijet models are both adequate fits and differ little in prediction. The QCD-Pomeron models results in a somewhat better fit, therefore it is used in this analysis.

The X_{max} resolution is about $45g/cm^2$ on an event by event basis. The systematic error with the absolute X_{max} is $20g/cm^2$. The systematic error of hadronic shower Monte-Carlo prediction is $10g/cm^2$, due to the approximations in the Monte-Carlo.

Figure 4 shows the X_{max} versus the primary energy for the experimental data and Monte-Carlo prediction of protons and iron nuclei. The elongation rate (X_{max} changes per decade of energy) below 0.3EeV is consistent with constant composition of $50g/cm^2$. For 0.3EeV to 10EeV, the X_{max} distribution is inconsistent with any pure composition and the elongation rate is $78.9 \pm 3.0g/cm^2$ which indicates a changing composition, a change from predominately heavy to predominately light.

The dip structure of stereo spectrum and the change of composition suggests a change of cosmic ray source. We consider a two component fit, one being a steeply falling power spectrum of predominantly iron flux, the other a flatter flux of protons. The best fit results from $10^{17.6}eV$ to $10^{19.6}eV$ are :

$$\text{iron flux}: \ \log[J(E)(m^2 \cdot s \cdot sr \cdot eV)^{-1}] = 33.185 - 3.496 \times \log(E)$$

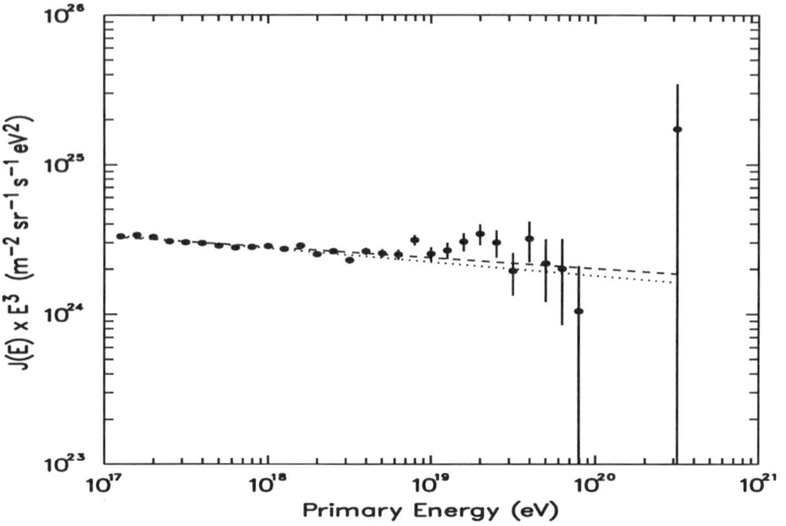

Figure 3: Fly's eye monocular differential spectrum multiplied be E^3. Points :data, Dashed line : Best fit of the total spectrum. Dotted line : best fit up to $10^{18.5}eV$

$$\text{proton flux : } \log[J(E)(m^2 \cdot s \cdot sr \cdot eV)^{-1}] = 16.782 - 2.610 \times \log(E)$$

The ratio of those two fluxes can be expressed as

$$\frac{\text{iron flux}}{\text{proton flux}} = 10^{-0.887 \times (\log(E) - 18.5)}$$

The prediction of spectrum and composition based on this two component fit are shown in figures 4 and 5. It fits the X_{max} observation very well. The predicted elongation rate is $10^{18}eV$ to $10^{19}eV$ is $81 g/cm^2$ which is consistent with the observed value $78.9 \pm 3 g/cm^2$.

5 Anisotropy

The uncertainty in the zenith angle $d\theta$ is about 1.7° for stereo data and 8.6° for monocular data. The resolution for the azimuth angle is similar to zenith angle for stereo data and slightly worse for monocular data. For better statistics, we use monocular data in previous papers [10]. We now have enough statistics in the stereo data. This enables us to use these much better angular resolution data to search for anisotropy.

The dependence of anisotropy on galactic latitude was analyzed in two ways, one in the galactic latitude gradient which searches for cosmic ray flux excess from the north/south galactic hemisphere and is formulated as

$$I(b)/I_0 = (1 + s\, b)$$

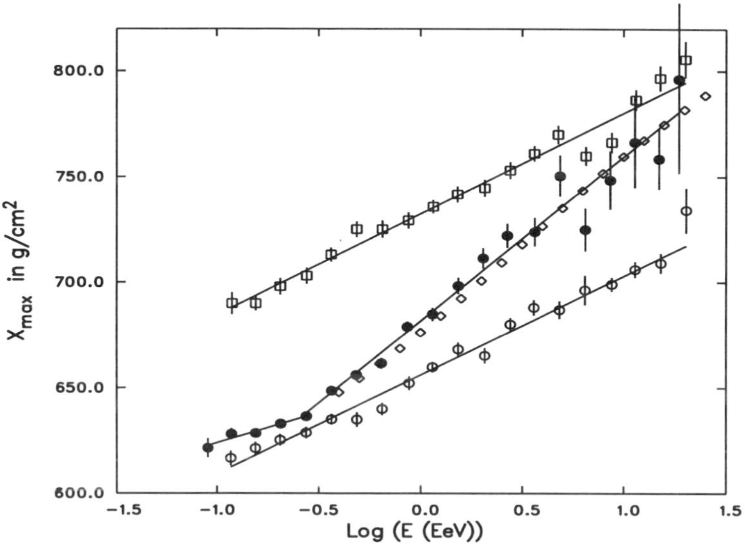

Figure 4: The X_{max} distribution of stereo Fly's Eye. The solid dots are the experimental data. The squares show the expectation of proton flux and the circles show the expecttion for iron flux. The dimonds show the expectation from the two component fit.

where b is the galactic latitude. The other is a search for galactic plane enhancement factor proposed by Wolfendale and Wdowczyk [11]

$$I(b)/I_0 = (1 - f + 1.402\, f\, e^{-b^2})$$

This analysis [12] use equal area bins of 10° in galactic longitude and 0.1743 in $\sin(b)$ which is about 10° near the galactic plane. The data has an angular uncertainty cut of $\leq 10°$. The flux ratio $I(b)/I_0$ is calculated from the ratio of the number of observed events and number of events predicted by an isotropic distribution. The result is shown in table 3 and no significant anisotropy is found. Figure 6 shows the galactic plane enhancement factor f from several groups.

The other large scale anisotropy study uses three major axes of galactic coordinates: Center-Anticenter, Forward-Backward, and North-South. The sky lobes analysis use 6 equal area lobes centered on each axi to search for excess/deficit from each axi. Previous analysis [13] [14] and recent analysis [16] show no consistent and significant deviation from isotropic prediction for several energy ranges starting from $10^{17.5}eV$.

Another analysis fits data to a second order Legendre polynomial to search for the angular dependence along these major axes. Again, the previous analysis [14] [15] and recent analysis [16] all show no significant deviation from isotropy. Notice in the fitting of the North-South angular dependence that the coefficient of the first Legendre polynomial has a similar meaning to the galactic latitude gradient, and the coefficient of the second

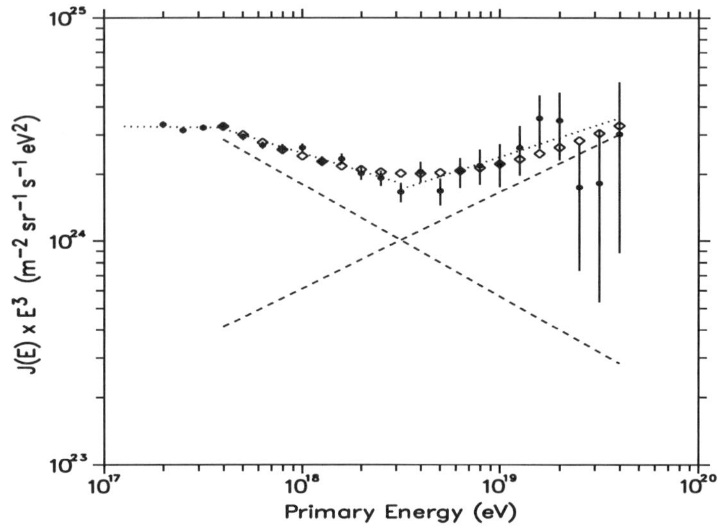

Figure 5: Two component fit to the stereo Fly's Eye energy spectrum. The diamond is the expectation from the two component fit.

Legendre polynomial has a similar meaning to the galactic plane enhancement. Those two analyses show consistent isotropy.

A harmonic analysis has recently been done [16]. There are three declination bands, $10°$-$30°$, $30°$-$50°$, and $50°$-$70°$. The first harmonic ranges from 0.69% (energy $10^{17.5}eV$ -$10^{18.0}eV$, $\delta = 10° - 30°$, phase=$234°$) to 26.2% (Energy $\geq 10^{19.0}eV$, $\delta = 30° - 50°$, phase=$26°$). The probability to have a first harmonic $\geq$ 26.2% is 0.53 for this bin. The minimum probability 0.12 occurred at Energy $10^{18.5}eV - 10^{19.0}eV$, $\delta = 30° - 50°$, and phase=$281°$. There is no consistently high first harmonic and phase.

For the small scale anisotropy, 4773 stereo events with energy $\geq 0.5EeV$ are used in this sky map analysis [17]. The arrival direction of each shower is represented by a probability function in the celestial coordinate which has a gaussian distribution and centered on its arrival direction and has a width according to the angular uncertainty. The expected density function is the average of simulated data sets produced by randomly choosing a sidereal time sampled from the collection of actual shower sidereal times. The excess/deficit of event density is calculated from

$$\sigma = (actual\ \ density - expected\ \ density)/(uncertainty\ \ of\ \ expected\ \ density)$$

the *uncertainty of expected density* is the RMS width of simulated data sets. The results for some candidate sources are shown in table 4. No significant excesses of flux were detected. Figure 6 plots the contour of the σ distribution. No obvious large scale pattern of excess/deficit are found in this figure. The distribution of σ can be compared with the simulated data sets. It follows the statistical fluctuation of isotropic distribution. [17]

Table 3: The result of galactic latitude dependent anisotropy, s is the galactic latitude gradient and f is the galactic plane enhancement factor. The δ is the uncertainty from the least χ^2 fit.

Energy (EeV)	# of events	$s \pm \delta_s$	σ	$f \pm \delta_f$	σ
0.3 - 1.0	5971	-0.0316 ± 0.0207	-1.52σ	0.0420 ± 0.0376	1.12σ
1.0 - 3.0	1772	-0.0757 ± 0.0384	-1.97σ	0.0551 ± 0.0688	0.80σ
3.0 - 10.0	363	0.1422 ± 0.0829	1.72σ	$0.0863 \pm 0.1397^*$	0.62σ
≥ 10.0	$63^\dagger$	$0.1019 \pm 0.1824^*$	0.56σ	$0.0686 \pm 0.4510^*$	0.15σ

† : This bin has only 63 events, large fluctuation should be expected.
* : Notice these number are updated values of [12].

Table 4: Results for candidate sources for energy $\geq 0.5EeV$

source	Expected density ($/deg^2$)	Percent excess	Chance probability	95% C.L. flux upper limit ($cm^{-2}sec^{-1}$)
Cyg X-3	0.233	-7.73%	0.60	7.0×10^{-18}
Her X-1	0.188	-13.9%	0.69	6.2×10^{-18}
Crab	0.205	-4.92%	0.56	8.0×10^{-18}

Assuming a $2.2\mu G$ galactic magnetic field and the two component fit of proton and iron flux, table 5 shows the gyroradius of proton, iron and the mean gyroradius at different energy. If we compare the gyroradius to the galactic magnetic disk thickness ($3kpc$) [18], the required energy is about $6EeV$ for protons and $160EeV$ for iron nuclei. It is thus reasonable to observe an isotropic distribution of iron flux of energies below $160EeV$. For the proton flux, the anisotropy should be very weak when proton energy $\leq 6EeV$. At $6EeV$, protons constitute approximately 57% of over-all flux, if the protons originate in galactic disk, the anisotropy should be clearly seen. Our analysis results show no significant anisotropy for either large or small scales in over-all flux. This fact suggests that the proton flux may not come from the galactic disk.

6 300EeV event

On Oct. 15, 1991, an $320^{+35}_{-40} \times 10^{18}eV$ event was detected by Fly's Eye I at universal time 7:34:16.443. The depth of shower maximum, $X_{max} = 815^{+45}_{-35}g/cm^2$, is consistent with a proton ($X_{max} = 850g/cm^2$ based on QCD Pomeron model). But we can not rule out a gamma ray induced shower. The geometry of this shower had been carefully checked. This shower hit the ground behind FE II, and therefore FE II did not record this event. This fact

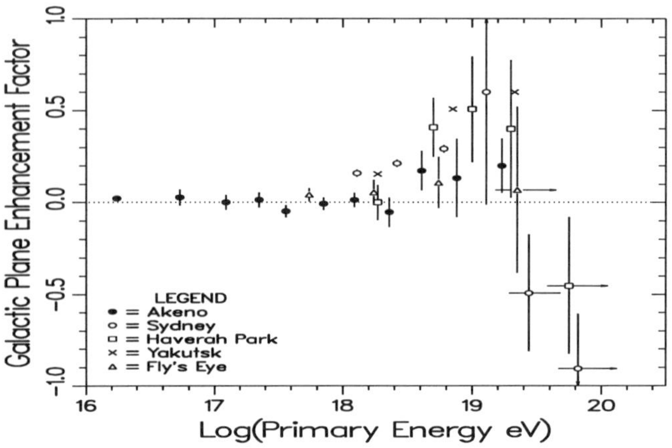

Figure 6: The galactic plane enhancement factor f for several groups.

puts a lower limit on the impact parameter of shower axi, which results in a lower limit of energy $150 EeV$. To check the shower reconstruction, we use a laser to simulate the shower axi at the best reconstructed shower impact point where shower hits the ground. The laser shot data has an uncertainty of $0.13°$ on the beam-detector plane which is consistent with the uncertainty of this event.

The arrival direction is 85.2 ± 0.2 degrees of right ascension and $48.0^{+3.2}_{-4.8}$ degrees of declination ($9.6°$ galactic latitude and $163.4°$ galactic longitude). This event almost comes from the direction near the anticenter of the galaxy. At this energy, the gyroradius is about 150kpc for protons in the galactic magnetic field of $2.2\mu G$. The angular deflection is less then $1°$ in the galactic magnetic fields [19]. Two nearby pulsars, the Crab and Geminga, are about $20°$ away. No other known energetic sources are near the arrival direction of this event. The source of this event is probably not in the galactic disk.

Table 5: The gyroradii of protons and iron nuclei at different energy. A $2.2\mu G$ galactic magnetic field and the two component fit of proton and iron flux was assumed.

	$\log_{10}(Energy(eV))$					
	17.5	18.0	18.5	19.0	19.5	20.0
proton flux ratio	0.11	0.26	0.50	0.74	0.89	0.96
R_H (kpc)	0.155	0.491	1.55	4.91	15.5	49.1
iron flux ratio	0.89	0.74	0.50	0.26	0.11	0.04
R_{Fe} (kpc)	0.006	0.019	0.060	0.189	0.597	1.888

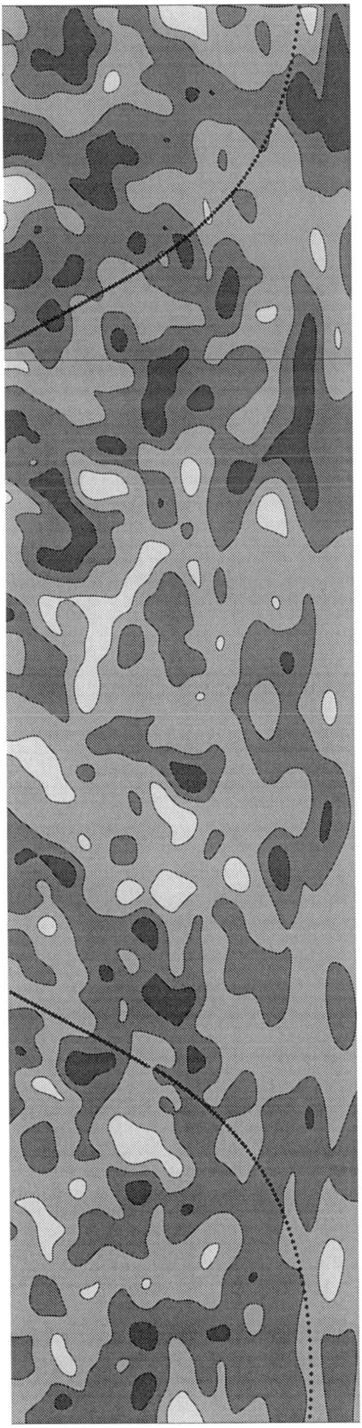

Fig. 7 This contour plot represents the statistical significance of density excesses and deficits as a function on the sky. Declination increases from -15 to 75 degrees from left to right. Right ascension increases from 0 to 360 degrees from bottom to top. Darker shades represent higher statistical significance. Contour lines are at -1, 0, +1. [17]

7 Conclusion

- The Fly's Eye monocular spectrum in the energy interval $10^{17.3} - 10^{19.9}$ is

$$J(E) = 10^{-29.55} \times (E/10^{18} eV)^{-3.07 \pm 0.01}$$
$$(m^2 \cdot s \cdot sr \cdot eV)^{-1}$$

- The stereo Fly's Eye observed a dip in the spectrum between $10^{17.6} - 10^{19.6} eV$. The spectral slope before the dip is -3.01 ± 0.06 and it steepens to -3.27 ± 0.02 at $10^{17.6}\, eV$ then flattens to -2.72 ± 0.10 at $10^{18.5} eV$.

- The composition of cosmic rays changes from predominately heavy below $10^{17.5} eV$ to predominately light above $10^{19} eV$.

- The stereo spectrum and composition can be explained by a two component fit.

$$\text{iron flux}: \log[J(E)(m^2\ s\ sr\ eV)^{-1}] = 33.185 - 3.496 \times \log(E)$$

$$\text{proton flux}: \log[J(E)(m^2\ s\ sr\ eV)^{-1}] = 16.782 - 2.610 \times \log(E)$$

The ratio of those two fluxes can be expressed as

$$\frac{\text{iron flux}}{\text{proton flux}} = 10^{-0.887 \times (\log(E) - 18.5)}$$

- The studies of anisotropy found no significant deviation from isotropy either on large or small scale. This is consistent with the previous two component fit. More statistics are needed to search for the anisotropy at an energy above 3 EeV.

- A $320^{+35}_{-40} \times 10^{18} eV$ event was detected. This is the highest energy ever detected and well beyond the Greisen energy cutoff.

Acknowledgement

We would like to acknowledge the earlier contributions of R.M. Baltrusaitis, H. Bergeson, R. Cady, R. Cooper, B. Fick, P.R. Gerhardy, D. Liebing, Y. Mizumoto and D. Steck to this experiment. In particular, we would like to thank G. Cassiday for his innumerable contributions to the development of the air fluorescent techniques. We are indebted to Colonels Frank Cox and James King and the staff of the Dugway Proving Ground for their continued cooperation and assistance. This work had been supported in part by the National Science Foundation (grants PHY-91-0021 at Utah) and the U.S. Department of Energy (grant FG-02-91-ER40677 at Illinois).

References

[1] R.M. Baltrusaitis *et al.*, Nucl. Instrum. and Methods, **A240**, 410, (1985)
[2] G.L. Cassiday *et al.*, Ann. Rev. Nucl. Part. Sci., **35**, 321, (1985)
[3] R.M. Baltrusaitis *et al.*, Phys. Rev. L., **54**, 16, 1875, (1985)
[4] clg:spe D.J. Bird *et al.*, procc. 23rd ICRC (Calgary), **2**, 34, (1993)
[5] D.J Bird *et al.*, Phys. Rev. L., **71-21**, 3401, (1993)
[6] D.J Bird *et al.*, Ap. J., **424**, 685, (1994)
[7] D.J. Bird *et al.*, procc. 23rd ICRC (Calgary), **2**, 38, (1993)
[8] T.K. Gaisser *et al.*, Phys. Rev. D, **47-5**, 1919, (1993)
[9] T. Stanev *et al.*, Comments Astrophys. **17-2&3**, 103, (1993)
[10] R.M. Baltrusaitis *et al.*, Astrophys. J., **303**, 313-315, (1986)
[11] X. Chi, *et al.*, J. phys. G., **18**, 1259, (1992)
[12] D.J. Bird *et al.*, procc. 23rd ICRC (Calgary), **2**, 55, (1993)
[13] G.L. Cassiday *et al.*, proc. ICRC (Adelaide), **3**, 196 (1990)
[14] G.L. Cassiday *et al.*,Ap. J., **351**. 454 (1990)
[15] R. Cooper *et al.*, "Astrophysical Aspect of the most Energetic Cosmic Rays", World Sci., Singapore, 34, (1991)
[16] D.J. Bird *et al.*, procc. 23rd ICRC (Calgary), **2**, 51, (1993)
[17] D.J. Bird *et al.*, procc. 23rd ICRC (Calgary), **2**, 30, (1993)
[18] D. Badhwar & S.A. Stephens, Ap.J. **212**, p494 (1977)
[19] P. Sommers, procc. Tokyo workshop on Techniques fro the study of Extremely High Energy Cosmic Rays, ICRR, univ. of Tokyo, 31 (1993)

THE HIGH RESOLUTION (HiRes) FLY'S EYE DETECTOR

D. J. BIRD (for the HiRes Collaboration)
Department of Physics
University of Utah
Salt Lake City
UT 84112
U.S.A.

ABSTRACT: In order to measure the primary cosmic ray flux beyond energies of 10^{19}eV in detail, a next generation Fly's Eye detector has been developed and is currently being constructed in the west desert of Utah. Details of this detector and preliminary results from a prototype of this detector investigating a lower energy regime of the primary cosmic ray flux will be discussed.

1. Introduction

The Fly's Eye detectors (Baltrusaitis *et al.* 1985, Bird *et al.* 1994) were operated from the late 1970's until late 1992 by the cosmic ray group of the University of Utah, observing the primary cosmic ray flux at particle energies greater than 0.3 EeV. The details of the results from these detectors are discussed in Bird *et al.* (1994), as well as the technique involved in detecting these highest energy cosmic rays (nitrogen fluorescence). The number of events detected above 10^{19}eV in a decade of operation is insufficient to provide statistically significant results on the chemical composition, energy spectrum and arrival direction distribution (anisotropy) of the primary cosmic ray flux above this energy.

To overcome this deficiency, a new detector, employing greatly improved computing and electronic capabilities has been developed to investigate this regime of the highest energy cosmic rays.

2. The High Resolution Fly's Eye Detector

The High Resolution Fly's Eye Detector (HiRes) has increased sensitivity compared to the original Fly's Eye detectors, due to both a reduction in the phototube aperture and an increase in the size of the mirror (see Table 1 for a comparison of the parameters of the two detectors).

The HiRes detector will consist of two sites (Five Mile Hill—the site of Fly's Eye 1 and CamelsBack Ridge), located 13 km apart, on Dugway Proving Grounds in the west

M. M. Shapiro et al. (eds.), Currents in High-Energy Astrophysics, 135–142.

desert of Utah. Each site will possess approximately fifty-four detectors. Each of these detectors consists of 256 photomultiplier tubes with a 1° aperture, giving a sky coverage of approximately 13° × 16°. Night-sky light is reflected from a two meter diameter spherical mirror through a UV-pass filter onto this array of closely-packed hexagonal photomultiplier tubes.

The resulting electronic signals are then processed by a local electronics system, located in a VME crate. If an event is deemed to have occurred, the relevant information is then relayed to a central computing facility over a thin-wire ethernet. Each detector is a stand-alone independent device, and this modular design has the advantage that it allows each individual detector to come on-line as soon as it is available.

The HiRes detector is being constructed in stages. Currently, a prototype detector has been built and operated for more than a year. Construction of the full HiRes detector has commenced at the CamelsBack site.

Detector Parameters	Fly's Eye	HiRes
PMT Aperture	5.5°	1°
Mirror Diameter	1.5m	2.0m
Time Resolution	50 ns	20 ns
Primary Energy Range	0.3 – 50 EeV	0.1 – > 200EeV
X_{max} Resolution at 1 EeV	50 – 70 gcm^{-2}	20 – 30 gcm^{-2}
Events per year > 10 EeV X_{max} Res < $30gcm^{-2}$	1	> 200
Electronics	Sample/Hold	Flash ADC

Table 1. Comparison of the Fly's Eye and HiRes Detectors

3. The HiRes Prototype Detector

A fourteen mirror HiRes prototype detector has been constructed on the site of the old Fly's Eye 1 detector. The arrangement and field of view of these mirrors is indicated in Figure 1. It has been oriented to overlook the Fly's Eye 2 (Baltrusaitis *et al.* 1985), CASA (Ong 1990) and MIA (Sinclair 1989) detectors located 3.4 km to the NNE and 150 m lower in altitude.

This detector has been fully operational since March 1993, and will remain operational during the construction of the full HiRes detector, ultimately being incorporated into one of the HiRes detectors.

4. Coincident Measurements

The prototype detector was constructed to test the hardware, electronic and software designs as well as to compare the observed background skynoise to that expected from Monte Carlo simulations.

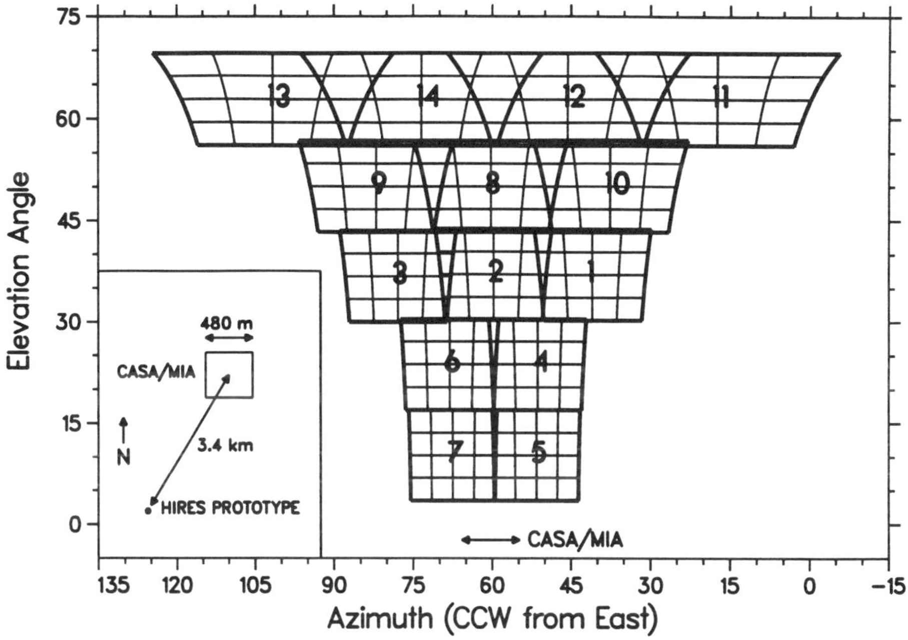

Figure 1. Sky Coverage of the HiRes Prototype Detector and position relative to CASA/MIA.

As a stand-alone detector, the Hires Prototype is very limited in its physics capabilities. This is due to both the lack of a significant sky coverage ($\sim 10\%$) and the complete lack of stereo observations. This implies that all cosmic ray showers detected need to be geometrically reconstructed through timing measurements alone, greatly reducing the accuracy of track reconstruction, and hence the physics obtainable from these events.

However, the prototype detector overlooks the neighbouring CASA (Ong 1990) and MIA (Sinclair 1989) detectors. CASA is an extensive air shower array, consisting of 1089 $1.5m^2$ particle detectors arranged in a 33×33 array with a separation of $15m$. MIA is an underground muon detector consisting of 16 widely spaced patches of 64 scintillation counters each buried $3m$ below CASA.

There is a limited overlap in sensitivity to extensive air showers between the prototype and these ground-based detectors, allowing the coincident measurement of these showers by the two different techniques. These observations will allow a significant investigation into the primary cosmic ray flux at energies between 50 and 1000 PeV.

It is expected that approximately 700 events will be observed per year in coincidence with the CASA/MIA detectors and a further 2000 per year will be observed in coincidence with the MIA detector alone. These rates appear to be consistent with those observed to date ($\sim 0.7hr^{-1}$ and $\sim 2.0hr^{-1}$ respectively since September 1993). A preliminary event

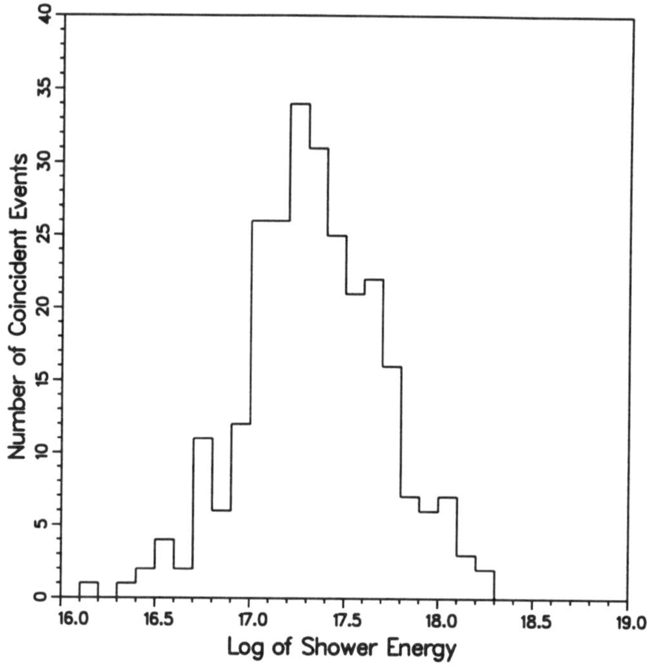

Figure 2. Preliminary Event Energy Distribution for events coincident with MIA. All energy estimates are uncertain by a factor of 2 (Fick 1994).

energy distribution for events coincident with MIA is presented in Figure 2 (all energy estimates have a factor of 2 uncertainty (Fick 1994)).

Coincidences are determined by both a match in event time (to within $\pm 2ms$ (WWVB)) and a flag set in the CASA/MIA data stream by a light signal sent from the HiRes detector to the CASA/MIA system after every trigger with a coincidence window of $50\mu s$. A post-acquisition merge of the data from the different detectors is carried out daily.

Figures 3a–3d are an example of a coincident event as viewed by the HiRes prototype and Figures 3e–3f are the same event as observed by the CASA/MIA detectors.

Cosmic ray showers observed by the prototype detector and with cores less than $500m$ from the center of CASA will trigger CASA/MIA and provide a shower direction and core with resolution of approximately 1.2° and $5m$ respectively. Knowledge of these parameters significantly improves the prototype geometric reconstruction, allowing a better estimate of the depth of shower maximum (X_{max}) (hence composition) and the number of particles at shower maximum (hence primary particle energy). These showers should also provide muon information from the MIA array and a comparison between the composition predicted by X_{max} and the number of muons (N_μ) should be possible, both overall and in some cases, on a shower by shower basis. If the composition of primary cosmic rays is mixed at these energies, then we should expect to see a correlation between X_{max} and N_μ.

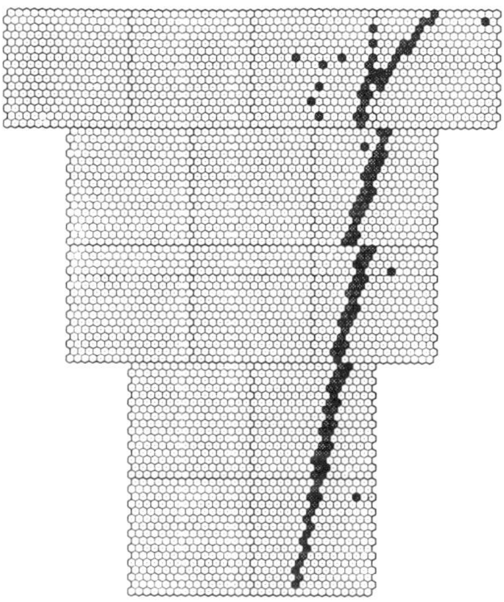

Figure 3a

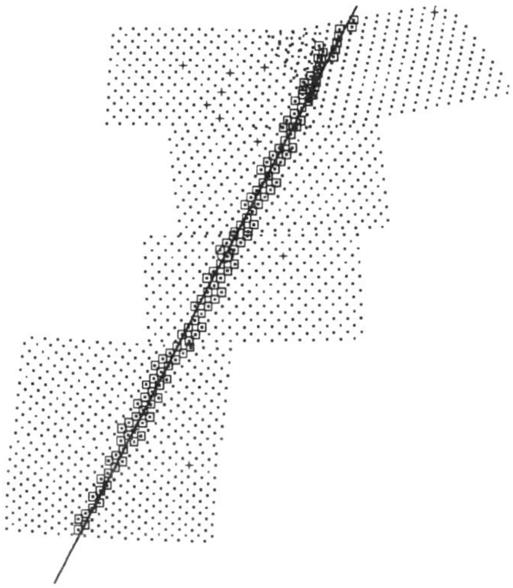

Figure 3b

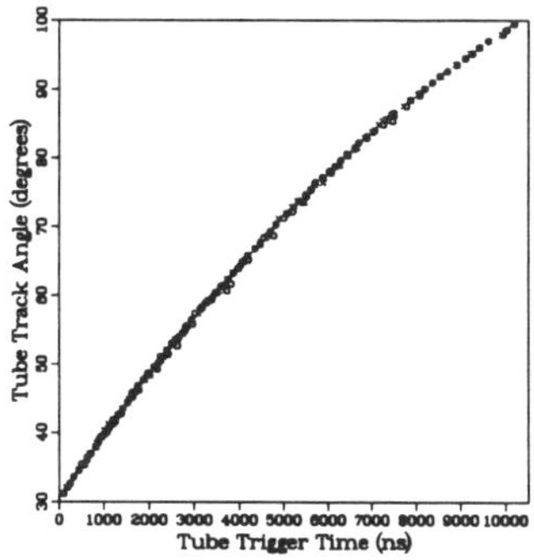

Figure 3c

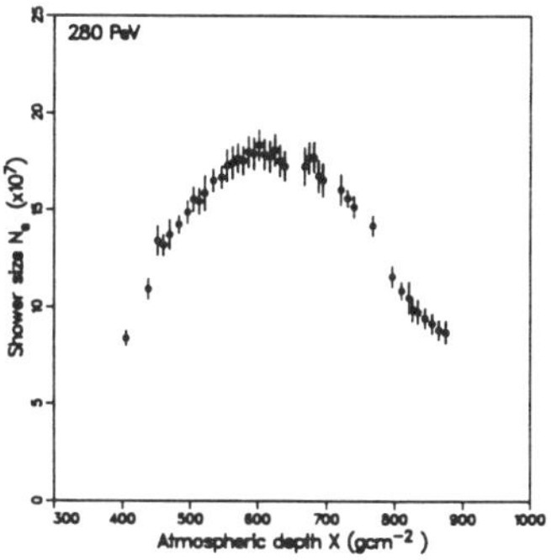

Figure 3d

Figure 3. Example of a Coincident event between the HiRes Prototype Detector and CASA/MIA

Figure 3a : Raw Tube Trigger Pattern
Figure 3b : Position in Sky of Tubes Triggered and best-fit shower plane
Figure 3c : Relative Tube Trigger Times (∘ observed times, × best-fit times)
Figure 3d : Longitudinal Shower Profile derived from timing fit

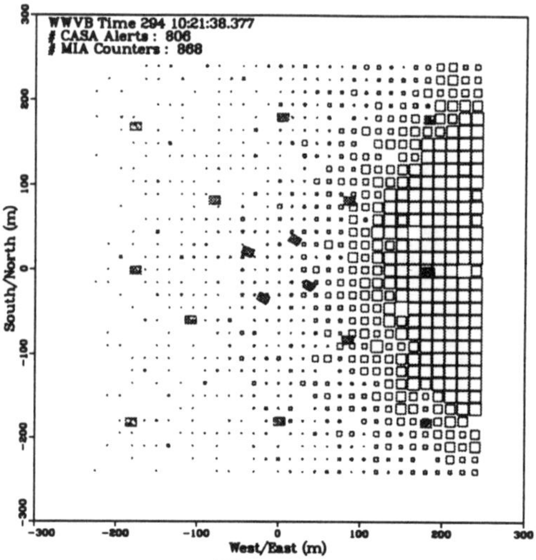

Figure 3e

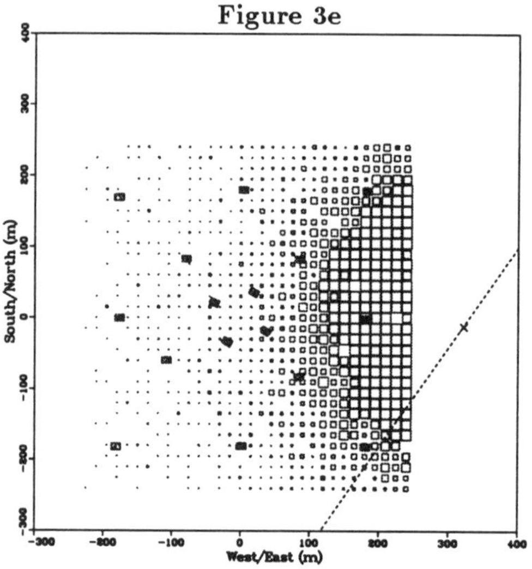

Figure 3f

Figure 3e: Distribution of particles detected by CASA and MIA. The boxes represent individual counters and their size is proportional to the number of particles detected with the maximum size corresponding the CASA saturation limit of $\sim 100m^{-1}$. The MIA patches are also indicated as are the individual counters in each patch which recorded a coincident muon.

Figure 3f: The dotted line indicates the projection of the HiRes determined plan of best-fit and the $\times$ indicates the core position determined from the timing fit. This core appears to be within $20m$ of the core predicted by CASA. In general, the timing-fit core position may be the order of a kilometer from the true core position.

Cosmic ray showers observed by the prototype detector and with cores less than $2km$ from the center of CASA will, in most cases, not trigger CASA, but due to the broader lateral distribution of muons, should still trigger MIA. While not as accurate as the core and directional information from CASA, the information provided by the muons still allows an improvement in the shower geometry reconstruction, and at a slightly lower energy threshold than those triggering CASA (30 PeV compared to 80 PeV). Monte Carlo simulations suggest that Fe-initiated showers should contain 75% more muons than that of a same-energy proton-initiated shower (Cassiday *et al.* 1990). It should be possible to investigate this effect using these coincident events as the longitudinal profile observed by the prototype detector provides a superior determination of the shower energy to any ground-based estimate (a good energy estimate has been lacking in previous investigations).

Using these coincident observations a number of other investigations into the primary cosmic ray flux in this energy regime are also possible.

4.1 X_{max}

To date, there is no information on the distribution of the X_{max} parameter below $3 \times 10^{17} eV$, which provides a measurement of the cosmic ray composition in this energy regime. Composition measurements in the coincident regime are almost non-existent. Satellite-based measurements extend to only about 1 PeV and the Fly's Eye estimates begin at 0.3 EeV. The Fly's Eye result on the energy spectrum (Bird *et al.* 1994) suggests a change in the composition below about $5 \times 10^{17} eV$ but there are insufficient statistics to support this hypothesis in this data alone.

4.2 $\rho(600)$

Large EAS arrays use the electron density at a distance of $600m$ from the core of the shower ($\rho(600)$) to determine the primary energy of the particle. Using these coincident observations, it will be possible to measure both $\rho(600)$ and the energy derived from the longitudinal profile providing a relative energy calibration between the two techniques. Such a calibration is necessary to understand the different energy spectra recorded by these different techniques.

4.3 MUON LATERAL DISTRIBUTION

There is evidence (Chiba *at al.* 1993) to suggest that the muon lateral distribution at distances greater than about $800m$ from the core of an EAS is not adequately explained by the Greisen formula (Greisen 1960). Using the coincidence measurements between the prototype and MIA should provide measurements of the muon lateral distribution to several kilometers and determine if this discrepancy is real.

5. The Future

The first four mirrors of the CamelsBack site are nearing completion and it is expected that they will be instrumented with a Flash ADC system by the end of the 1994. Ground has also been broken for the next twenty-four mirrors at this site, and they are expected to be on-line summer 1995.

6. Acknowledgements

The continued assistance and cooperation of Colonels Frank Cox and James King and the staff of the Dugway Proving Ground is acknowledged. This work has been supported in part by the National Science Foundation and the U.S. Department of Energy.

7. References

Baltrusaitis R.M. *et al.*: 1985, Nucl. Instr. Meth., **A240**, 410.
Bird D.J. *et al.*: 1994, these proceedings.
Cassiday G.L. *et al.*: 1990, NSF Proposal to Construct a High Resolution EYE (HiRes) Detector.
Chiba N. *et al.*: 1993, Proc. 23rd I.C.R.C. (Calgary), **4**, 307.
Fick B.E.: 1994, private communication.
Greisen K.: 1960, Ann. Rev. Nucl. Sci., **10**, 63.
Ong R.A.: 1990, Nucl. Phys. B. (Proc. Suppl.), **14A**, 273.
Sinclair D.: 1989, Nucl. Instr. Meth., **A278**, 583.

THE HEGRA EXPERIMENT AT LA PALMA

V. Fonseca
Dpto Física Atómica
Facultad Ciencias Físicas
Universidad Complutense
28040 Madrid, Spain

ABSTRACT. The HEGRA experiment is a ground based experiment located on the Canary Islands (Spain) for the measurement of the extensive air showers produced in the Earth's atmosphere by very high energy cosmic rays. It was originally proposed as a scintillation counter matrix and at present the experiment has a combination of different techniques aimed to cover a wide energy range, to decrease the energy threshold and to obtain the maximum information about the measured parameters of the showers from the combined measurements of the several sub-arrays.

1 Introduction

Cosmic rays, discovered by Victor Hess in 1912, have been, since then, extensively studied but nevertheless, they still pose many unanswered questions. In fact their origin is unclear and their nature and composition at the higher energies is still unknown. Cosmic rays are the most energetic radiation impinging on the Earth's atmosphere with energies ranging from 1 MeV ($= 10^6$ eV) up to the highest observed energies of 100 EeV ($= 10^{20}$ eV). Most of the cosmic radiation consists of hadrons (protons and nuclei), a small fraction (1%) are electrons and even a smaller fraction (0.001%) are gamma rays. Cosmic radiation contributes a significant amount to the energy density of the Universe. The local galactic cosmic ray energy density is about 1 eV cm^{-3} which is of the same order of magnitude as the local energy density of starlight. When very high energy cosmic particles penetrate the Earth's atmosphere, a cascade of relativistic secondary particles is produced that can be detected at ground level with ground based arrays, if the energy of the primary particle is large enough. In order to get reliable information about the original cosmic particle, as many parameters as possible of the air shower front should be measured. This concept has guided the design of the HEGRA experiment as an array built up of several sub-arrays located at the same site, each sub-array being capable to detect, in coincidence with the others, different parameters of the same shower (1).

M. M. Shapiro et al. (eds.), Currents in High-Energy Astrophysics, 143–155.

In fact the idea of simultaneous observations of cosmic phenomena in different energy ranges is being applied to combined measurements made by extraterrestial detectors on satellites and ground based detectors (scintillation arrays, telescopes, etc.) (2). Correlated observations of satellites and ground based detectors distributed around the world allow measurements of emissions in real time from celestial objects in a broad wavelength region (IR, optical, UV, X rays etc.).

In the last years many important discoveries in the field of High Energy Astrophysics have been made with detectors placed on board satellites (3) and with ground based experiments (see 4, 5 and references therein). New galactic objects have been discovered as strong gamma ray emitters in the very high energy region, as the Crab Nebula (6), others are extragalactic objects, as Mrk421 (7), other sources of gamma rays are isotropically distributed as the gamma ray bursters (GRB) (8), etc. which opens a new window in high energy astrophysics research. These discoveries have encouraged the construction of new and challenging experiments around the world with energy thresholds ranging from a few hundred GeV (9) to the highest energy region (10^{20} eV) (10, 11).

In order to extract the information about the origin, composition and nature of the cosmic rays powerfull gamma/hadron separation criteria based on new approaches are needed. In what follows an overview of the present status of the HEGRA experiment is given. Several publications describing in detail the different parts of the HEGRA array are in preparation.

2 What HEGRA measures in Air Showers?

The Hegra (High Energy Gamma Ray Array) experiment, a ground based array originally proposed by the Kiel group of the Hegra collaboration, started taking data in 1988 at La Palma, (one of the Canary Islands, Spain) as a small scintillation matrix (37 counters). Since then, the HEGRA detector has been thoroughly extended and at present a major upgrading program is taking place. The experiment is operated by an international collaboration formed by groups from the Universities of Hamburg, Kiel, Madrid and Wuppertal, the Max-Planck Institutes from Munich and Heidelberg and the Yerevan Physics Institute (Armenia). The experiment, located on the Roque de los Muchachos at 2240 m above sea level covers a square area of 40000 m^2 (geographical coordinates 28.8^o N, 17.9^o W).

In figure 1 a layout of the experimental area is shown. At present the HEGRA experiment combines four experimental techniques arranged in sub-arrays in the same area to study different parameters of the showers initiated high up in the atmosphere by high energy cosmic rays. What can be measured with the actual HEGRA configuration can be summarized as follows:

- Measurement of the electromagnetic component (mostly electrons, positrons and photons) of the shower front with a scintillation array that samples the cosmic ray shower tail reaching the ground level. This array is operated continuosly day and night.

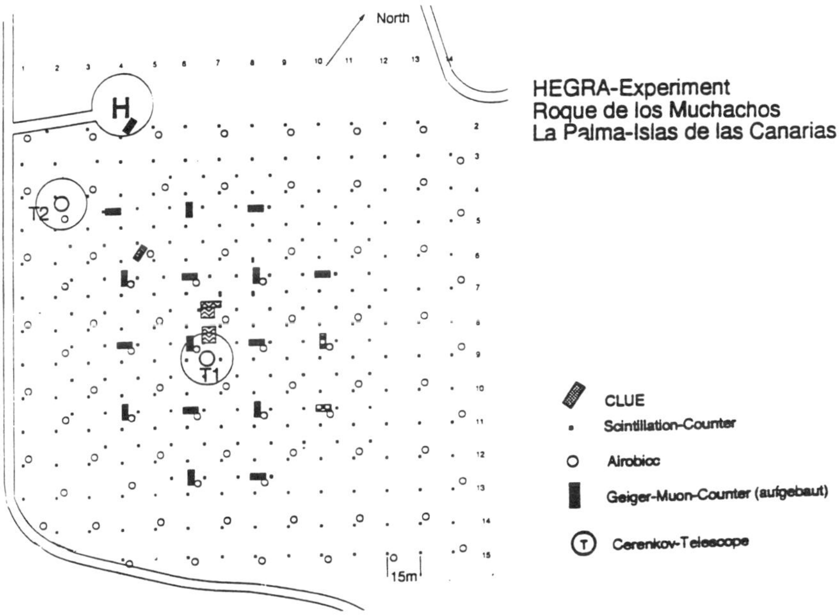

Figure 1: Layout of the HEGRA experiment (Spring 1994).

- Measurement of the muon component with muon detectors distributed in the experimental area at several positions to measure the muon content of the shower disk and the direction of the primary cosmic ray in order to improve the gamma/hadron separation power. The muon content of the cosmic gamma and hadron induced showers differs at least by an order of magnitude if the production cross sections can be scaled in the conventional way. Additionally, the HEGRA muon counters can also be used to sample the electromagnetic component of the shower. It is planned to improve the sensitivity of the muon counters in order to increase further the sensitivity of the HEGRA detector to the electromagnetic component of the shower.

- Measurement of the Cerenkov light disk with a combination of several imaging Cerenkov telescopes for the observation of the Cerenkov light flashes emitted by the charged particles of the atmospheric showers. The imaging air Cerenkov telescope technique is the one that measures cosmic radiation at ground level with the lowest energy threshold ($E_{th} > 0.1$ TeV) and with a very high gamma/hadron separation power, but telescopes have limited angular acceptance, thus restricting their observation window to only one source at a time. The actual Cerenkov telescopes operate only during clear and moonless nights (duty cycle 10 - 15 %).

New moon-blind Cerenkov telescopes able to measure the ultraviolet component of the Cerenkov light (below 240 nm) have been developed and are under test (CLUE collaboration) at the HEGRA site.

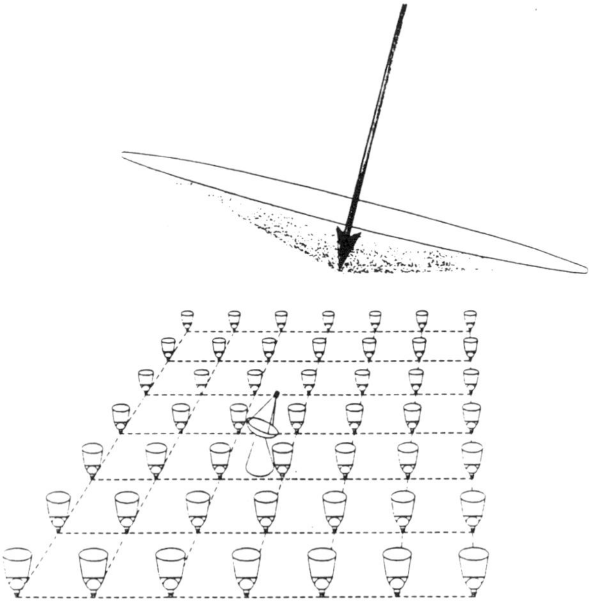

Figure 2: Artistic picture of the Cerenkov light cone of an air shower approaching the AIROBICC matrix from the direction of the arrow.

- Measurement of the Cerenkov light disk with an air Cerenkov counter matrix (named AIROBICC array) to measure the Cerenkov light of the showers. A picture of the method is shown in figure 2.

This is a new concept developed by the HEGRA collaboration to obviate the need for tracking and to measure many sources at a time (12, 13). The basic principle is similar to the principle used for the detection of the particles of the shower front by an array of scintillation counters, but in this case we measure the much more numerous Cerenkov photons produced in the shower as it develops deeper in the atmosphere. The multiple sampling of the light disk provides a good measurement of the energy and of the fluctuations of the light distribution. This array concept allows measurement not only of the lower part of the shower, as is done with the scintillation array, but also the longitudinal development of the shower in the atmosphere, thus improving the power to distinguish primary photon from primary hadron induced showers. The caveats of the actual AIROBICC concept are the restriction to measure only during clear and moonless nights and the integration of the night sky background light

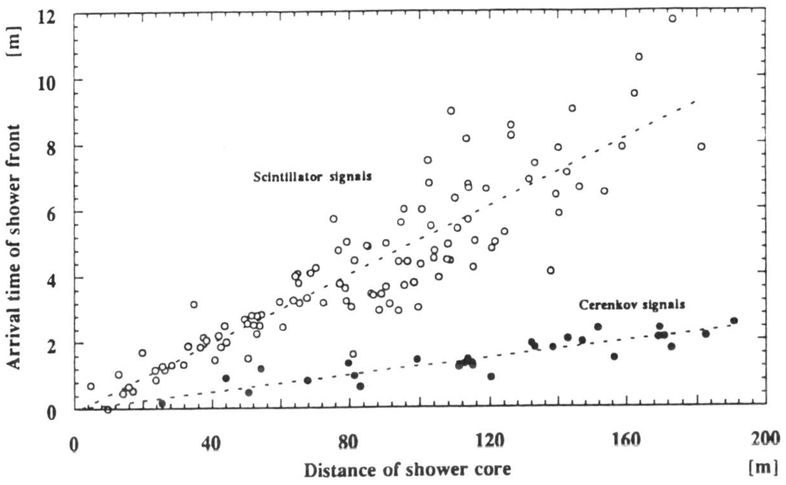

Figure 3: Simultaneous measurement of the shower front of an extensive air shower in the HEGRA scintillator matrix and in the air Cerenkov matrix AIROBICC

front is better defined (Cerenkov photons have a delay of about 5 ns/100 m with respect to the shower core) than the particle disk of the same shower (mainly formed by e^+ , e^- and γ particles with a measured delay of 15 ns/100 m). Also, the fluctuations on the arrival time of the Cerenkov photons are smaller than those of the particle disk components.

In figure 4, the angular resolution of the AIROBICC matrix is compared with the angular resolution of the scintillation matrix. The angular resolution (defined as $\sigma_{63\%}$) is obtained using the standard checker board approach, and it is shown as a function of the shower size which was determined from the number of electrons measured by the scintillation array. As can be seen from figure 4, the angular resolution of the AIROBICC array is about five times better than the angular resolution of the scintillation array.

Additional AIROBICC counters (36 in all) are ready to be installed at the HEGRA site to form an inner region with a higher density of Cerenkov counters in order to improve the sensitivity and the angular resolution of the array further.

3.3 THE MUON ARRAY

It consists of 20 muon tracking detectors of 6×2.8 m^2 with an angular resolution of 1°. The detectors consist of high density absorbers ($\rho = 0.8$ g/cm^3) interleaved with layers of geiger tubes (6 layers). There are 4.5 radiation lengths (Xo) of Pb located over a large angular window (35° angular acceptance), which limits the threshold.

3 The HEGRA Instruments

A brief description of the instruments of the HEGRA sub-arrays is given here. During summer 1993 a shutdown occurred to replace the data taking system from CAMAC to FASTBUS electronics for the scintillator and AIROBICC arrays (the telescopes use CAMAC for the time being).

3.1 THE BASIC SCINTILLATION MATRIX

The basic detector consists of a 1 m^2 plastic scintillator viewed from below by two photomultipliers, one provides a fast timing signal for the determination of the shower direction while the other one measures the multiplicity of particles from which the energy of the primary particle is estimated. Each counter is covered with a 5 mm layer of lead. From 1988 to June 1992 the HEGRA complex was operated only as a scintillation array with an increasing number of counters (37 counters in 1988-89, 100 in Nov 1990, 169 in June 1992) (14). The trigger rate of the '169' counter array was 7.8 Hz ($\geq$ 14 huts with signal); the average angular resolution was $\leq 0.8^o$; and the energy threshold was E_{thr}= 50 TeV. In summer 1992 the inner part of the array was increased by 52 additional stations of the same type (using only one photomultiplier) in the inner area of the array in order to enhance the sensitivity and to improve the precision of the core determination. The trigger rate increased to 14 Hz.

3.2 AIROBICC (AIR SHOWER OBSERVATIONS BY ANGLE INTEGRATING CERENKOV COUNTERS)

This array consists of 49 individual stations arranged in a 7×7 matrix with a 30 m grid spacing (12, 16). Its area matches closely that of the scintillation matrix. Each individual counter consists basically of a 20 cm diameter photomultiplier that views the night sky with a 1 sterad angular acceptance. In order to increase the light collection area, the photomultiplier is surrounded by a mirrored Winston cone. A filter restricts the spectrum sensitivity to a range of 300-470 nm to reduce the background from the diffuse night sky light. The angular acceptance is 35°; the energy threshold is 35 TeV for hadrons (vertical incidence); the average angular resolution 0.25^o for hadrons; and the uptime about 15%/year. The actual trigger condition for the scintillation and AIROBICC arrays requires that at least 14 scintillation counters or 6 AIROBICC counters are fired in 150 ns. The trigger rate with both arrays is 18-22 Hz, and the AIROBICC threshold is 5000 photons/m^2.

The detection of a few TeV hadron shower is presented for both the scintillation and the AIROBICC arrays for comparison in figure 3.

Figure 3 shows that the time structure of the Cerenkov light disk of the shower

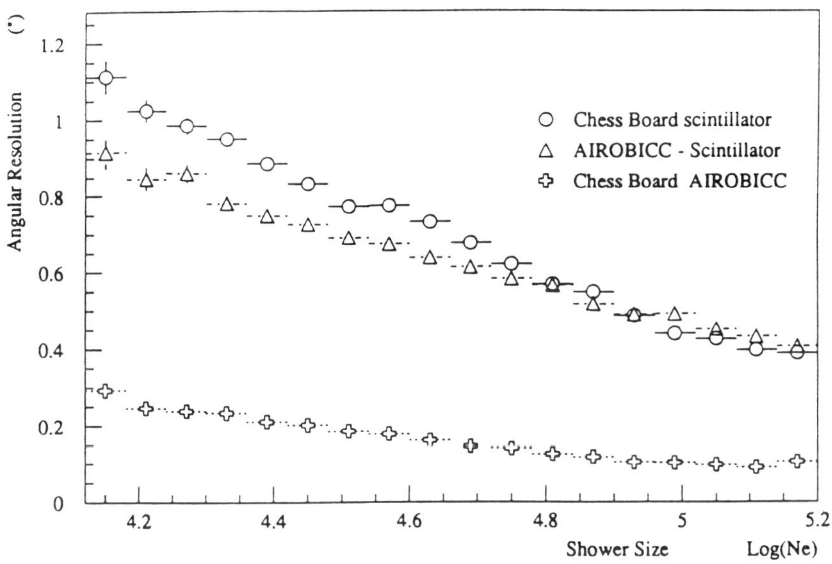

Figure 4: Angular resolution of the scintillator array and AIROBICC matrix (using the checker-board method) as a function of shower size.

at the first and second planes of the towers to absorb the low energy electromagnetic component of the shower front and to add information to the scintillator array. The muons are identified by tracking (one projection) in 6 layers after 9 Xo. The minimum energy of muons to cross all 6 layers is about 160 MeV (15). The measurement of muons is important to know the nature of the primary particle and for the detection of the abnormally muon-rich showers from point sources (16). The muon detectors work in combination with other HEGRA sub-arrays.

3.4 THE IMAGING AIR CERENKOV TELESCOPE MATRIX

The complete HEGRA air Cerenkov telescope subarray will consist of 6 units. The first telescope located at the center of the array became operational in 1992, and this is the prototype unit with a mirror of area 5 m^2, a 37 pixel camera, a trigger rate of 0.5 Hz and an energy threshold of 2 TeV. In 1993 a second telescope was installed at a distance of about 92 m from the first telescope. It was equipped with a mirror of 8.3 m^2 and a 61 pixel camera and the trigger rate is 2 Hz. By Autumn 1994 a third telescope with the same size as the previous one will be installed with a 5° camera (271 pixel), aimed to search for gamma emissions from extended sources. By the end of next year, it is expected to have the full telescope array ready, which in its final form will consist of: 5 telescopes of 8.3 m^2 each being a unit equipped with a

271 pixel camera and the other ones equipped with 127 pixel cameras. The sixth telescope, the prototype, will be used as a test unit for new detector developments. The expected energy threshold of this 5 telescope array is of the order of 500 GeV.

The telescopes can either track sources individually or follow one source with a very high gamma/hadron separation power. While the individual telescopes would mainly track specific sources in the energy range of 1 - 2 TeV, it is expected that enough data can be collected for a subset of events that overlaps with the data from the other sub-arrays.

At the moment of writing this paper, two UV telescopes from the CLUE collaboration are operating in the HEGRA array in a test mode and are taking data with the general trigger of the scintillation array. These single mirrored telescopes (2 m diameter mirror) are equipped with photosensitive proportional chambers with a 256 pixel pad readout. The filling photosensitive gas, TMAE, has an upper spectral cutoff in its sensitivity around 230 nm and a quantum efficiency of about 15%.

4 Results

In this section a brief description of the results obtained with the HEGRA experiment is presented. Full description will be available in the coming months.

A search for point γ ray sources has been carried out taking the data collected with the scintillation array. A mean angular resolution of 0.8^o and a proton energy threshold of 50 TeV was considered in the source search. These parameters are obtained using 4 independent methods: 1) a full Monte Carlo simulation of air showers and the detector response (17), 2) the checker-board method based on actual data, 3) an analysis of the cosmic ray obscurations by the moon and the sun, 4) a comparison of air showers observed simultaneously by the scintillation and the AIROBICC arrays.

Neither steady nor pulsed gamma ray emission from the known sources accesible to the HEGRA array has been detected. Therefore only flux upper limits have been obtained for the scanned sources, and these flux values are in agreement with those obtained by other experiments (18 - 22). Based on the BATSE gamma-ray burst catalogue, a search for GRB candidates has been carried out using the scintillation data sample. No significant excess has been found from any of the possible candidates.

A similar search for γ point sources has been done using the data obtained with the Cerenkov light information given by the AIROBICC array (13, 23). In the first search done using the data obtained with the CAMAC electronics and no cuts derived from Monte Carlo studies for gamma/hadron separation or from the muon tower data, no significant excess coming from the well established gamma ray sources has been detected.

Intensive Monte Carlo studies are in progress to study the gamma/hadron separation as well the chemical composition of the cosmic rays based on the shower measurements obtained with the scintillation and the air Cerenkov arrays as a first approach.

In figure 5, an example is shown of the distribution of the characteristic parameter 'radius' of the Cerenkov light distribution of the air showers as measured from AIROBICC and as calculated from Monte Carlo simulations for proton and iron showers. The 'radius' parameter is defined as the perpendicular distance between the shower axis and the shower front point where the Cerenkov light intensity of the shower front decreases by a factor 1/e. This simple example is an indicator that Monte Carlo studies combining the information left by the atmospheric showers on several detectors can greatly help to understand the hard problem of the cosmic ray chemical composition using data from ground based detectors.

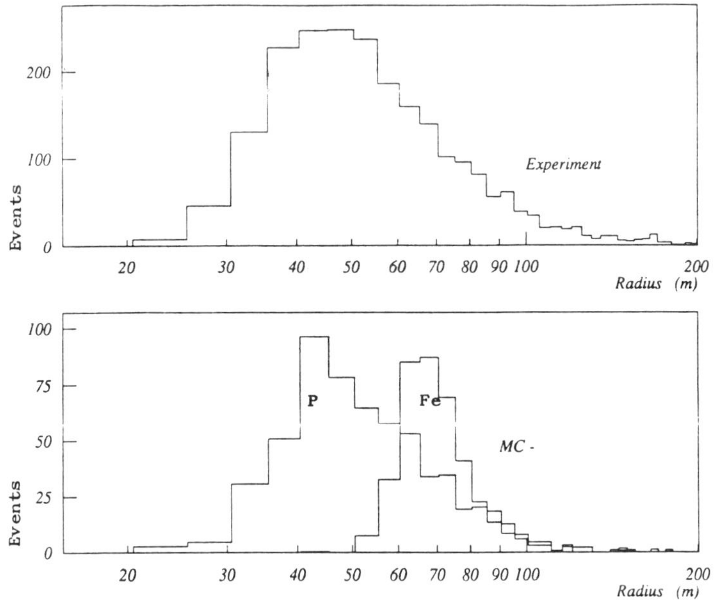

Figure 5: Distribution of the measured 'radius' for the air showers compared with the Monte Carlo calculations for initiated proton and iron showers.

The upper limit of the flux values obtained for many sources in the field of view of the AIROBICC and the scintillation arrays, agree well with those obtained by other experiments. Our flux values (90% CL) for the Crab Nebula of $1.78*10^{-13}$ $cm^{-2}s^{-1}$ (γ-E_{th} = 24 TeV) and for Mrk421 of $1.24*10^{-13}$ $cm^{-2}s^{-1}$ (γ-E_{th} = 25 TeV) (13) are already about a factor of two below the extrapolated values from lower energy regions and from those obtained by other experiments in higher energy regions (24-33), due to the improved sensitivity and angular resolution of the combined array.

Concerning the analysis of the data obtained with the Cerenkov telescopes of the HEGRA array, the gamma ray point-source search has been sucessful (34) (the azwidth parameter (35) has been used in the analysis). From the analysis of the

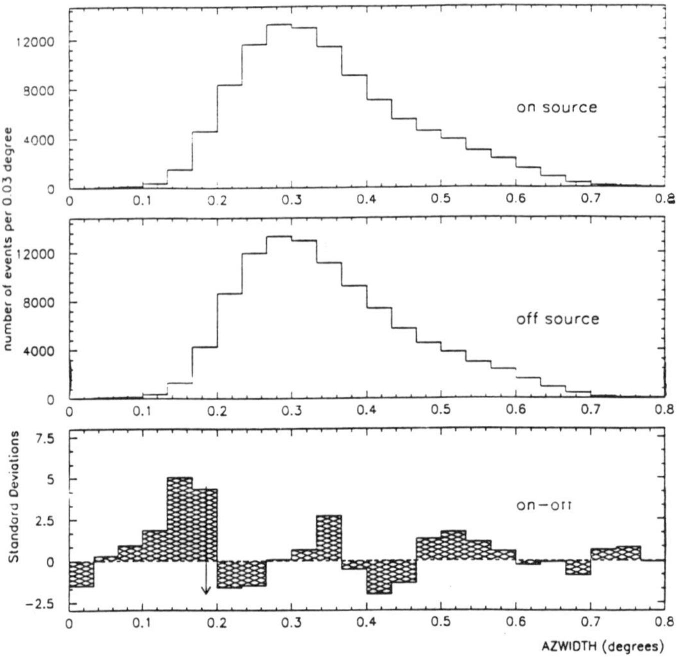

Figure 6: Azwidth distributions for the "on" (upper) and "off" (middle) source measurements and significance distribution (lower) for the differential excess from the Crab. The arrow indicates the Monte Carlo prediction for the azwidth cut.

data taken with the first telescope during September 92 until October 93, a 6.2 σ excess is visible from the Crab Nebula during a 70 hours on/off observation period (see figure 6). From the analysis of the data taken with the second telescope during last winter, a 4.6 σ excess is visible from the Crab Nebula during a 14 hours on/off observation period.

5 Summary and Conclusions

HEGRA is one of the most powerfull tools for the search of cosmic rays in the energy region (0.5 TeV - 10 PeV) in the North Hemisphere. The lack of steady or sporadic gamma ray sources, as obtained from the source search done up to now using the data obtained with the scintillation array alone, seems to indicate that the actual scintillation technique is not sensitive enough to detect gamma rays coming from point sources. The upgrade program of the HEGRA detector aimed to decrease the energy threshold, to improve substantially the angular resolution by sampling the Cerenkov light of the shower front, and to measure the muon component, has proved to be very successful. An improvement of the angular resolution of the AIROBICC array by a factor five, in comparison with the angular resolution of the scintillation array, has already been obtained. Also, the analysis of the data taken with the two first telescopes has shown a clear gamma ray signal from the direction of the Crab Nebula in the TeV energy range.

This unique combination of different techniques placed in the same experimental area with its cross-calibration possibilities opens new prospects in gamma-ray astronomy.

6 Acknowledgements

I want to thank Dr. M. M. Shapiro for inviting me to attend the 1994 Erice Summer School.

References

[1] Aharonian, F. et al., The HEGRA collaboration. 1993, Proceedings of the 23rd International Cosmic Ray Conference (I.C.R.C.), 4, 291.

[2] Lorenz, E. et al., 1994, A proposal for the Compton Gamma Ray Observatory Cycle 4.

[3] Kanbach, G., 1992, EGRET status report. Contribution Workshop: Towards a Mayor Atmospheric Cerenkov Detector, Palaiseau.

[4] Cronin, J. W., Gibbs, K. G. and Weekes, T. C., 1993, Annu. Rev. Nucl. Part. Sci., 43, 883.

[5] Lorenz, E., 1993, Nuclear Physics B (Proc. Suppl.) 33A,B, 93.

[6] Weekes, T. C., 1988, Phys. Rep., 160, 1.

[7] Punch, M. et al., 1992, Nature, 358, 477.

[8] Meegan, C. A. et al., 1992, Nature, 355, 123.

[9] Mirzoyan, R. et al., 1992, Imaging Cerenkov multitelescope approach of the HEGRA collaboration. Contribution Workshop: Towards a Mayor Atmospheric Cerenkov Detector, Palaiseau.

[10] Cronin, J. W., 1992, Nuclear Phys. B (Proc. Suppl.), 28B, 213.

[11] Busch, J. et. al., 1992, Nuclear Phys. B (Proc. Suppl.), 28B, 163.

[12] Bott-Bodenhausen, M. et al., 1992, Nucl. Inst. and Methods, A315, 236.

[13] Karle, A., 1994, Ph.D. thesis, University of Munich, Germany.

[14] Merck, M., 1993, Ph.D. thesis, University of Munich, Germany.

[15] Just, F. et al., 1993, The HEGRA collaboration. 1993, Proceedings 23rd I.C.R.C., 4, 203.

[16] Samorski, M. and Stamm, W., 1983, Ap. J., 268, 117.

[17] Martinez, S. et al., 1993, The HEGRA collaboration. 1993, Proceedings 23rd I.C.R.C., 4, 742.

[18] Amenomori, M. et al., 1993, The Tibet Asγ collaboration. 1993, Proceedings 23rd I.C.R.C., 1, 412, 342.

[19] Alexandreas, D. E. et al., 1993, The Cygnus collaboration. 1993, Proceedings 23rd I.C.R.C., 1, 416.

[20] Borione, A. et al., 1993, The CASA-MIA collaboration. 1993, Proceedings 23rd I.C.R.C., 1, 424.

[21] Merck, M. et al., 1993, The HEGRA collaboration. 1993, Proceedings 23rd I.C.R.C., 1, 361.

[22] Prahl, J. et al., 1993, The HEGRA collaboration. 1993, Proceedings 23rd I.C.R.C., 4, 734.

[23] Karle, A. et al., 1993, The HEGRA collaboration. 1993, Proceedings 23rd I.C.R.C., 4, 235.

[24] Nolan, P. L., 1993, Ap. J., 409, 697.

[25] Weekes, T. C. et al., 1989, Ap. J. 342, 379.

[26] Vacanti, G. et al., 1991, Ap. J., 377, 467.

[27] Akerlof, C. W. et al., 1990, Proc. 21st ICRC, 1, 135

[28] Baillon, P. et al., 1993, Astroparticle Physics, 1, 341

[29] Lewis, D. A. et al., 1993, Proc. 23rd ICRC, 1, 279.

[30] de Jager, O. C. and Harding, A. K., 1992, 396, 161.

[31] Lin, Y. C. et al., 1992, Ap. J. Lett., L49, 490.

[32] Mohanty, G. et al., 1993, Proc. 23rd ICRC, 1, 440.

[33] Schubnell, M., 1993, Proc. 23rd ICRC, 1, 409.

[34] Krennrich, F. et al., 1993, Proc. 23rd ICRC, 1, 251.

[35] Hillas, A. M., 1985, Proc. 19th ICRC, 3, 445.

THE EXTENSIVE AIR SHOWER EXPERIMENT IN TIBET

M. NISHIZAWA
Institute for Cosmic Ray Research, University of Tokyo
3-2-1 Midori-cho, Tanashi 188, Japan
For the Tibet ASγ Collaboration [1]

ABSTRACT. The Tibet air shower array has operated at an altitude of 4300 m above sea level in Tibet. The results obtained by the observation during the period of 1990 June through 1992 September are summarized in this paper. We have searched for continual emission of 10 TeV gamma rays from the Crab Nebula, X-ray binaries, pulsars, AGNs and other active objects. No *DC* excess is found from any source. Flux upper limits above 10 TeV and above 30 TeV are obtained for the each source. This is the first observation, with the air shower array, in 10 TeV energy region.

The shadows of the Sun and Moon in the 10-TeV cosmic ray flux were clearly observed at the 7.1σ level for the Moon, while the shadow of the Sun was found at the direction significantly away from the apparent solar position, under the influence of the solar and interplanetary magnetic fields. We also observed that the shadows by cosmic rays coming from the away and toward field sectors in the interplanetary space are shifted in opposite directions from each other according to its polarity of

[1]**The Tibet ASγ Collaboration:** M.Amenomori[1], Z.Cao[2], L.K.Ding[2], Z.Y.Feng[3], K.Hibino[4], N.Hotta[5], Q.Huang[3], A.X.Huo[2], H.G.Jia[3], G.Z.Jiang[3], S.Q.Jiao[3], F.Kajino[6], K.Kasahara[4], Labaciren[7], D.M.Mei[7], L.Meng[2], X.R.Meng[7], Mimaciren[7], K.Mizutani[8], J.Mu[9], H.Nanjo[1], M.Nishizawa[10], Nusang[7], A.Oguro[8], M.Ohnishi[10], I.Ohta[5], T.Ouchi[10], J.R.Ren[2], To.Saito[11], M.Sakata[6], Z.Z.Shi[2], M.Shibata[12], T.Shirai[4], H.Sugimoto[13], X.X.Sun[2], K.Taira[13], Y.H.Tan[2], N.Tateyama[4], S.Torii[4], H.Wang[2], C.Z.Wen[3], Y.Yamamoto[6], X.Y.Yao[9], G.C.Yu[3], P.Yuan[2], T.Yuda[10], J.G.Zeng[3], C.S.Zhang[2], H.M.Zhang[2], L.Zhang[9], Zhasang[7], Zhaxiciren[7] and W.D.Zhou[9]
[1] Department of Physics, Hirosaki University, Hirosaki 036, Japan
[2] Institute of High Energy Physics, Academia Sinica, Beijing 100039, China
[3] Department of Physics, South West Jiaotong University, Chengdu 610031, China
[4] Faculty of Engineering, Kanagawa University, Yokohama 221, Japan
[5] Faculty of Education, Utsunomiya University, Utsunomiya 321, Japan
[6] Department of Physics, Konan University, Kobe 658, Japan
[7] Department of Mathematics and Physics, Tibet University, Lhasa, China
[8] Department of Physics, Saitama University, Urawa 338, Japan
[9] Department of Physics, Yunnan University, Kunming 650091, China
[10] Institute for Cosmic Ray Research, University of Tokyo, Tanashi 188, Japan
[11] Tokyo Metropolitan College of Aeronautical Engineering, Tokyo 116, Japan
[12] Faculty of Education, Yokohama National University, Yokohama 240, Japan
[13] Shonan Institute of Technology, Fujisawa 251, Japan

M. M. Shapiro et al. (eds.), Currents in High-Energy Astrophysics, 157–175.

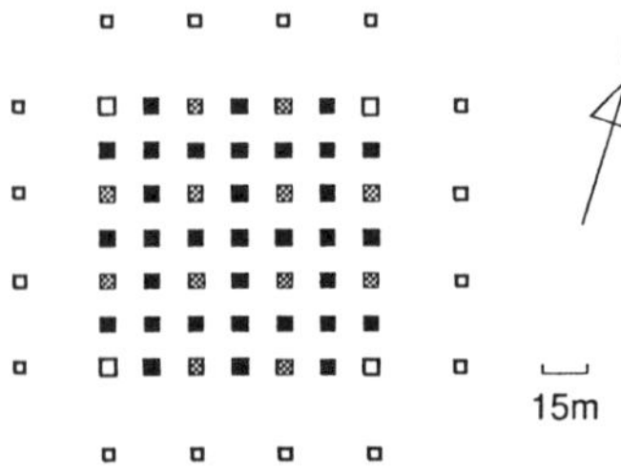

Figure 1: Layout of scintillation detectors in the Tibet air shower array. Closed squares and open squares stand for the FT- and Density-detectors, respectively.

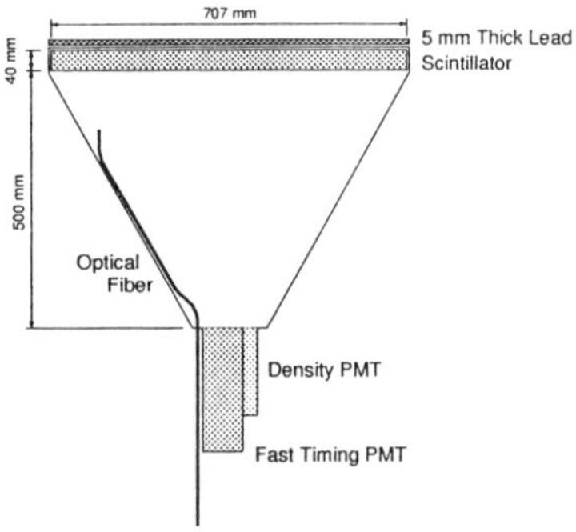

Figure 2: Sectional view of the FT-detector.

the fields. This is the first direct observation of the influence of the interplanetary magnetic field on the shadow's displacement. The results imply that observation of the cosmic ray shadow by the Sun can produce new information on the relation between a time variation of the large-scale structure of the solar and interplanetary magnetic fields and the cycle of solar activity. Finally, the Tibet-II experiment is briefly introduced.

1 Experiment and Data Analysis

1.1 TIBET AIR SHOWER EXPERIMENT

The Tibet air shower array has been operating from January 1990 at Yangbajing. It is located at an altitude of 4300 m a.s.l. (90.53°E and 30.11°N), corresponding to an atmospheric depth of 606 g/cm^2. As shown in Fig 1, the fast-timing (FT) detector array consists of 45 scintillation detectors of 0.5m^2 each, which are placed in a grid pattern of a 15m spacing. These FT-detectors are equipped with 2" ϕ fast response photomultiplier tubes (PMTs)(Hamamatsu H1949) with a transit time spread less than 0.55 ns to measure the arrival time of shower particles. Among these, furthermore, 16 detectors are equipped with 1.5" ϕ PMTs (Hamamatsu H3178) to measure the particle densities in a wide dynamic range.

This FT-array is surrounded by 20 density detectors to obtain a good core location for each shower event, as shown in Fig 1. All the detectors are set up in an inverted pyramidal style, and a lead plate of 5 mm thickness is put on top of each detector for converting gamma rays in the air showers into electron pairs. The inside of the box is painted by VH white enamel and an optical fiber is installed for monitoring a timing, PMT gain and of its linearity. A sectional view of the FT-detector is shown in Fig 2.

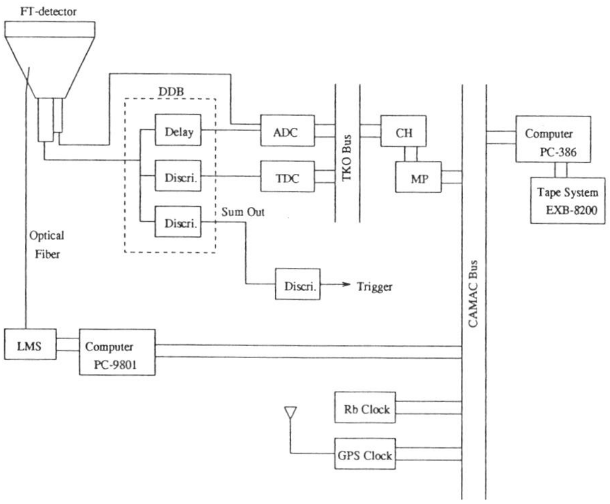

Figure 3: Block diagram of the data acquisition system.

1.2 DATA ACQUISITION SYSTEM

A TKO and CAMAC system is employed for data acquisition of the signals from the scintillation counters. The CAMAC bus is controlled by a computer (Epson PC-386) using i386-20MHz via a crate controller (Toyo CC/7000) and an interface board (Toyo CC/9800). A signal from each FT-detector is divided into three signals. One of them is used to generate a trigger pulse, and others are used to measure the values of an arrival time and a charge of signal from PMT, respectively. The timing signals are measured by a time-to-digital converter (TDC) with a 250 psec resolution and the charge signals are measured by an analog-to-digital converter (ADC) with a 0.25 pC resolution. A block diagram of the data acquisition system is shown in Figure 3.

A gain of each PMT and a time offset value of each channel are calibrated by using a laser monitoring system [1] at the beginning of every data set segment (RUN). All the TDCs are also calibrated at the start of every RUN by using a TDC tester module. Here, one RUN records about 4×10^5 events and it takes about six hours at a trigger rate of about 20 Hz under the condition of any 4-fold coincidence in the 45 FT detectors within a 300 ns time interval. The discrimination level to make a trigger pulse for each detector is set to a value equivalent to single-particle incidence. The signals from the density detectors are recorded only when a trigger pulse is generated by the FT-detector array. A Rubidium clock and GPS (Global Positioning System) provide information on the recording time of each event with sufficient accuracy. The data are recorded in 8 mm video-tape using Exabyte EXB-8200 drivers with a 2.3 Gbytes/tape capacity in the format of 320-byte-fixed-length binary. About 5 ms dead time, in an average, is lost to record one event in our system, resulting in about 10% of the whole running time at the trigger frequency of 20 Hz.

1.3 EVENT SELECTION

Since June of 1990 our system has triggered the events at a rate of about 20 Hz under any 4-fold coincidence in the FT detectors. About 8.5 $\times 10^8$ events have been recorded during the period from June 18 1990 through September 29 1992. The effective running time for this period was 598.6 days. These data are used not only to search for gamma-rays from point sources, but also to measure the cosmic ray shadows by the Sun and the Moon.

The event selection was made imposing the following two conditions. First, each of the four FT detectors should give a signal more than 1.25 particles per 0.5 m^2. Second, among the four detectors which record the highest particle densities, two or more should be inside the central 5×5 detector matrix. It is found that about 50% of the total number of trigger events pass through these selection criteria. These selected events are used to examine the shadowing of cosmic rays by the Sun and the Moon.

In the case of gamma-ray point source search, we further imposed a selection criterion to reject the out-array events. Since gamma induced showers are expected to have narrower lateral spread than those of background cosmic rays, this selection was made by imposing the condition of $\langle r_i \rangle = \sum_{\mathrm{ALL}} r_i \rho_i / \sum_{\mathrm{ALL}} \rho_i \leq 25$ m, where r_i is a distance of the i-th channel from estimated core location and ρ_i is a number of particle density (m^{-2}) at the i-th detector. About 25% of the total events are picked up as the "contained events" by these procedures. It is estimated by the simulation that about 74% of gamma-ray induced showers remain in the contained events and the fraction of mimic events, *i.e.* truly being outside of the array, is about 27%. We thus selected about 2.2×10^8 contained events from all the events recorded.

1.4 ARRAY PERFORMANCE BY THE MONTE CARLO SIMULATION

In order to examine the performance of the array, a Monte Carlo simulation has been done by taking account of the experimental conditions such as, observation altitude, detector shape and arrangement, output pulse shape and time jitter, trigger condition and so on. In this simulation, a subroutine package, GENAS[2], is employed. Figure 4 shows the energy spectra of proton-induced (A) and gamma-induced (B) showers to be observed with the Tibet array, respectively. The differential primary energy spectrum of the form $E^{-\gamma}$, is assumed, where the power index γ is taken to be 2.0 for gamma rays while being 2.73 at $E \leq 2 \times 10^{15}$ eV and 3.00 at $E > 2 \times 10^{15}$ eV for protons. The selection of the simulated showers is done under the same criteria as in the experiment. The upper histogram in this figure denotes the spectrum of protons producing the showers satisfying the selection criteria mentioned above, and the lower one is that of protons generating those with $\sum \rho_{\mathrm{FT}} > 100$, where ρ_{FT} denotes the shower particle density per square meter in each detector and the summation is taken over all the FT detectors. From this figure, the mode energy of primary protons is estimated to be 7 TeV (the median energy is about 17 TeV) for all the shower events and 35 TeV (56 TeV) for those with $\sum \rho_{\mathrm{FT}} > 100$.

The correlation between the sum of particles, $\sum_{\mathrm{FT}} \rho_i$, detected by 45 FT-detectors and the primary gamma-ray energy, E_γ, is examined for the contained events by the simulation, which is expressed as $E_\gamma(\mathrm{TeV}) \cong 0.3 \sum_{\mathrm{FT}} \rho_i$. This relation is used to estimate each primary gamma-ray energy.

The coordinates of each air shower core on the observation plane of the array are

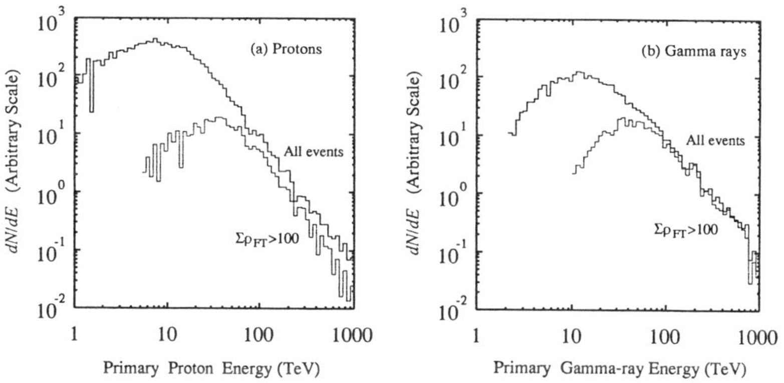

Figure 4: Energy spectra of protons(a) and gamma-ray(b) producing air showers to be observed with the Tibet array, obtained from the simulation. The upper histogram denotes the spectrum of protons producing air showers satisfying the selection criteria of all events (see text). The lower one is that of protons(a) and gamma-ray(b) producing the showers with $\sum \rho_{\rm FT} > 100$.

estimated by using a formula

$$(x, y) = \left(\frac{\sum_{\rm ALL} \rho_i^\alpha x_i}{\sum_{\rm ALL} \rho_i^\alpha}, \frac{\sum_{\rm ALL} \rho_i^\alpha y_i}{\sum_{\rm ALL} \rho_i^\alpha}\right),$$

where x_i and y_i are the coordinates of i-th detector respectively and ρ_i the number density ($\rm m^{-2}$) of detected particles. A weight parameter α is optimized to get a better core location and is given as follows,

$$\begin{aligned} \alpha(\textstyle\sum_{\rm FT} \rho_i) &= -1.54 \log_{10} \textstyle\sum_{\rm FT} \rho_i + 5.54 \quad \text{for} \quad \textstyle\sum_{\rm FT} \rho_i < 200 \quad \text{and} \\ \alpha &= 2.0 \quad \text{for} \quad \textstyle\sum_{\rm FT} \rho_i \geq 200. \end{aligned}$$

The error of core location at the median is estimated to be 6.6 m for proton showers and 6.1 m for gamma showers.

An arrival direction of individual air shower event is estimated by using the fast-timing data from the FT-detectors and by assuming that the air shower front forms a cone with a half angle 89° at the estimated shower core position. For each Monte Carlo shower, we can estimate its arrival direction with the same method as used in the experiment. The opening angle between the estimated direction and the given one is calculated for each of simulated showers. This opening angle distribution was used to estimate the angular resolution. Figure 5 shows the dependence of the opening angle on the shower size $\sum \rho_{\rm FT}$ for proton-induced (A) and gamma-ray induced (B) events. The simulation data are fitted by a straight line for each case, and they stand for the opening angles in which 20%, 50% and 80% of events are contained at each shower size bin, respectively. It is seen that the opening angle decreases as $\sum \rho_{\rm FT}$

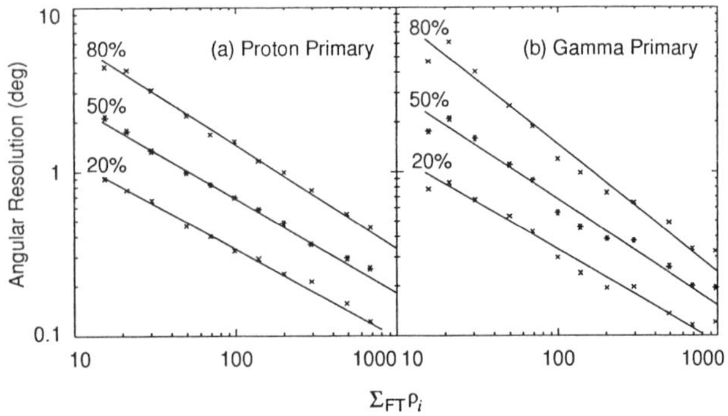

Figure 5: Dependence of the opening angle on the air shower size $\sum \rho_{FT}$ for shower events induced by protons. Simulation results are fitted by a straight line for each, and they stand for the opening angles in which 20%, 50% and 80% of events are contained at respective

increases. The overall angular resolution of the Tibet array, defined to be the half-angle of the cone containing 50% of events coming from a source direction, is estimated to be 1.15° for all proton-induced events and 0.49° for those with $\sum \rho_{FT} > 100$. In the case of gamma-ray showers, the overall angular resolutions are 0.8 ° and 0.4° for all and $\sum \rho_{FT} > 100$, respectively.

2 Shadows of the Moon and the Sun

High sensitivity air shower arrays specially designed to search for gamma ray point sources have a capability of detecting the shadowing of cosmic rays by the Sun and Moon as a collimator with the apparent angular diameter of about 0.5°. This was first suggested by Clark [3] and recently several groups [4][5][6] have observed the shadows of the Sun and Moon to confirm the angular resolution of the air-shower array using cosmic rays with energies upwards of about 50 TeV.

Almost all the cosmic rays are charged and propagate through the interplanetary space while arriving at the earth. Therefore, these particles will measurably bend under the influence of solar and interplanetary magnetic field, as well as geomagnetic field, if their energies are the order of 10 TeV or lower. Thus, observation of the cosmic ray shadow by these objects at these energies may give some new information on the solar magnetic field as well as the interplanetary magnetic field (IMF) [3].

The Tibet air shower array [7][8], being successively operated from January of 1990, has a good angular resolution better than 1° at 10-TeV energies and is just suitable for examining this interesting anticipation. Using this array, actually we observed for the first time the Sun shadow shifted from the apparent solar position to the west-southwest by about 0.9° [8]. We showed that this observed displacement can not be explained by statistical deviations nor by systematic pointing errors of the array, and then concluded that it is caused by the combined effect of the magnetic fields between the Sun and the Earth. In another paper [9], we further studied the shadowing of cosmic rays by the Sun in order to examine how the shadow was influenced by the solar and interplanetary magnetic fields. Here, we summarized our results about the shadows of the Sun and Moon.

2.1 EVENT ANALYSIS

In order to measure a small decrease in event density around the object (Sun and Moon), the background was carefully estimated using the events which come from the following eight off-source positions. These fake objects are located on the same zenith angle as the object, but apart by $\pm 5°, \pm 10°, \pm 15°$ and $\pm 20°$ in the azimuth angle. The average of the event densities around these eight off-source positions was taken to be the background while some appropriate choices were done depending on the zenith angle. The background density distribution was then obtained for each of the Sun and Moon. The cosmic ray deficits around these objects were examined as follows. First, the area of $4° \times 4°$ centered on each object was divided into 32×32 cells each size being a $0.125° \times 0.125°$ in real angle, and a two-dimensional histogram of event densities per cell was obtained. Same histogram was also made for the background obtained above, while it was once smoothed out by taking a moving average on the event densities over 401 cells within a circle of radius 1.5° from each concerning cell. We then subtract this smoothed background from the event histogram. The deficit event density obtained at each cell was again smoothed out over 69 cells within a circle of radius 0.6°. The weight of the deficit event density at each cell is defined by $(E - B)/\sqrt{E}$, where E and B are the event and background densities, respectively.

2.2 MOON SHADOW AND ARRAY PERFORMANCE

The contour map of the weights of deficit event densities around the Moon, thus obtained, is shown in Fig. 6 Contour lines start from 0σ deficit with a step of 1σ. The shadowing of cosmic rays is clearly perceived on this contour map, while its profile seems to be on the whole shifted westward. The maximum-likelihood method was used to estimate the angular resolution. To use this method, *a priori* angular resolution function $\mathrm{P_r}(\theta, R)$ is assumed to be a linear combination of three two-dimensional Gaussian distributions including a scaled angular resolution parameter R [8]. The likelihood function is the product of the differential probability as $dP/dz \propto \{N_{\mathrm{BG}}(z) - \pi\theta_{\mathrm{M}}^2 P_r(\theta(z), R)\}$ for each event in the sample, where z is a scaled variable defined as $z = (1 - \cos\theta)/(1 - \cos\theta_{\mathrm{max}})$ with $\theta_{\mathrm{max}} = 4.0°$, $N_{\mathrm{BG}}(z)$ the normalized background function and θ_{M} the angle radius of the Moon. The likelihood is computed numerically for many trial R's for the Moon data set ; its natural logarithm is plotted against trial R in Fig. 7. The curves become maximum at $R = 0.88°$ for all events and at $R = 0.52°$ for events with $\sum \rho_{\mathrm{FT}} > 100$, respectively. From this figure, the angular resolution of our array is estimated to be $0.88°{}^{+0.11°}_{-0.09°}$ for all events and $0.52°{}^{+0.09°}_{-0.07°}$ for those with $\sum \rho_{\mathrm{FT}} > 100$. If we use a single two-dimensional Gaussian distribution, however, the resolution becomes $0.67°{}^{+0.09°}_{-0.07°}$ for all events, though the statistical significance is 6.2σ for the shadow of the Moon. Using the maximum likelihood method, the maximum deficit position of the shadow profile is found at $0.16°{}^{+0.11°}_{-0.09°}$ to the west and $0.02°{}^{+0.07°}_{-0.09°}$ to the south when the angular resolution of the array is 0.88°.

As a statistical significance, the shadow at the maximum deficit is calculated to be 7.1σ level. The shadow profile and its westward displacement become more clear than the previous result [8] with increasing statistics. The geomagnetic field will bend

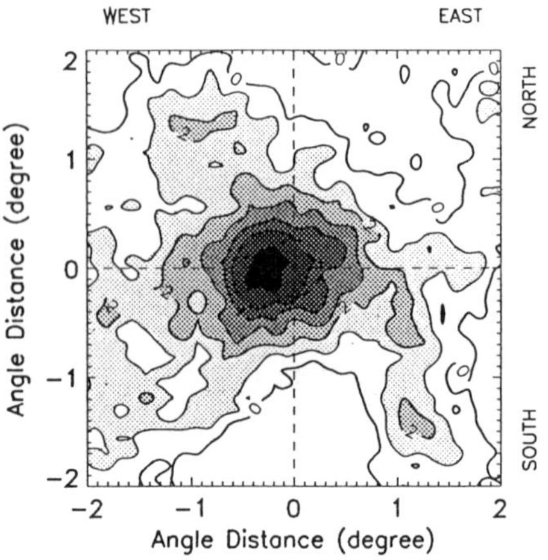

Figure 6: Contour map of the weights of deficit event densities around the Moon in the area of $4° \times 4°$ centered on the Moon (dotted circle). The contour lines are drawn from a level of no deficit, 0σ, with a step of 1σ.

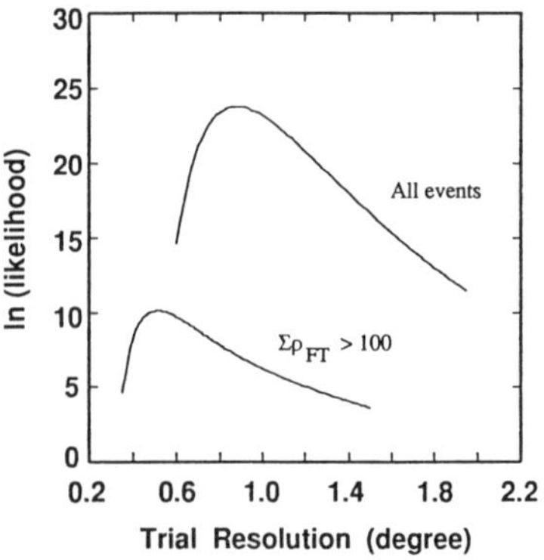

Figure 7: Logarithm of the likelihood vs. trial angular resolution R for the Moon data set. The upper curve is for all events and the lower one for those with $\sum \rho_{\rm FT} > 100$.

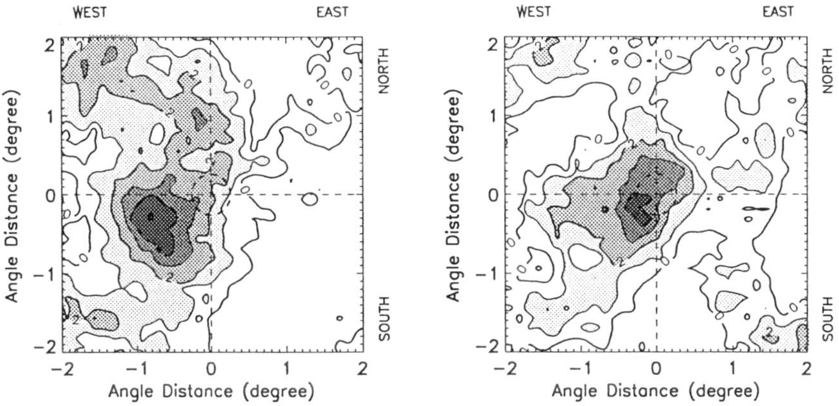

Figure 8: (a) Contour map of the weights of the event densities around the Sun, as shown in Fig. 6, for all showers. (b) Same as (a), but for showers with $\sum \rho_{\rm FT} > 100$.

incident positively charged cosmic rays to make their apparent arrival directions shift to the west. Using a Monte Carlo method, the displacement of the Moon shadow is calculated to be 0.15° for all showers, which is comparable with the observed one. Moreover, the shadow by high energy cosmic rays with $\sum \rho_{\rm FT} > 100$ was ascertained to fall almost on the Moon, as expected. Thus, the observed displacement can be attributed to the geomagnetic effect. Taking these into consideration, the overall pointing error of the Tibet array is estimated to be less than 0.1°. The results obtained above are used as reference standards in the following.

2.3 SUN SHADOW

To begin with, we show the contour map of the deficit event densities around the Sun in Fig. 8 for all showers. A large displacement of the shadow from the Sun direction is evident in this figure.

The profile is slightly disordered when compared with that for the Moon, and this may be partly attributed to a complex structure of the solar and interplanetary magnetic fields. The primary composition will also complicate this somewhat. The maximum likelihood analysis shows that the most probable position of the center of the deficit is at $0.70^{\circ}{}^{+0.10^{\circ}}_{-0.10^{\circ}}$ to the west and $0.40^{\circ}{}^{+0.09^{\circ}}_{-0.11^{\circ}}$ to the south when the angular resolution of the array is 0.88°. The logarithm of likelihood is calculated to be 10.6 at this position, while it becomes –6.0 at the true Sun position. With a ratio of likelihoods between observed and true Sun positions and a trials factor of 1024 (total number of cells), the probability of chance occurrence of this displacement is given as 6.3×10^{-5}. This value is still so small that the shadow's displacement is highly significant. The significance of the maximum event deficit is also calculated to be 4.6σ.

For cosmic rays with $\sum \rho_{\rm FT} > 100$, the Sun shadow was also clearly observed, but its maximum deficit position is calculated to be $0.25^{\circ}{}^{+0.15^{\circ}}_{-0.15^{\circ}}$ W and $0.10^{\circ}{}^{+0.15^{\circ}}_{-0.15^{\circ}}$ S as shown in Fig. 8(b)(The significance of the shadow is 4.3σ). The displacement is about one third of that for all showers. When the displacement of the shadow is mainly caused by the magnetic field effects, this decreasing tendency for high energy showers is quite reasonable, since the median primary energy for these showers is about three times higher than that for all showers.

After subtracting the displacement due to the geomagnetic field, the resultant center of the Sun shadow for all events is given at $0.54° \pm 0.10° \pm 0.10°$ W and $0.38° \pm 0.10° \pm 0.10°$ S, where the second errors are from the positioning error of the center of the Moon shadow. This net displacement can be attributed to the effect of the solar and interplanetary magnetic fields.

2.3.1 Effects of the Interplanetary Magnetic Field

As well known, the configuration of the solar and interplanetary magnetic fields considerably changes with solar cycle. Note that almost all the data used for this analysis were taken in the period between June 1990 and July 1992 when the solar activity was near maximum or rather at the decreasing phase. At that time, the solar magnetic field had large deviations from a pure dipole, i.e., small dipole moment with large contributions from higher multipoles, and the dipole field axis was almost sideways especially in the period from 1990 to 1991 just after the reversal of the polar fields [10]. Thus, the structure of the solar magnetic field at this time is too complicated to give a quantitative estimate of its effect on the shadow.

On the other hand, the effect of IMF may be directly examined as follows. It is known that the IMF has a sector structure with the field direction reversing across the sector boundary [11][12]. This sector structure, with the "away" and "toward" field directions, varies with the phase of solar activity cycle and near maximum it is highly modified, while keeping almost the two-sector structure [13]. During the declining phase, however, the IMF structure rather simplifies and becomes rather stable [14]. According to the solar geophysical data [10], the IMF had almost the two-sector structure in the period when the data were taken, while gradually changing to the four-sector structure in the last period after April 1992.

Because of the opposite polarity of the fields, cosmic ray particles arriving through the away and toward sectors will bend in opposite directions to each other. If the effect of IMF is significant, then the Sun shadows by cosmic rays from respective sectors should be shifted oppositely each other with respect to the shadow center obtained above. Shown in Fig. 9 are the deficit maps of cosmic rays for respective sectors. Although the statistics becomes worse for each case, the expected behavior is well found. The most probable position of each shadow center is calculated to be at 0.57° W and 0.48° S for the toward-sector cosmic rays and at 0.86° W and 0.29° S for the away-sector cosmic rays, using the maximum likelihood method. After removing the geomagnetic displacement, these centers are given as (0.41° W, 0.46° S) and (0.70° W, 0.27° S), respectively. The angular distance between two centers is found to be about 0.34°.

Figure 10 shows a summary of these results. This well confirms that the IMF gives a large contribution to the displacement of the Sun shadow. This figure also suggests that a weak magnetic field component facing nearly north-west-northward steadily exits to make the apparent arrival directions of cosmic rays shift to the southwest. A displacement of the Sun shadow due to IMF will generally occur in the north-south direction, since the sector fields are almost parallel to the plane of the ecliptic due to the high-speed solar wind blowing from the rotating sun [15]. This component may be unexpectedly produced by a large-scale disordered structure of the solar and interplanetary magnetic fields in the period near solar maximum. The solar cycle is

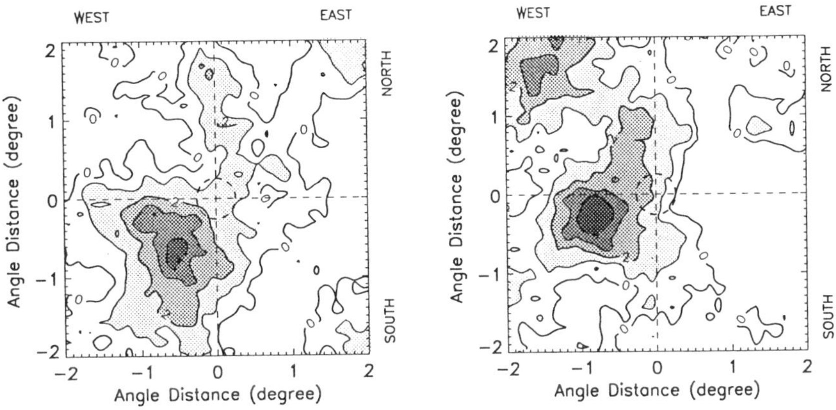

Figure 9: (a) Contour map of the weights of deficit event densities for the cosmic rays coming from the toward sector. (b) Same as (a), but for those from the away sector.

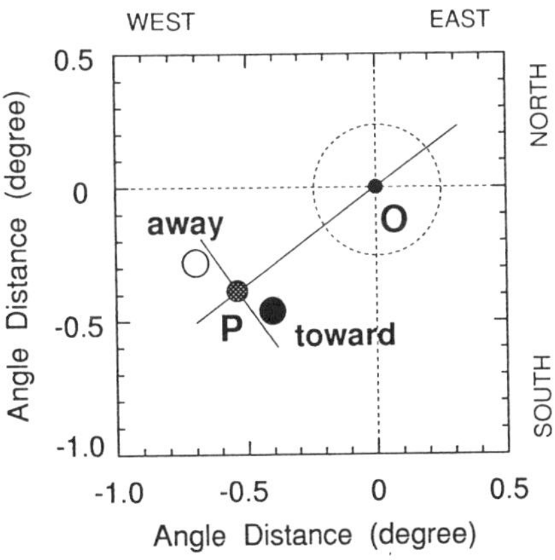

Figure 10: Most probable positions of the shadow center for each case of the toward and away sectors in IMF, respectively. The center of the shadow for all events is shown by a mark **P**, and **O** denotes the apparent center of the Sun.

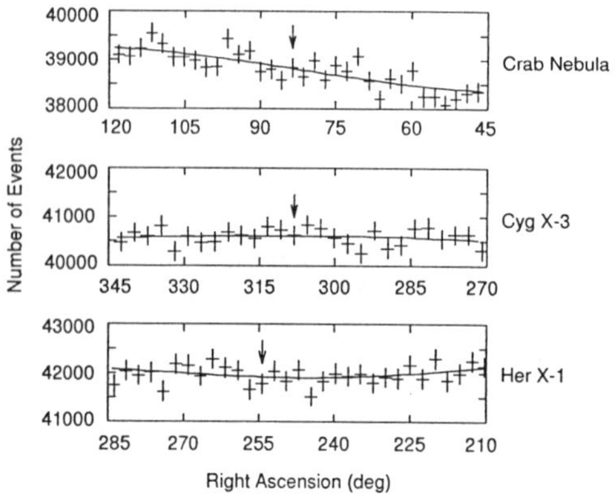

Figure 11: Right ascension scan. The solid line on each source indicates the expected average level of cosmic-ray background.

now coming into a quiet phase. From this point of view, it is very interesting and important to compare these results with those which will be obtained in the near future, and this may provide a new step in the study of the solar and interplanetary magnetic fields.

3 Gamma-ray source search

3.1 CYGNUS X-3, HERCULES X-1 AND THE CRAB NEBULA/PULSAR

The most promising candidates radiating VHE/UHE gamma rays in the northern sky are Cygnus X-3, Hercules X-1 and the Crab Nebula/Pulsar. The Whipple group [16] observed a steady emission of TeV-gamma rays from the Crab Nebula with the high-resolution atmospheric Čerenkov camera and a signal was detected at the 20σ level or more, and the observed flux is seemingly in favor of the synchrotron-self-Compton model by De Jager and Harding [17]. This model suggests that gamma-ray signals may be detected in the 10-TeV energy region with a high-resolution surface array at high altitude. The results of the search for gamma-ray sources from this experiment were already reported elsewhere [7][18].

Figure 11 shows the right ascension scans for the Crab Nebula, Cygnus X-3 and Hercules X-1. The number of events coming from the region within a circle of apparent radius 1.0° is plotted as a function of right ascension at each declination of these sources. The background contribution is estimated from the events falling in the five windows adjacent to the source window on both sides at the same declination. This size of search window is chosen to maximize the ratio $N_S/N_B^{1/2}$, where N_S is the number of signals and N_B the number of background events, and to contain about 50% of the signals. We obtain 38822 events for the Crab Nebula, 40617 events for Cygnus X-3 and 41940 events for Hercules X-1. The results of searching at the locations of the Crab Nebula, Cygnus X-3 and Hercules X-1 are excesses of 0.01σ, -0.06σ and -0.75σ, respectively. We conclude that no evidence has been found for steady emission from these sources in our observation period.

Upper limits on the excess number of events at the 95% confidence level are obtained: 433 for the Crab Nebula, 423 for Cygnus X-3 and 348 for Hercules X-1

Table 1: Flux upper limits above 10 TeV and 30 TeV at the 95%C.L.

	F ($\geq$ 10 TeV) [$cm^{-2}s^{-1}$]	F ($\geq$ 30 TeV) [$cm^{-2}s^{-1}$]
Crab Nebula	9.1×10^{-13}	1.7×10^{-13}
Cygnus X-3	7.4×10^{-13}	1.3×10^{-13}
Hercules X-1	5.8×10^{-13}	1.4×10^{-13}

based on the Protheroe procedure [19]. Then, flux upper limits for these sources are calculated using the Monte Carlo simulation. Showers from gamma rays are generated from a differential-power-law spectrum of the form $E^{-\gamma}$ with $E_{\rm min} = 1$ TeV. The power index γ is assumed to be 2.0 both for Cygnus X-3 and Hercules X-1 and 2.4 for the Crab Nebula. The results obtained are summarized in Table 1

In Fig. 12 we compare our upper limit fluxes for the Crab Nebula with other results [20][21][22] [23][24][25]. Our data is located between the flux levels extrapolated from the Whipple [22] spectrum and Themistocle [25] one. The Themistocle data seem to contradict with ours. For Cygnus X-3, our limits are far below the levels obtained by extrapolating of the Kiel flux [26] back to 10 TeV regions, using an integral spectrum of E^{-1}, and our upper limit flux at 30 TeV is lower than the level of the CYGNUS experiment [20]. It is well established that Cygnus X-3 is no longer a steady source emitting VHE/UHE gamma rays at the Kiel flux level or has now fallen below the level of detectable emission. For Hercules X-1, we also place stringent flux upper limits on the steady emission of gamma rays at energies above 10 TeV and 30 TeV.

3.2 AGNs AND OTHER SOURCES

Recent gamma-ray observations [27][28] from the active galaxy Mrk421 at GeV and TeV energies may also suggest that active galactic nuclei (AGNs) can be promising candidates capable of supplying cosmic rays above 10^{14} eV [29], while gamma rays from such an extragalactic source are strongly attenuated at energies above several hundred GeV by interactions with extragalactic starlight [30]. At present, unfortunately, we have no anomalously bright sources emitting UHE gamma rays. It is, however, very important to search for individual point sources of neutral radiation which would signal the presence of a powerful accelerator of protons to solve the mystery of the origin of UHE cosmic rays.

From this point of view, we also searched for gamma rays from other several potential sources including extragalactic objects: Geminga, Cas A, Cyg X-1, Cyg X-2, 3C273, 3C279, M87 and Mrk 421, in the northern sky [31](the newest results for AGNs are reported in [32]). The RA scans are shown for Geminga, Cyg X-1 and Mrk421 in Fig. 13. The results of searching at the location of Geminga, Cas A, Cyg X-1, Cyg X-2, 3C273, 3C279, M87 and Mrk421 are summarized in Table 2. We conclude that no evidence has been found for steady emission from any source. Upper limits on the excess number of events at the 95% confidence level and flux upper limits above 10 TeV are also shown in Table 2. Then, flux upper limits for these sources are calculated using the Monte Carlo simulation. Showers from gamma

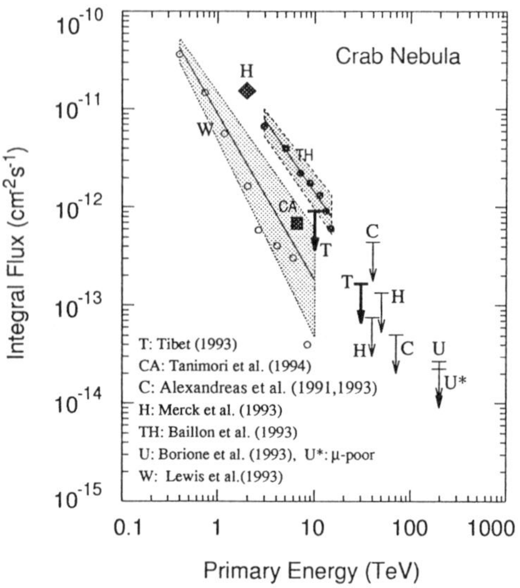

Figure 12: A comparison of the experimental results for Crab fluxes and flux upper limits.

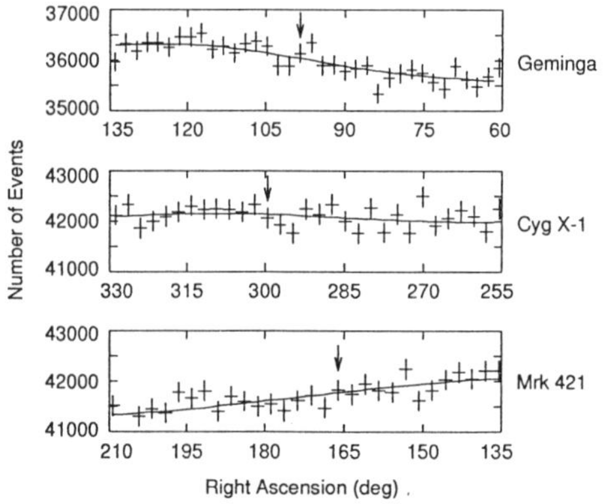

Figure 13: Right ascension scans for Geminga, Cyg X-1 and Mrk421 as examples. The solid line on each source indicates the expected average level of

Table 2: Flux upper limits above 10 TeV and 30 TeV at the 95%C.L.

	Declination	Energy	N_B	Excess	N_S	F (cm^{-2}s^{-1})
Geminga	+17.8°	$\geq$ 10 TeV	36047	$+0.44\sigma$	< 456	$< 9.1 \times 10^{-13}$
		$\geq$ 30 TeV	1685	-0.66σ	< 70	$< 1.2 \times 10^{-13}$
Cas A	+58.8°	$\geq$ 10 TeV	25402	-0.91σ	< 247	$< 6.7 \times 10^{-13}$
		$\geq$ 30 TeV	1227	-1.16σ	< 53	$< 1.2 \times 10^{-13}$
Cyg X-1	+35.2°	$\geq$ 10 TeV	42158	-0.50σ	< 370	$< 6.3 \times 10^{-13}$
		$\geq$ 30 TeV	1958	$+0.88\sigma$	< 124	$< 1.9 \times 10^{-13}$
Cyg X-2	+38.3°	$\geq$ 10 TeV	41561	-0.55σ	< 357	$< 6.1 \times 10^{-13}$
		$\geq$ 30 TeV	1926	$+2.24\sigma$	< 171	$< 2.8 \times 10^{-13}$
3C273	+2.1°	$\geq$ 10 TeV	19483	$+1.74\sigma$	< 489	$< 18 \times 10^{-13}$
		$\geq$ 30 TeV	931	$+1.07\sigma$	< 87	$< 2.6 \times 10^{-13}$
3C279	−5.7°	$\geq$ 10 TeV	11190	$+1.89\sigma$	< 386	$< 22 \times 10^{-13}$
		$\geq$ 30 TeV	538	$+0.14\sigma$	< 50	$< 2.3 \times 10^{-13}$
M87	+12.4°	$\geq$ 10 TeV	30775	$+0.26\sigma$	< 395	$< 9.2 \times 10^{-13}$
		$\geq$ 30 TeV	1427	$+0.24\sigma$	< 85	$< 1.7 \times 10^{-13}$
Mrk421	+38.2°	$\geq$ 10 TeV	41737	$+0.48\sigma$	< 501	$< 8.6 \times 10^{-13}$
		$\geq$ 30 TeV	1930	$+0.41\sigma$	< 105	$< 1.7 \times 10^{-13}$

rays are generated from a differential-power-law spectrum of the form $E^{-\gamma}$ with E_{min} = 1 TeV. The power index γ is assumed to be 2.0 for every source. The results are summarized in Table 2 and flux upper limits above 10 TeV are shown in Fig. 14.

It has been discussed that because of strong attenuation of TeV photons by the interaction with extragalactic starlight, signals from objects such as 3C273, 3C279 and Mrk421 could not be detected at very high energies, if any. In other words, observations of VHE/UHE gamma rays from these objects could provide a constraint on the intergalactic background infrared photon field [30]. Even with the advent of the high sensitive Tibet array, we have still failed to confirm the bright sources continuously radiating UHE gamma rays which were reported in the 1980's, and the flux limits obtained are well more than an order of magnitude lower than the earlier observations. It should be noted, however, that the threshold of sensitivity for detecting gamma-ray sources with this array is nicely approaching to that by Čerenkov techniques.

4 Tibet-II Experiment

4.1 TIBET-II ARRAY

The Tibet air shower array will be enlarged by a factor of about four in 1994. As schematically shown in Fig 15, the fast-timing (FT) detector array consists of 185 scintillator detectors of 0.5m^2 each. For new detectors (not filled), plastic scintillators of 3cm thickness each (Bicron BC408A) are used and 2" ϕ PMTs (Hamamatsu H1161) are equipped. They are placed in a grid pattern of a 15m spacing. Moreover, each

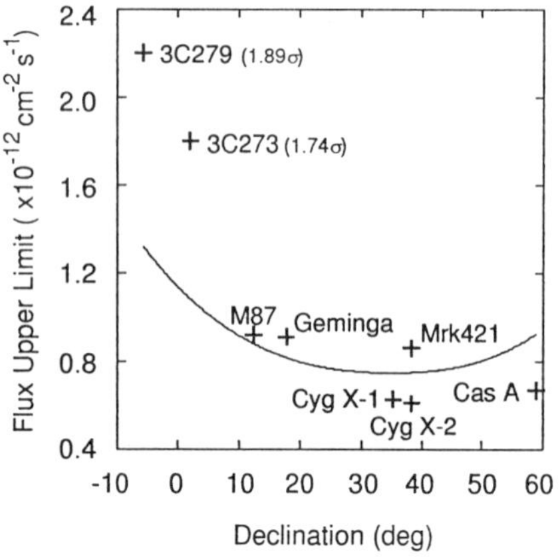

Figure 14: Flux upper limits above 10 TeV at the 95% confidence level. The solid curve shows the case of no excess events from the source position.

of 52 FT-detectors contains a 1.5" ϕ PMT (Hamamatsu H3178) for a wide-range particle density measurement.

This FT-array is surrounded by 32 density detectors of 0.5 m^2 each to obtain a good core location for individual shower event. The overall structure is almost same as that shown in Fig. 2. The effective area for detecting the contained events is estimated to be about 8 times as large as the old one, and a trigger rate is estimated to be about 200Hz.

4.2 DATA ACQUISITION SYSTEM

In order to achieve a high speed and low dead time data acquisition, a FASTBUS system is employed as shown in Fig. 16. Event buffers and zero-suppress system will work well to reduce a dead time and it is estimated to be about 5% at a trigger rate of 200Hz.

Gain of each PMT and time offset value of each channel are calibrated by a newly developed LED and mismatched-reflection monitoring system at the beginning of every data set segment. All the TDCs and ADCs are also calibrated by using a CAT module in the FASTBUS system. These data are recorded using the 8 mm tape drivers (Exabyte EXB-8505) with about 10 Gbytes/tape capacity in the format of variable-length binary (400 byte in average).

4.3 EXPECTATION

Using this new Tibet-II array, the sensitivity to detect gamma rays will increase by a factor of about three compared with the present array. For the Crab-nebula, this will almost correspond to the Cangaroo point [24] or flux level extrapolated from the Whipple [22] spectrum, resulting in substantially detectable positive signal. Our data could then provide reference standards to the energy calibration of Čerenkov observation. In the case for Mrk421, we could also obtain two or three times lower upper bound than that extrapolated from the Whipple [28] spectrum. This will

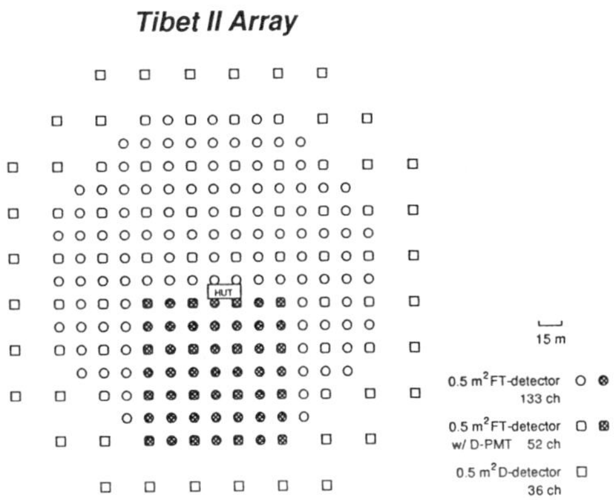

Figure 15: Layout of the Tibet-II scintillation counter array.

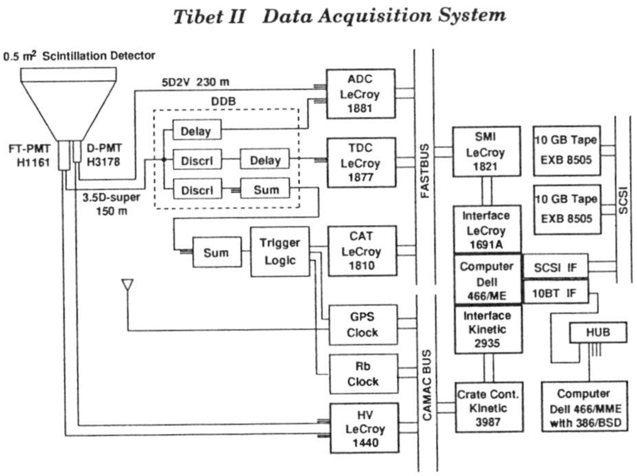

Figure 16: Block diagram of the Tibet-II data acquisition system.

provide new information about the absorption of gamma rays by the intergalactic infrared or far infrared background field [30].

For the Sun's shadow, sufficient data allow us to estimate the displacement of the shadow at every one-two months. Then, we could get new information on the relation between a time variation of the large-scale structure of the solar and interplanetary magnetic fields and the solar activity, which may considerably contribute to the study of solar terrestrial physics.

ACKNOWLEDGMENTS. This work is supported in part by Grants-in-Aid for Scientific Research and also for International Scientific Research from the Ministry of Education, Science and Culture, in Japan and the Committee of Natural Science Foundation and Academy of Sciences in China.

References

[1] M. Nishizawa *et al.* *Nucl. Inst. and Methods*, **A285**, 532, (1989).
[2] K. Kasahara and S. Torii. *Comput. Commun.*, **64**, 109, (1991).
[3] G. W. Clark. *Phys. Rev.*, **108**, 450, (1957).
[4] D. E. Alexandreas *et al.* *Phys. Rev. D*, **43**, 1738, (1991).
[5] B. E. Fick *et al.* in *22nd Int. Cosmic Ray Conf., Dublin, Ireland, 1991*, volume 2, 728, (The Dublin Institute for Advanced Studies, Dublin, Ireland, 1991).
[6] A. Karle *et al.* in *22nd Int. Cosmic Ray Conf., Dublin, Ireland, 1991*, volume 4, 460, (The Dublin Institute for Advanced Studies, Dublin, Ireland, 1991).
[7] M. Amenomori *et al.* *Phys. Rev. Lett.*, **69**, 2468, (1992).
[8] M. Amenomori *et al.* *Phys. Rev. D*, **47**, 2675, (1993).
[9] M. Amenomori *et al.* *Astrophys. J.*, **415**, L147, (1993).
[10] NOAA/USAF. in *Solar Geophysical Data*, (Space Environmental Information Center, 1990, 1991 and 1992).
[11] J. M. Wilcox and N.F Ness. *J. Geophys. Res.*, **70**, 5793, (1965).
[12] J. M. Wilcox. *Space Sci. Rev.*, **8**, 258, (1968).
[13] J. M. Wilcox and P. H Scherrer. *Journ. of Geophys. Res.*, **77**, 5385, (1972).
[14] J. T. Hoeksema. in *The Sun and the Heliosphere in Three Dimensions*, 241, (ed. by R. G. Marsden, D. Reidel Publishing Company, 1986).
[15] E. N. Parker. in *Interplanetary Dynamical Processes*, (Interscience Pub. Co., N. Y., 1963).
[16] G. Vacanti *et al.* *Astrophys. J.*, **377**, 467, (1991).
[17] O. C. De Jager and A. K Harding. *Astrophys. J*, **396**, 161, (1992).
[18] M. Amenomori *et al.* in *23rd Int. Cosmic Ray Conf., Calgary, Canada, 1993*, volume 1, 342, (International Union of Pure and Applied Physics, 1993).
[19] R. J. Protheroe. *Astron. Expr.*, **1**, 33, (1984).
[20] D. E. Alexandreas *et al.* *Astrophys. J.*, **383**, L53, (1991).
[21] D. E. Alexandreas *et al.* *Astrophys. J.*, **405**, 353, (1993).
[22] A. Borione *et al.* in *23rd Int. Cosmic Ray Conf., Calgary, Canada, 1993*, volume 1, 286, (International Union of Pure and Applied Physics, 1993).
[23] M. Merck *et al.* in *23rd Int. Cosmic Ray Conf., Calgary, Canada*, volume 1, 290, (International Union of Pure and Applied Physics, 1993).
[24] T. Tanimori *et al.* *Astrophys. J.*, (1994). in press.

[25] M. Bailon *et al.* in *23rd Int. Cosmic Ray Conf., Calgary, Canada*, volume 1, 271, (International Union of Pure and Applied Physics, 1993).
[26] M. Samorski and W. Stamm. *Astrophys. J.*, **268**, L17, (1983).
[27] Y. C. Lin *et al. Astrophys. J.*, **401**, L61, (1992).
[28] M. Punch *et al. Nature*, **358**, 447, (1992).
[29] R. J. Protheroe *et al. Phys. Rev. Lett.*, **69**, 2885, (1992).
[30] F. W. Stecker *et al. Astrophys. J.*, **390**, L49, (1992).
[31] M. Amenomori *et al.* in *23rd Int. Cosmic Ray Conf., Calgary, Canada, 1993*, volume 1, 412, (International Union of Pure and Applied Physics, 1993).
[32] M. Amenomori *et al. Astrophys. J.*, **429**, (1994), in press.

PRODUCTION AND ACCELERATION OF ULTRA HIGH ENERGY PARTICLES BY BLACK HOLES AND STRINGS

C.SIVARAM
Indian Institute of Astrophysics
Bangalore
India
and
INFN Sezione di Bologna
Bologna, Italy

ABSTRACT. Some possible new mechanisms based on physical processes associated with objects such as black holes and strings are outlines to account for the production of extremely energetic cosmic ray particles ($> 10^{20}$ eV) such as the 300 EeV event.

Keywords: UHE particles, decaying cosmic superstrings, evaporating black holes.

Conventional supernova remnant models cannot accelerate cosmic ray particles to much above energies $\sim 10^{14} eV$. As is well known, this is essentially bceause the interstellar magnetic field is only $\sim 5\mu G$. However, the cosmic ray spectrum is observed to continue through the 'knee' to much higher energies. Pulsars are possible sources or sites for acceleration to higher energies. Even here there are limits ($\sim 10^{18} eV$) or at most $\sim 10^{19} eV$ for iron nuclei.

Moreover acceleration near neutron stars is difficult because of the high energy loss rates of charged particles in intense magnetic fields. Consequently in some models the acceleration sites are located further into the relativistic wind region. In short none of these processes can account for particles of $\sim$ 300 EeV energy. A proton with such energy will pass straight through the galaxy. The Larmor radius of a 300 EeV proton in the galactic magnetic field is $\sim$ 100 Kpc and the estimated maximum propagation length is $\sim$ 30 Mpc.

As acceleration in pulsar magnetic and electric fields is clearly inadequate by a large factor to produce such energetic particles one can consider more compact objects than neutron stars, i.e. black holes. As an illustration of the potential (in a literal sense!) of such objects to produce ultra high energy (UHE) particles we next discuss the acceleration of charged particles by electrically charged black holes. Such black holes are predicted to exist and for a black hole of mass M and charge Q, the radius of the event horizon[1] is given by:

$$r = GM/c^2 \pm [\frac{G^2M^2}{c^4} + \frac{GQ^2}{c^4}]^{1/2} \tag{1}$$

M. M. Shapiro et al. (eds.), Currents in High-Energy Astrophysics, 177–182.

(G is the gravitational constant and c the velocity of light). The maximal electric charge that a black hole of mass M can have is given by:

$$Q = G^{1/2}M \tag{2}$$

This implies that a particle with charge e, approaching such a black hole would have at the horizon, i.e. on reaching $r_s = GM/c^2$, an energy given by: (i.e. maximum possible energy)

$$E_{max} \simeq \frac{G^{1/2}Me}{GM/c^2} = \frac{c^2 e}{G^{1/2}} \simeq 10^{25}\ eV \tag{3}$$

This is a number independent of the black hole mass. This is potentially the maximal energy (allowed by general relativity) to which a proton can be accelerated to by a charged black hole. In general the equation of motion is: (for particle of mass m)

$$\ddot{x} = \frac{G^{1/2}Me}{mx^2} - \frac{e^2}{3c^3}\dddot{x} + \ higher\ order\ terms \tag{4}$$

where x is the position co-ordinate.

Electrons would lose energy at a far more rapid rate, as loss rate is proportional to $(\ddot{x})^2$ which in turn is proportional to $1/m^2$. So electrons as usual will not be accelerated to very high energies.

Solution[2] of eq.(4) shows that a proton deflected at $r \simeq 3r_s$, can leave the hyperbolic trajectory with a maximal energy $\sim 10^{23}eV$. Below $r < 3r_s$, as is well known, orbits are unstable.

It is also well known[3] that charged black holes can have a magnetic dipole moment (indeed for a rotating charged black hole, the gyromagnetic ratio is 2, the same as for a Dirac particle). Such a black hole can thus also interact with a particle having a magnetic moment. The interaction energy in this case is given by:

$$E_{int} \simeq \frac{{}^{\mu}BH\,{}^{\mu}P}{r^3} \times \ curved\ space\ factors \tag{5}$$

Here ${}^{\mu}$BH and ${}^{\mu}$P are the magnetic dipole moments of the black hole and the particle respectively.

For ${}^{\mu}P \approx {}^{\mu}B$ (the Bohr magneton) and for a maximally charged hole (cf. eq.(2)) this gives a maximal energy (at $r = r_s$) of:

$$E_{max} \simeq \frac{{}^{\mu}Bc^4}{MG^{3/2}} \tag{6}$$

For a 10^{17} gm. primordial Hawking black hole (to be discussed later) this gives (note E α 1/M)

$$E_{max} \simeq 10^{23}eV$$

The equations of motion can also be written down for this case as in the earlier example (cf. eg.(4)). If the black hole is embedded in a magnetic field such high energy particles accelerated by the hole can also emit ultra high frequency gamma radiation (suppressed by $1/m^4$). However, this turns out not to be significant.

We next consider the acceleration of particles by cosmic strings and fundamental superstrings. Superstrings are produced near the Planck scale (energy E_{pl} or $M_{pl} \sim 10^{19}\ GeV$). They are characterized by a tension $T_{pl} \sim c^2/G$ (mass per unit length). $T_{pl} \sim 10^{28} gcm^{-1}$ strings produced by symmetry breaking at any other energy (mass scale) $\sim M$ have a tension given by:

$$T_s \simeq \frac{c^2}{G}\left(\frac{M}{M_{pl}}\right)^2 \tag{7}$$

In addition one can have conducting cosmic strings which are essentially topological line defects.

There are some nice analogies between vortex lines in a Type II superconductor (carrying a quantized flux $\hbar c/2e$) and conducting cosmic strings. For instance, the field vanishes everywhere in a superconductor (Meissner effect), i.e. $F_{ab} = 0$, everywhere except along Abrikosov vortex lines carrying a confined quantized flux $\hbar c/2e$. Inside a superconductor we have the Landau equations:

$$\Delta^2 \bar{B} + \lambda^2 \bar{B} = \bar{J} \tag{8}$$

The vanishing of the field inside a superconductor is an effect of the Landau-Ginzburg theory where we have the Maxwell field coupled to a scalar field as:

$$D_\mu D^\mu \phi = a\phi(\mid \phi \mid^2 - \lambda^2) \tag{9}$$

$\mid \phi \mid \simeq \lambda$ near the broken symmetric state.
Far from the flux tube:

$$D_\mu \phi = (\delta_\mu + ieA_\mu)\phi = 0$$

and

$$[D_\mu, D_\nu]\phi = ieF_{\mu\nu}\phi = 0$$

So either ϕ or $F_{\mu\nu}$ must vanish. This has the solution:

$$A_\mu = -(1/e)\delta_\mu \phi, \phi = \lambda e^{i\theta}$$

The Higgs field responsible for these defects is described by a relativistic version of the Landan-Ginzburg model and consequently it can be shown that conducting strings also carry a flux[5,6].

$$\phi = n\hbar c/e \tag{10}$$

The flux can be shown to give rise to an electric field[5] given by:

$$V \simeq cT_s GM_{pl}^2/e\hbar$$

$$\simeq \frac{Gc}{e\hbar}.T_s.M_{pl}^2 \tag{11}$$

Thus charged particles can be accelerated to a maximal energy given by: (corresponding to a critical current[6]):

$$E \simeq ecT_s^{1/2} \;\; G^{1/2}M_{pl} \tag{12}$$

For a string tension, corresponding to a GUT scale $M \simeq 10^{15}GeV$, (the corresponding tension being given by eq.(7)):

$$E \simeq 10^{21}eV \tag{13}$$

A higher string tension T_s gives rise to a higher value of E. For a GUTs scale $M \sim 10^{16}GeV$, $E \approx 10^{22}$ eV.

So far we have considered production of UHE particles by acceleration by black holes and cosmic strings. However, it must be noted that UHE particles can be spontaneously generated by Evaporating black holes (EBH).

An important consequence of attempts to link gravity with quantum physics is the prediction[7] that black holes must spontaneously emit radiation and decay. This effect is however significant only for primordially formed black holes with masses $<< M_\odot$. The lifetime for decay is:

$$t_H \approx \frac{G^2M^3}{\hbar c^4} \tag{14}$$

The occurrence of $\hbar$ shows that this is a quantum effect. As $\hbar \to 0$, $t_H \to \infty$, i.e. a classical black hole stay for ever. For t_H comparable to the Hubble time scale $\sim 1/H(\sim 10^{10}yrs)$, we have a horizon size $GM/c^2 \approx 10^{-13}$ cms (1 fermi) which is a typical elementary particle Compton length. If the horizon size is comparable to Compton wavelength $\hbar/mc$ of particles of mass m, then the quantum uncertainty principle implies that virtual pairs of this particle would be produced near the horizon.

The rate of energy emission[8] by spontaneous creation of these virtual pairs is:

$$dE/dt \;\; \simeq \;\; \frac{mc^2}{\hbar/mc^2} \;\; \simeq \;\; \frac{m^2c^4}{\hbar},$$

and as $\frac{GM}{c^2} \simeq \frac{\hbar}{mc}$,

$$\frac{dE}{dt} \simeq \frac{\hbar c^6}{(GM)^2} \tag{15}$$

The rate of energy emission maximal for the heaviest elementary particles one can conceive. The energy of the particles spontaneously emitted by the EBH is given by:

$$E \simeq \frac{\hbar c^3}{GM}$$

$$\simeq 10^{20}eV\,[\frac{1kg}{M}] \tag{16}$$

Equations (15) and (16) imply that when the evaporating black hole reaches one kilogram mass, it can emit $\sim 10^{16}$ particles of $\sim 10^{20}eV$ energy each in a time scale (i.e. burst) of $\sim 10^{-20}s$. With a density of primordial black holes in a given mass range, one can get the background flux of such spontaneously emitted UHE particles by integrating over all red shifts[9]. From known upper limits on the flux of such UHE particles (i.e. with energies $> 10^3 EeV$), one can estimate that the background density of such priomordial evaporating black holes is $< 10^9/Kpc^3$ (assuming they cluster, eg. in Virgo). The Hawking-Page bound from the diffuse gamma ray background is interestingly of same order. Such holes emit TeV $\gamma - rays$, when their mass becomes $\sim 10^8 kg \sim 10^2$ Tons. At lower masses they emit, pairs of particles, so that antiparticles of the same energy (blackhole decay is expected not to violate (CPT invariance) are also emitted.

Strings produced at GUTs phase transition, i.e at energies around 10^{15} or 10^{16} GeV, also decay spontaneously into high energy particles with energies ranging from 10^{21} to 10^{25} eV; with a rate depending on the string tension T_s. The spectrum of high energy particles produced by the string is of the form:

$$N(m) \approx Am^{-n}exp[m/KT_{Hag}] \tag{17}$$

T_{Hag} is the Hagedorn temperature; which is related to the string tension and for superstrings is $\sim 10^{27}eV$. n $\simeq$ 3 for the heterotic string (a favourite candidate for ToE). For m<< T_{Hag} which is the case for UHE particle energies m between 10^{20} to 10^{25} eV,

$N(E) \sim AE^{-3}$, not inconsistent with FLY's EYE results. Black holes emitting UHE particles in their terminal stages have a similar E^{-3} spectrum.

Thus in conclusion both evaporating black holes in their terminal stages and decaying strings, both produced in the early universe, are capable of producing UHE particles of energies $> 10^{20}$ eV with an approximate E^{-3} spectrum.

However, there are some very definite signatures of such processes which are testable in future experiments.

Some of these are:

1. One expects also equal numbers of antiparticles, (i.e. for eg. antiprotons) at such high energies. Thus one predicts $\bar{P}/P \rightarrow$ one, at $> 10^{21}eV$.
2. One does not expect, any heavier elements i.e. nothing other than protons. If one sees oxygen or Fe nuclei at energies $> 10^{21}eV$, then clearly these are not the mechanisms to produce such particles.
3. In the case of decaying strings, one expects a cut off around 10^{25} eV or less (these are based on constraints from inflation theories which fix the scale M<< M_{pl}). However for evaporating black holes the spectrum can continue till 10^{28} eV, the Planck scale! What about the possibility that we have some neutrinos with UHE $> 10^{21}$ eV? Such UHE neutrinos can interact with the thermal cosmic background neutrinos which are expected to have a temperature around $2^\circ K$, through processes like $\nu + \bar{\nu} \rightarrow e^+ + e^-$ etc. In principle this can cause a cut off in the UHE ν flux. However, the cross sections are low[10]. There is a sharp enhancement of the C.S. at the Z° resonance which is $\approx 10^{-32}$cr corresponding to a cut-off energy around $\sim 10^8$. Such UHE

neutrinos interacting with the galactic neutrino halo (assuming all the DM in the halo is due to neutrinos with mass of a few electron volts) have a mean free path $\sim$ halo radius. Thus if the DM density is distributed as $\rho \sim \rho_o/(1 + a^2/r^2)$, then $\rho_o \sim 10^{-25} gcm^{-3}$ gives a background number density $n_\nu \sim 10^9 cm^{-3}$, which with a C.S $\sim 10^{-32} cm^2$, implies a $mfp \sim 10^{23} cm \sim$ few Kpc.

In principle future experiments should be able to discern such dips in the UHE neutrino cosmic rays[10].

References

1. V.Frolov and I.Novikov: The Physics of Black Hole, Pergamon Oxford 1989.
2. C.Sivaram 1994 (to be published).
3. See for eg. MTW Gravitation, Freeman 1973 (chapter on black holes) also C.Sivaram et al. Phys. Rev. D **16**, 1975 (1977).
4. For a review: A Vilenkin Phys. Rep. **121**, 263 (1985).
5. E.Witten: Nucl. Phys. B **249**, 557 (1985).
6. C.Sivaram: Proc. of the 20th ICRC (Moscow, 1987) HE5; Nature **327**, 108 (1987).
7. S.Hawking: Nature **246**, 30 (1974).
8. C.Sivaram: Amer. J. Phys. **51**, 277 (1983).
9. C.Sivaram: Proc. 23rd ICRC (Calgary), (July 1993).
10. C.Sivaram and R.Gandhi: (1994) to be published.

COSMIC RAY NONLINEAR EFFECTS IN SPACE PLASMA, 1. GENERAL CHARACTERISTICS AND DYNAMIC GALACTIC HALO

L.I. DORMAN
Cosmic Ray Dept. of IZMIRAN, Troitsk 142092, Moscow reg., RUSSIA
Cosmic Ray Res. Center of MIGAL, Qazrin 12900, POB 3593, ISRAEL

ABSTRACT. Effects of cosmic ray pressure on space plasma dynamics and effects of kinetic stream instability of non isotropic cosmic ray fluxes on the magnetic structure of space plasma are considered. These effects are important in our Galaxy and other galaxies, in the outer heliosphere, in cosmic ray sources, in the processes of cosmic ray acceleration by shock waves and in regions of magnetic field reconnection. Here we will consider general characteristics of cosmic ray nonlinear effects and its application to the dynamic processes in the galactic halo.

1. Introduction

At the foundation of cosmic ray nonlinear effects in space plasma are two phenomena (Berezinsky et al.,1990): 1) influence of cosmic ray pressure on plasma dynamics - it was first considered by Axford and Newman (1965) for solar wind propagation, and then by Dorman and Dorman (1968, 1969), Dorman (1975), Babayan and Dorman (1977, 1979, 1990); 2) kinetic stream instability of anisotropic cosmic rays and generation of Alfven turbulence - it was first considered by Ginzburg (1965) for cosmic ray propagation in the interstellar medium, and then by Wentzel (1974), Cesarsky (1980), Babayan et al. (1987), Zirakashvili et al. (1991, 1993).

The cosmic ray nonlinear effects are important:

i. in our Galaxy and other galaxies (galactic wind driven by cosmic rays and influence on cosmic ray propagation, chemical composition and energy spectrum formation);

ii. in the outer heliosphere (dynamics effects of galactic cosmic ray pressure on solar wind and interplanetary shock waves propagation, on the formation of terminal shock wave and the boundary of the heliosphere, Alfven turbulence generation by kinetic stream instability of non isotropic cosmic ray fluxes and its influence on cosmic ray propagation and modulation);

iii. in cosmic ray sources (influence of cosmic ray pressure and stream instability of escaping particles on acceleration efficiency and formation of energy spectrum and chemical composition of escaping particles);

iv. in the processes of cosmic ray acceleration by shock waves and in regions of magnetic field reconnection (inverse influence of pressure and stream instability of accelerated particles on the structure and propagation of shock waves, on processes of reconnection, on formation of accelerated particles energy spectrum and chemical composition).

We will consider here general characteristics of two cosmic ray nonlinear effects (pressure and stream instability) and their possible application to the physics of galactic halo as well as to the problems of cosmic ray propagation and modulation in the Galaxy.

M. M. Shapiro et al. (eds.), Currents in High-Energy Astrophysics, 183–191.

2. Effects of Cosmic Ray Pressure

The generally used approximation of cosmic ray motion in electro-magnetic fields is not valid in general. In many cases when cosmic ray density energy is comparable to the energy density of magnetic fields and plasma moving - the inverse influence of cosmic rays on space plasma dynamics and electro-magnetic field structure are important. As was shown by Ptuskin (1984) on the basis of cosmic ray kinetic equation that in diffusion approximation the pondermotive force from cosmic ray particles on plasma with stochastic magnetic field is determined by the pressure of these particles (Berezinsky et al., 1990):

$$P_{CR} = \frac{4p}{3}\int dp\, V p^3 f_0(p) \tag{2.1}$$

where V and p are the velocity and momentum of cosmic ray particles and $f_0(p)$ is the cosmic ray distribution function. The cosmic ray pressure P_{CR} is connected with cosmic ray total energy density

$$W_{CR} = 4\pi \int dp\, p^2 E(p) f_0(p), \tag{2.2}$$

where $E(p)$ is the total energy of particle with momentum p. As was shown by Ptuskin (1984)

$$\frac{\partial P_{CR}}{\partial W_{CR}} = \frac{1}{3}\left\{1 - \frac{1}{3(W_{CR} + P_{CR})}\sum_i m_i c^2 \int_{m_i c^2}^{\infty} dE \frac{2E^2 + m_i^2 c^4}{E^3} N_i(E)\right\} \tag{2.3}$$

where

$$N(E) = \left(4\pi p^2 / V\right) f_0(p) \tag{2.4}$$

is the differential energy spectrum and index i shows the sort of cosmic ray particles. For

$$N(E) \propto E^{-\gamma}, \tag{2.5}$$

equations (2.1) - (2.3) give

$$P_{CR} = \frac{2}{3\gamma} W_{CR}\ ; \qquad \frac{\partial P_{CR}}{\partial W_{CR}} = \frac{2\,(5\gamma + 6)}{3\,(\gamma + 2)(3\gamma + 2)}\ . \tag{2.6}$$

For $\gamma = 2.7$ it gives $P_{CR} \approx 0.25\, W_{CR}$ and $\partial P_{CR} / \partial W_{CR} \approx 0.27$. For ultra relativistic gas $\left(E >> m_i . c^2\right)$ it will be $P_{CR} = (1/3)\, W_{CR}$ and $\partial P_{CR} / \partial W_{CR} = 1/3$.

Let us note that if the gradient of cosmic ray pressure is zero then the pondermotive forces on the elemental volume of space plasma from all directions will be the same and therefore the resultant pondermotive force acting on the elemental volume will be zero. Cosmic ray pressure gradient influenced on the dynamic of space plasma according to the set of hydrodynamics equations:

$$\frac{\partial \rho}{\partial t} + \nabla(\rho\, \mathbf{u}) = 0\ ; \qquad \rho\left(\frac{\partial\, \mathbf{u}}{\partial t} + (\mathbf{u}\, \nabla)\, \mathbf{u}\right) = -\nabla P_G - \nabla P_{CR} + \frac{1}{4\pi}\left[\nabla \times \mathbf{H},\, \mathbf{H}\right]; \tag{2.7}$$

$$\frac{\partial\, \mathbf{H}}{\partial t} = \nabla \times \left[\mathbf{u},\, \mathbf{H}\right]; \quad \nabla\, \mathbf{H} = 0\ ,$$

where $\mathbf{u}$ is the velocity of matter $\left(|\mathbf{u}| << c\right)$ and P_G is the gas-dynamical pressure of plasma.

3. Effects of Cosmic Ray Kinetic Stream Instability

The importance of cosmic ray kinetic stream instability effects for cosmic ray acceleration and propagation was first noted by Ginzburg (1965), and then by Wentzel (1974) and Cesarsky (1980). This type of instability is well known in plasma physics (Tsytovich, 1971; Vedenov and Rjutov, 1972; Akhiezer et al., 1974; Artsimovich and Sagdeev, 1979). The cosmic ray kinetic stream instability generates a broad spectrum of waves in space plasma, but generation of high-frequency waves (Langmuir and whistler types) is not effective because of a big damping of these waves; moreover these waves are not effective for cosmic ray particle scattering. Another situation is for magneto hydrodynamic waves that are effective for cosmic ray scattering. The growth rate is the largest for waves propagated along magnetic field. The main interaction between cosmic ray particles and these waves is based on cyclotron resonant on the first harmonic; the growth rate will be determined by equation (Berezinsky et al., 1990):

$$\Gamma_{CR}(K) = \frac{Z^2 p^2 e^2 V_A^2}{2\, e^2} \,*$$

$$*\iint p^2 dp\, d\mu\,(1-\mu^2)\, V\left\{\frac{\partial f(p,\mu)}{\partial p} + \left(\frac{K V}{\omega_A(K)} - \mu\right)\frac{1}{p}\frac{\partial f(p,\mu)}{\partial p}\right\}\delta\left(KV\mu \pm \frac{ZeHc}{E}\right) \quad (3.1)$$

Here μ is cosine of particles pitch-angle, $\omega_A(K)$ is the frequency of Alfven wave with wave number $K = 2\pi/\lambda$. Let us suppose that cosmic ray distribution function is characterized by isotropic part $f_0(p)$ and some small anisotropy with amplitude A which implies:

$$f(p,\mu) = f_0(p)(1 + A\mu) = f_0(p) - \frac{U_0}{V} p \frac{\partial f_0(p)}{\partial p}\,\mu\,. \quad (3.2)$$

In (3.2) the value

$$U_0 = -(V f_0(p) A)/(p\partial f_0(p)/\partial p) \quad (3.3)$$

is the effective velocity of observer relative to the system of coordinate with cosmic ray isotropic distribution function $f_0(p)$ - to obtain anisotropy A.

If we take into account only resonance scattering with

$$p_{\mathrm{RES}} = \left|\frac{ZeH}{cK}\right| = \left|\frac{Zm_i\,\omega_{Hj}}{Z_i K}\right| \,, \quad (3.4)$$

where Z_i , ω_{Hi} are the charge and ion gyro frequency of background plasma then according to (3.1) and (3.2) it will be

$$\Gamma_{CR}(K) = \frac{\gamma-1}{\gamma}\frac{\pi}{4}\left|\frac{Z\omega_{Hi}}{Z_i}\right|\frac{N_{CR}(>p_{\mathrm{RES}})}{n_i}\left(\frac{U_0}{V_A} - 1\right), \quad (3.5)$$

where

$$N_{CR}(>p) = 4\pi\int_p^\infty p^2 dp\, f_0(p) \propto p^{-\gamma+1} \quad (3.6)$$

is integral cosmic ray momentum spectrum and n_i is the plasma density. Eq. (3.5) shows that $\Gamma_{CR}(K) > 0$ for $U_0 > V_A$. This condition is necessary for development of cosmic ray stream instability and generation of Alfven turbulence.

For the spectrum (3.6) $\Gamma_{CR}(K) \propto K^{\gamma-1}$, but if the spectrum has a maximum at $p = p^*$ then for $p < p^*(K > K^*)$ $\Gamma_{CR}(K) \propto 1/K$. Here it is assumed that $\Gamma_{CR} \ll \omega(K)$ i.e. that $AW_{CR} \ll H_0^2/4\pi$. Usually this condition is valid (in the Galaxy, in the Heliosphere, in the processes of particle acceleration by shock waves and in regions of magnetic field reconnection). Maybe only in very powerful compact sources of cosmic rays the accelerated particle anisotropy and energy density can be so high that it necessary to consider a stronger approximation.

4. Cosmic Ray Propagation in Dynamic Model of the Galaxy

Cosmic rays can give important information on galactic wind. The extended dynamical halo (galactic wind) was take first into account in cosmic ray propagation in the convection - diffusion model by Bulanov et al. (1972). Dogiel et al. (1980) extended this model, taking into account the adiabatic losses. Bloemen et al. (1993) investigated in great detail cosmic ray

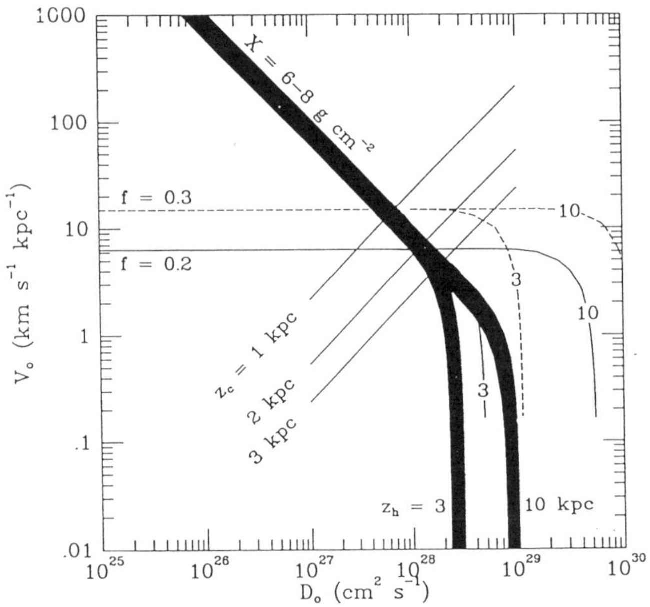

Fig. 1. Constraints on the relation between V_0 and D_0 in a convection - diffusion model from the observed cosmic ray grammage **X** and the observed abundance **f** of radioactive isotope ^{10}Be. Results are shown for different extent of the halo z_h. According to Bloemen et al. (1993).

propagation in the Galaxy, taking into account the galactic wind (diffusion - convection processes and adiabatic energy losses in extended halo). It was supposed that the velocity of galactic wind U increased proportional to the distance z from the equatorial plane of disk $(U = V_0 z)$. The problem is that we don't know exactly the coefficient V_0 and effective diffusion coefficient in the disk and halo D_0. In the frame of this model it was calculated expected grammage **X** of cosmic

ray crossing of matter for the time of living and relative contents **f** of radioactive isotope ^{10}Be. Fig.1 shows the results of calculation Bloemen et al. (1993) of expected relations $V_0 - D_0$ for $\mathbf{X} = 6-8\ g\ cm^{-2}$ (that fits experimental data on cosmic ray chemical composition) and for $\mathbf{f} = 0.2-0.3$ (that fits experimental data on relative contents of radioactive isotope ^{10}Be).
Fig. 1 shows that the best fit of V_0, D_0 that can explain simultaneously $\mathbf{X} = 6-8\ g\ cm^{-2}$ and $\mathbf{f} = 0.2-0.3$ is $D_0 \approx 10^{28}\ cm^2 s^{-1}$, $V_0 \approx 10\ km\ s^{-1}\ kpc^{-1}$. At a distance 30 kpc from the disk the velocity of galactic wind is expected to be 300 km/s. This result shows that the galactic wind plays an important role for cosmic ray propagation and formation of the chemical composition. Moreover this result can be considered as some additional evidence of existence of galactic wind.

5. Galactic Wind Driven by Cosmic Rays

5.1. POSSIBLE GEOMETRY OF GALACTIC WIND

The model of galactic wind driven by cosmic rays was proposed by Ipavich (1975). Recently we have some radioastronomical evidence for the existence of galactic wind (Reich and Reich, 1988; Hummel et al., 1988; Hummel and Dettmar, 1990; Pohl et al., 1991). In Sect. 4 we considered the importance of the existence of galactic wind for cosmic ray propagation and formation of chemical composition. One-dimensional cartesian geometry of galactic wind (that is valid near the center and for distances $z \leq 10 kpc$ from the disk) was considered by Breitschwerdt et al.(1991) and Fichtner et al. (1991). On the other side, one-dimensional spherical symmetric geometry considered by Ipavich (1975) and Zank (1989) is applicable only for very large galactocentric distances $\geq 15 kpc$. Multi-dimensional model for ellipsoidal geometry (contains planar and spherical regimes as asymptotic cases) was considered by Fichtner et al. (11991), Vormbrock & Fichtner (1993).

5.2. EXPECTED DISTRIBUTION OF GALACTIC WIND VELOCITY AND COSMIC RAY DENSITY IN THE HALO (ELLIPSOIDAL GEOMETRY MODEL).

Vormbrock & Fichtner (1993) considered ellipsoidal geometry for the model of galactic wind driving by cosmic rays for our Galaxy and for NGC 4631 with the same mass $2.75*10^{11} M_\Theta$ and same dimensions of disk $r_d = 15\ kpc$, $h_d = 1\ kpc$. The basic set of hydrodynamical equations is:

$$\nabla(\rho\ \mathbf{u}) = 0\ ;\ \rho\ (\mathbf{u}\nabla)\mathbf{u} = \mathbf{f} - \nabla P_G - \nabla P_{CR}\ ;\ \nabla\left(\rho\ \mathbf{u}\left(\frac{1}{2}u^2 + \frac{\gamma_G}{\gamma_G - 1}\frac{kT}{m}\right)\right) + \qquad (5.1)$$

$$+ \frac{1}{\gamma_{CR} - 1}(\mathbf{u}\nabla)P_{CR} - \mathbf{uf} - \Lambda_1 + \Lambda_2 = 0;\ \nabla(\ \mathbf{u}\,\gamma_{CR}\,P_{CR} - D\nabla P_{CR}) - (\gamma_{CR} - 1)(\mathbf{u}\nabla)P_{CR} = 0$$

where **f** is the gravitational force, Λ_1 reflects the heating processes (Coulomb interaction, ionization by cosmic rays), Λ_2 takes into account cooling processes (bremsstrahlung, recombination, collision induced line emission). Fig. 2 shows expected distribution of directions and values of galactic wind velocity in assuming that it has only one cause : driving by cosmic rays. The expected full mass loss rate is $\approx 0.8\, M_\Theta/year$, in good agreement with observations.

Let us note that the calculated values for galactic wind must be considered as a lower limit because there are at least several additional sources (supernova explosions, stellar winds etc.). In Fig. 3 the expected distribution of cosmic ray pressure P_{CR} in the dynamical halo is shown (according to Vormbrock & Fichtner, 1993) .

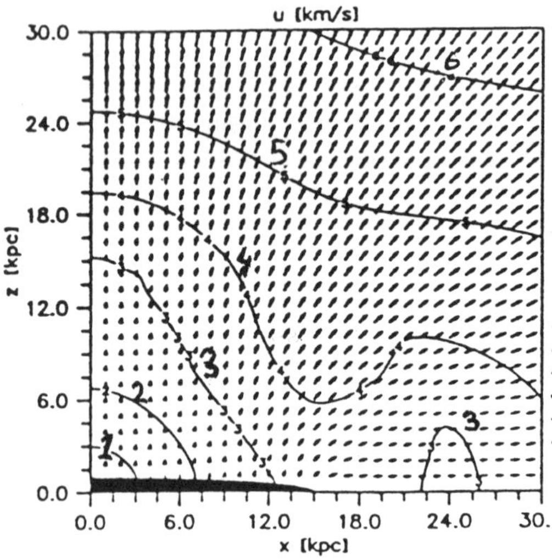

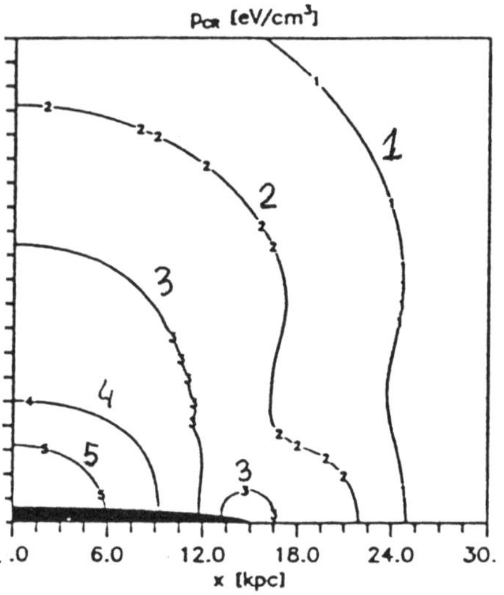

Fig. 2. Expected galactic wind velocity field: 1 - 25, 2 - 75, 3 - 150, 4 - 225, 5 - 300 and 6 - 350 km/s .
Fig. 3. Expected distribution of cosmic ray pressure: 1 - 0.001, 2 - 0.003, 3 - 0.015, 4 - 0.075, and 5 - 0.2 eV/cm^3 .

5.3. SELF-CONSISTENT PROBLEM FOR DYNAMIC HALO IN ROTATING GALAXY.

5.3.1. *Solution for galactic wind and magnetic field.* The self-consistent problem of cosmic ray propagation in the extanded halo, taking into account the rotation of the Galaxy was considered by Zirakashvili et al. (1993), Ptuskin and Zirakashvili (1993). In the frame of axisymmetrical model it was supposed that plasma moved along some surface of rotation S (according to Weber and Davis, 1967; Yen, 1976). The vectors **u** (galactic wind velocity) and **H** (frozen magnetic field) are coplanar to S. It is supposed that $\partial/\partial s$ is derivative in meridional direction (symbolized by ') and the cross section of the tube is $B(s,r_0)$. The basic steady-state MHD equations which include cosmic ray pressure P_{CR} (and under the assumption of the smooth transition of the solution through the possible critical points) will be:

$$B\rho\, u = const;\; BH = const;\; \rho\,\left(u' - u_r u_\Phi^2/ur\right) = -\left(P_G + P_{CR}\right)' - \left(r^2 H_\Phi^2\right)'/8\pi r^2 + \rho\, f'; \tag{5.2}$$

$$B^{-1}\left(B\left(\rho\, u\left(\frac{u^2}{2}+\frac{u_\Phi^2}{2}+\frac{\gamma_G P_G}{\rho\,(\gamma_G-1)} - f\right) - \frac{\Omega\, r\, H H_\Phi}{4\pi}\right)\right)' + (u+V_A)P'_{CR} = 0; \tag{5.3}$$

$$B^{-1}\left\{B\left[\gamma_{CR}\,(u + V_A)P_{CR} - DP'_{CR}\right]\right\}' - (\gamma_{CR} - 1)(u + V_A)P'_{CR} = 0 \;. \tag{5.4}$$

Here the gravitational potential f is taken as $f = f_0(1+s/s_1)^+$, $f_0 = 1.9.10^{15}\ cm^2/s^2$, and $s_1 = 45$ Kpc . The shape of flux tube is taken in the form

$$B(s) = B_0\left(1+(s/s_2)^2\right),\ r(s) = r_0\left(1+(s/s_2)^2\right)^{1/2},\ s_2 = 15\ \text{Kpc} . \qquad (5.5)$$

The calculations are made for boundary condition at $z_0 = 3$ Kpc : $P_{CR\,0} = 2.10^{-13}\ erg/cm^3$, $H_0 = 10^{-6}$ G , $T_0 = 4.10^5$ °K , $n_0 = 10^{-3}\ cm^{-3}$, $u_{\Phi 0} = 250$ km/s , and $\gamma_G = 1.6$ was used. It was found that the meridional initial wind velocity $u_0 = 28.4$ km/s. The results of expected changes of magnetic field H, temperature T, meridional u and azimuthal u_Φ components of galactic wind velocity with z up to $z = 75$ Kpc are shown on the Fig. 4 (the critical points with smooth transition are located at 5.3 Kpc - slow magnetosonic , at 8.1 Kpc - Alfvenic and at 30 Kpc - fast magnetosonic; generated by Galaxy rotation azimuthal magnetic field falls as s^{-1}).

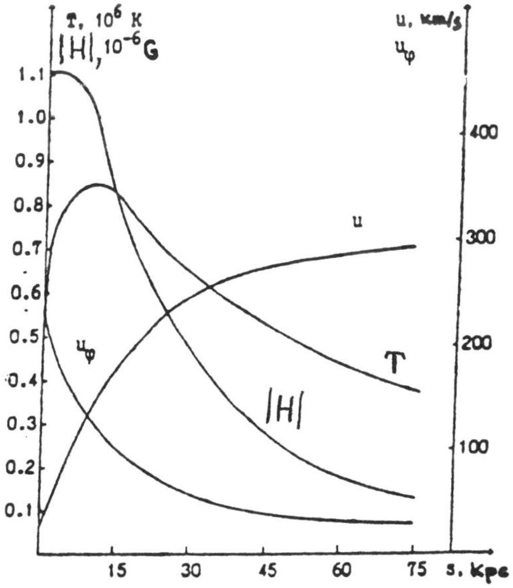

Fig. 4. Expected distribution of magnetic field $|\mathbf{H}|$, meridional u and azimuthal u_Φ galactic wind velocities and gas temperature T in the rotating Galaxy (according to Zirakashvili et al., 1993)

5.3.2. *Solution for cosmic ray propagation.* On the basis of the solution shown on Fig. 4 Ptuskin and Zirakashvili (1993) investigated cosmic ray propagation in the Galaxy governed by the diffusion - convection processes:

$$-B^{-1}\left(BDf'\right)' + \left(u+V_A\right)f' - B^{-1}\left(B\left(u+V_A\right)\right)'\frac{p}{3}\frac{\partial f}{\partial p} = 2Q(p)\delta(s) , \qquad (5.6)$$

where $f(p,s)$ is the cosmic ray distribution function and cosmic ray sources concentrated in disk are assumed to have a power law spectrum $Q \propto p^{-\gamma_s}$. The diffusion coefficient

$$D = D_{||}\cos^2\alpha ,\ D_{||} = Vr_L H^2/6\pi^2 K_{RES} W(K_{RES}) ,\ K_{RES} = 1/r_L \qquad (5.7)$$

and r_L is the Larmor radius. The growth rate of cosmic ray streaming instability for Alfven waves generation

$$\Gamma_{CR} = -6\pi^3 e^2 V_A \left(K_{RES}\ c^2\right)^{-1} \int_{p_{RES}}^{\infty} dp\, p\left(1 - p_{RES}^2/p^2\right) D_{||} \cos\alpha\ f'\ , \tag{5.8}$$

where $p_{RES} = e|H|/cK_{RES}$. The nonsaturated Landau damping of Alfven waves will be (according to Volk and Cesarsky, 1982; Fedorenko, 1992):

$$\Gamma_{NL} = 2\pi^2 \beta^{1/2} V_A\ K \int dK\, W(K)/H^2\ ;\ \beta = 4\pi n\, T/H^2 \tag{5.9}$$

The diffusion coefficient will be

$$D_{||} = \frac{\gamma_s(\gamma_s - 2)}{9\pi^2(\gamma_s - 4)} \beta^{1/2} c^3 \omega_{Hi}^{-1} H^2 \mathrm{E}^{-1} (p/mc)^{\gamma_s - 3} B(s) \cos\alpha\ , \tag{5.10}$$

where E is the full energy . This model is in a good agreement with data on mean matter thickness $<X> = 10(V/c)\ \mathrm{g/cm^2}$ for $pc/Ze < 5$ GV and $<X> \propto p^{-0.55}$ for $pc/Ze > 5$ GV, but predicts too large anisotropy $A = -3DV^{-1}(\ln f)' \approx 10^{-4}(p/mc)^{0.55}$ that gives 5% at energy $10^{14}eV$ (measured amplitude is about 0.05 % at $10^{12} - 10^{14}eV$ and about 3 % at $10^{17}\ eV$). Ptuskin and Zirakashvili (1993) note that this discrepancy can be caused by peculiarities of local conditions near the Sun, that do not affect the energy spectrum and chemical composition, but are very important for the observed cosmic ray anisotropy .

6. Conclusion.

Cosmic ray nonlinear effects (pressure and stream instability) considered here can play a vital role in many astrophysical objects. In our Galaxy these nonlinear effects are expected to be mainly responsible for formation of dynamic halo (galactic wind) and for significant influence on cosmic ray propagation in the Galaxy and formation of chemical composition, energy spectrum and anisotropy. It is expected that these nonlinear effects (pressure and stream instability) can play important role also in dynamics of other galaxies.

ACKNOWLEDGMENTS. I would like to thank Prof. Maurice Shapiro, Prof. John Wefel and Prof. Rein Silberberg for the possibility to attend the school of cosmic ray astrophysics, for very interesting discussions and fruitful comments, and the staff of Ettore Majorana Center for very kind hospitality. I would like to thank also Prof. N. Iucci and Dr. G. Villoresi for their kind invitation to work in Rome and stimulating discussions. This work was partly supported by 3-rd Rome University.

7. References

Akhiezer I.A., Polovin R.V., Sitenko A.G., Stepanov K.N., 1974. *Plasma Electrodynamics*, Nauka, Moscow.

Artsimovich M.A and Sagdeev R.Z.,1979. *Plasma Physics for Physisists*, Atomizdat, Moscow.

Axford I.W and Newman R.C., 1965. Proc. 9-th ICRC, London, Vol.1, p. 173.

Babayan V.Kh. and Dorman L.I., 1977. Proc. 15-th ICRC, Plovdiv, Vol. 3, p. 107, p. 113.

Babayan V.Kh. and Dorman L.I., 1979. Proc. 16-th ICRC, Kyoto, Vol. 3,p. 117; p. 123.

Babayan V.Kh. and Dorman L.I., 1990. In Physics of the Outer Heliosphere, ed. S.Grzedzielski and D.E.Page, Pergamon Press, p. 204.
Babayan V.Kh., Dorman L.I., Ptuskin V.S., 1987. Proc Intern. Conf. on Plasma Physics, Kiev, Vol. 4, p. 228.
Berezinsky V.S., Bulanov S.V., Ginzburg V.L., Dogiel V.A., Ptuskin V. S., 1990. *Cosmic Ray Astrophysics* (Ed. V.L.Ginzburg). Fizmatgiz, Moscow.
Bloemen J.B.G.M., Dogiel V.A., Dorman V.L., Ptuskin V.S., 1993. Astr. & Astrophys., Vol. 267, p. 372.
Breitschwerdt D., McKenzie J.F., Volk H.J.,1991.Astronomy&Astrophysics,Vol. 245, p.79.
Bulanov S.V., Dogiel V.A., Syrovatskii S.I., 1972. Space Research (Moscow), Vol. 10, p.532, p.721.
Cesarsky C.J., 1980. Ann. Rev. Astron. Astrophys., Vol. 18, p. 289.
Dogiel V.A., Kovalenko V.M., Prischep V.L., 1980. Letters in Astron. J. (Moscow), Vol.6, p. 696.
Dorman I.V. and Dorman L.I., 1968. Geomagnetism and Aeronomy, Vol. 8, p. 817; p.1008.
Dorman I.V. and Dorman L.I., 1969.Izvestia Ac.of Sci.of USSR, ser. Phys., Vol. 33, p.1908.
Dorman L.I., 1975. *Variations of Galactic Cosmic Rays.* Moscow University Press, Moscow.
Drury L.O'C. and Volk H.J., 1981. Astrophys. J., Vol. 248, p. 344.
Fedorenko V.N., 1992. Astrophys. Space Phys. (Sov. Sci. Rev., ed. R.A. Sunyaev), Vol. 8, p. 1.
Fichtner H. et al. 1991a. Nuovo Cimento, Vol. 106B, p. 909.
Fichtner H. et al. 1991b. Astrophys. J., Vol. 371, p. 98.
Ginzburg V.L., 1965. Astron. J. (Moscow), Vol. 42, p. 1129.
Hummel E. et al., 1988. Astronomy&Astrophysics, Vol. 197, p. L29.
Hummel E. and Dettmar R.J., 1990. Astronomy&Astrophysics, Vol. 236, p. 33.
Ipavich F.M., 1975. Astrophys. J., Vol. 196, p.107.
McKenzie J.F. and Webb G.H., 1984. J. Plasma Physics, Vol. 31, p.375.
Pohl M., Schlickeiser R., Hummel E., 1991. Astronomy&Astrophysics, Vol. 250, p. 302.
Ptuskin V.S., 1984. JETP, Vol. 86, p. 483.
Ptuskin V.S. and Zirakashvili V.N., 1993. Proc. 23- th ICRC, Calgary, Vol. 2,p. 290.
Reich P. and Reich W., 1988. Astronomy&Astrophysics, Vol. 196, p. 211.
Tsitovich V.N., 1971. Theory of Turbulent Plasma, Atomizdat, Moscow.
Vedenov A.A. and Rjutov L.D., 1972. In "Problems of Plasma Theory", Atomizdat, Moscow, Vol. 6, p. 3.
Volk H.J. and Cesarsky C.J., 1982. Z. Naturforsch. Vol. 37a , p. 809.
Vormbrock N. and Fichtner H., 1993. Proc. 23-th ICRC, Calgary, Vol. 2, p. 283.
Weber E.J. and Davis L. Jr., 1967. Astrophys. J., Vol. 148, p. 217.
Wentzel D., 1974. Ann. Rev. Astrophys., Vol. 12. p. 71.
Yen T., 1976. Astrophys. J., Vol. 206, p. 776.
Zank G.P., 1989. Astronomy&Astrophysics, Vol. 225, p. 37.
Zirakashvili V.N., Breitschwerdt D., Ptuskin V.S., Volk H.J., 1993. Proc. 23-th ICRC, Calgary, Vol. 2, p. 287.
Zirakashvili V.N., Dorman L.I., Ptuskin V.S., 1991. Proc. 22-th ICRC, Dublin, Vol. 3, p. 585.

COSMIC RAY NONLINEAR EFFECTS IN SPACE PLASMA, 2. DYNAMIC HELIOSPHERE

L.I. DORMAN
Cosmic Ray Dept. of IZMIRAN, Troitsk 142092, Moscow reg., RUSSIA
Cosmic Ray Res. Center of MIGAL, Qazrin 12900, POB 3593, ISRAEL

ABSTRACT. Effects of cosmic ray pressure on solar wind and interplanetary shock waves propagation and effects of kinetic stream instability of non isotropic cosmic ray fluxes on the magnetic structure of plasma in the Heliosphere are considered. These effects are important in the outer Heliosphere, where the cosmic ray pressure become comparable with dynamical pressure of solar wind and shock waves.

1. Introduction

In the first part of this paper we considered general characteristics of cosmic ray nonlinear effects and application to the dynamic processes in the galactic halo. We note that at the foundation of cosmic ray nonlinear effects in space plasma there are two phenomena :
1) influence of cosmic ray pressure on plasma dynamics - it was first considered by Axford and Newman (1965) for solar wind propagation, and 2) kinetic stream instability of anisotropic cosmic rays and generation of Alfven turbulence - it was first considered by Ginzburg (1965) for cosmic ray propagation in the interstellar medium. Nonlinear effects of local cosmic rays generated in the disk are important for formation and dynamic processes in the halo (galactic wind driven by galactic cosmic rays); effects of extragalactic cosmic rays are expected negligible. We have an opposite situation in the Heliosphere: local cosmic rays (generated on the Sun, in magnetospheres of planets and in interplanetary space) play negligible role in dynamics of the Heliosphere (excluding some short periods of prominent events), but nonlinear effects of galactic cosmic rays from ouside the solar system play a vital role in dynamical processes in the Heliosphere (dynamic effects of galactic cosmic ray pressure on solar wind and interplanetary shock waves propagation, on the formation of terminal shock wave and the boundary of the heliosphere; Alfven turbulence generation by kinetic stream instability of non isotropic cosmic ray fluxes and its influence on cosmic ray propagation and modulation). As the first step only dynamical effects of galactic cosmic ray pressure have been investigated (Axford and Newman , 1965; Dorman and Dorman, 1968, 1969; Dorman, 1975; Babayan and Dorman, 1977, 1979a,b, 1981, 1990). We point out that in the Heliosphere galactic cosmic rays have a big solar anisotropy increasing with distance from the Sun. This should be important for generation of HMD turbulence by cosmic ray kinetic stream instability what is sufficient for cosmic ray propagation, acceleration and modulation as well as for dynamical processes in the Heliosphere (Babayan et al.,1987, Dorman et al., 1990, Zirakashvili et al., 1991).

M. M. Shapiro et al. (eds.), Currents in High-Energy Astrophysics, 193–208.

2. Interplanetary Cosmic Ray Modulation, Possible Structure of the Heliosphere and Expected Nonlinear Effects

2.1. COSMIC RAY HYSTERESIS EFFECTS AND DIMENSION OF THE HELIOSPHERE.

The studies of the neutron component data have made it possible to find the hysteresis character of the relationships between the variations in solar activity and in cosmic ray intensity (Simpson, 1963; Dorman and Dorman, 1967; Dorman, 1974, 1975a,b). This effect arises from the delay of the interplanetary processes (responsible for cosmic ray modulation) with respect to the initiating solar processes which correspond to some effective velocity of solar wind and shock waves propagation (see Fig. 1 for $r_0/u = 10$ months and Fig. 2 for $r_0/u = 20$months, where r_0 is an effective radius of modulation region and u is an effective radial velocity of solar wind and shock waves).

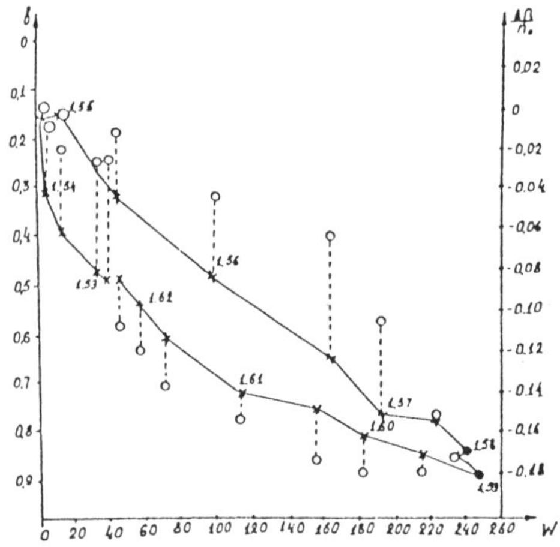

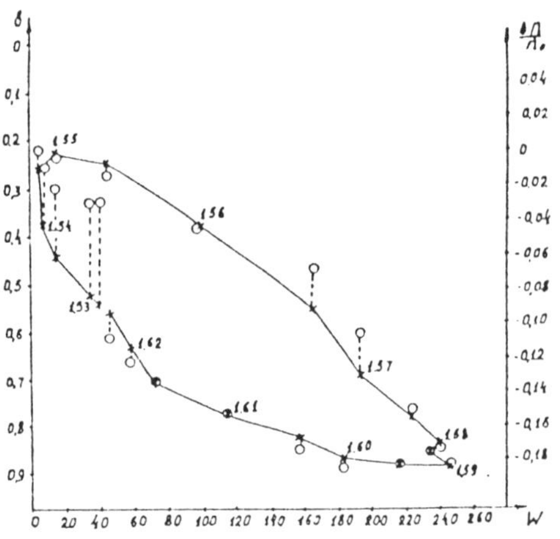

Fig. 1. Expected modulation (curve, left scale) for $r_0/u = 10$ months (which corresponds to $r_0 = 67AU$ at $u = 4.10^7\ cm/s$) as function of sunspot number W flattened over the 12-month period and comparison with observation data of the neutron component in Chicago, also flattened over the 12 - month period (circles connected with the corresponding theoretical values by the dashed vertical lines, the right scale).
Fig. 2. The same as in Fig. 1 but for $r_0/u = 20$ months (which corresponds to $r_0 = 133AU$).

This effect shows that the size of region of cosmic ray effective modulation for energy of few GeV is expected about 100 AU, but for high energy particles detected in equatorial station Huancayo is expected much smaller. This result is very easy to understand if we take into account that at large distances from the Sun solar wind magnetic field becomes so weak that it does not influence on high energy cosmic rays.

2.2. LONG - TERM COSMIC RAY SPECTRUM MODULATION IN THE HELIOSPHERE.

The investigation of long-term cosmic ray modulation in the interplanetary space gives important information on dimensions of the Heliosphere (from hysteresis effect, see Sect. 4.1) as well as on dependence of transport path Λ versus particle rigidity R (from long - term cosmic ray rigidity spectrum of modulation). The time-dependence of the primary variations of cosmic rays with effective rigidities R = 2, 5, 10 and 25 GV was found by Belov, Dorman et al. (1988, 1990) on the basis of ground measurements (muon and neutron components) as well as measurements in stratosphere on balloons and in space on satellites and space-craft. Results are shown on Fig. 3.

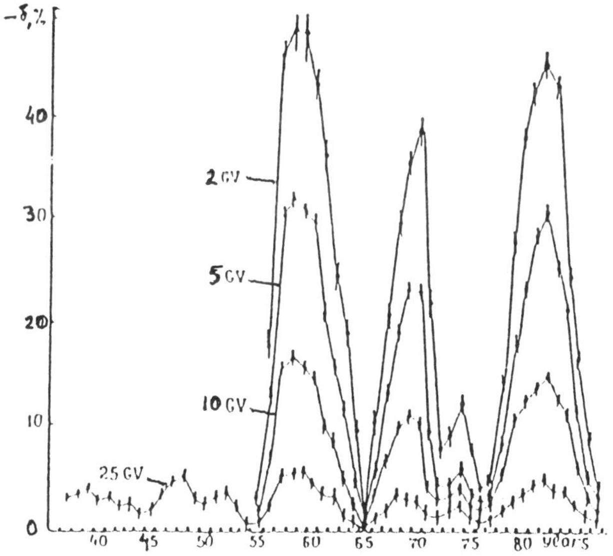

Fig. 3. Time dependence of primary cosmic rays with rigidity R = 2, 5, 10 and 25 GV (on the basis of ground level registration of muon and neutron components and measurements on balloons and satellites),

The smaller modulation in 1965-1975 solar activity cycle reflects the influence of reversal general magnetic field of the Sun (drift effects). Fig. 4 shows the observed rigidity spectrum of long-term modulation relative to minimum of solar activity in 1965 as well as residual modulation spectrum in 1965. The residual modulation is 6.0 ± 1.2 % at $R = 10$ GV (what is in good agreement with results on Fig. 2 for hysteresis effect in Chicago neutron monitor data). The slope of the residual spectrum gets steeper with increasing rigidity and the spectral index:

$\beta \approx 0.4$ at $R = 2-5$ GV, $\beta \approx 1.1$ at $R = 5-10$ GV and $\beta \approx 1.6$ at $R = 10-25$ GV . (4.1)

In the first approximation the effective transport path in the interplanetary space will be

$$\Lambda \propto R^{\beta} \quad (4.2)$$

where β is determined by (4.1). The tendency of increasing β with increasing rigidity is seen also for cosmic ray propagation in the interstellar space for much higher rigidity region $10^{14} - 10^{17}$ V (Berezinsky at al., 1990) and for solar cosmic rays in the solar atmosphere and in interplanetary space for the much lower energy region $10^6 - 10^{10}$ eV (Dorman and Miroshnichenko, 1976; Dorman, 1978; Dorman and Venkatesan, 1993). These intervals in the first approximation correspond to the product of magnetic field strength on the characteristic scale of turbulence in the space where cosmic rays propagate.

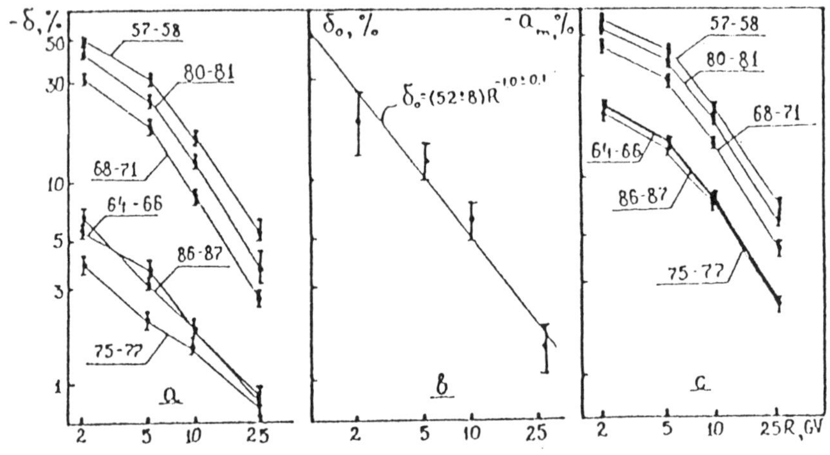

Fig.4. Rigidity spectra: of observed long - term cosmic ray variations (a), of residual modulation (b) and of total cosmic ray modulation (c) during the various time intervals indicated on the curves.

2.3. COSMIC RAY ANISOTROPY IN THE HELIOSPHERE.

For the problem of kinetic stream instability in the Heliosphere the information on possible types of cosmic ray anisotropy in the interplanetary space and their dependence on heliolatitude and radial distance as well as on the level of solar activity - is very important .Cosmic rays penetrate inside the Heliosphere and propagate in extended solar wind with frozen in rotated regular spiral magnetic field with inhomogeneities. Fig. 5 (from Moraal, 1993) shows several mechanisms of cosmic ray anisotropy formation in the meridional plane: convection - diffusion, cosmic ray density gradient drift, terminal shock wave drift, polar drift and neutral current sheet drift. The convection anisotropy is determined by the velocity of solar wind and directed radially from the Sun; diffusion anisotropy is mainly along spiral magnetic field and directed to the Sun. Resulting anisotropy increased with the radial distance (in the first approximation - proportionally) and does not depend on the direction of spiral interplanetary magnetic field. The direction and the value of the gradient density drift anisotropy depends on the directions of gradient and magnetic field and is proportional to the product of their particle Larmor radius and cosmic ray density gradient. The drift along terminal shock wave gives important particle acceleration. The neutral current sheet drift. is very important for cosmic ray long - term modulation; its direction changed every 11 years with changing of the sign of the Sun's general magnetic field - it is a main cause of 22 - year cosmic variation. In the equatorial plane there is also a very important convection - diffusion mechanism that gives some average anisotropy perpendicular to radial direction and directed as the Sun rotated (the amplitude of this type of anisotropy is expected to increase with increasing radial distance - it is important for cosmic ray kinetic stream instability.

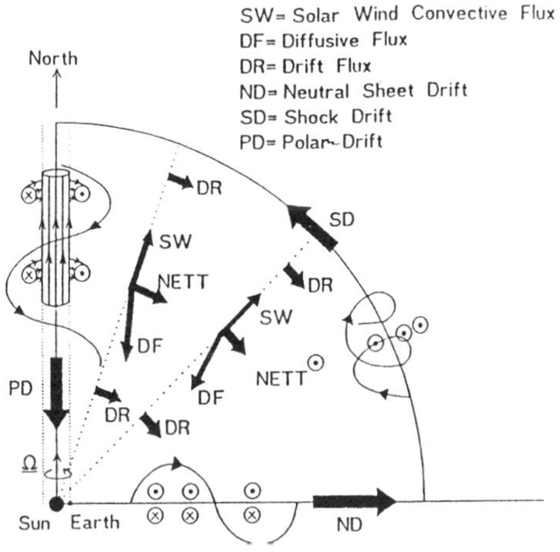

Fig. 5. Meridianal projection of a quarter Heliosphere showing the major cosmic ray transport processes and mechanisms of anisotropy formation (according to Moraal , 1993).

As was shown by Belov, Dorman et al. (1991, 1993) from statistical analysis for many years of the behavior of the first harmonic of cosmic ray anisotropy, there are amplitude - phase interrelation. This interrelation can be explained in the frame of convection - diffusion and drift anisotropy and can be find the expected distribution of radial and transverse cosmic ray gradients (see Fig. 6 for 1974-1977 and Fig. 7 for 1984-1987).

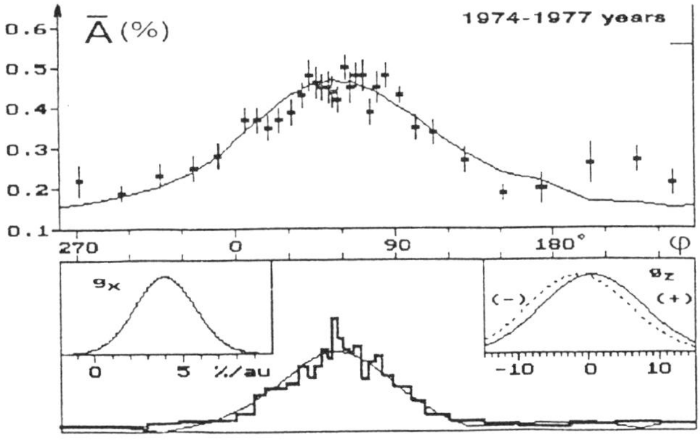

Fig. 6. The amplitude - phase dependence (top) and phase distribution (bottom) of the first harmonic of cosmic ray anisotropy near the minimum of solar activity in 1974 - 1977. Crosses and step line - experimental data, smooth line - model. On the inserts - used distributions of radial g_X and latitudinal g_Z cosmic ray gradients. 0° - direction from the Sun , $(+)$ and $(-)$ indicate the direction of IMF in the North hemisphere.

Dorman et al. (1994) show that observed anisotropy is mixed, produced by several mechanisms with different properties and different rigidity spectra. Some types of anisotropy reflect real cosmic ray fluxes and determine cosmic ray density distribution in space and energy balance (convection, anisotropy diffusion, neutral sheet drift, curvature drift, drift in inhomogeneous magnetic field, drift along shock wave front with energy change); other types give real anisotropy that can be measured, but don't affect cosmic ray flux (for example, the drift perpendicular to magnetic field and to cosmic-ray density gradient).

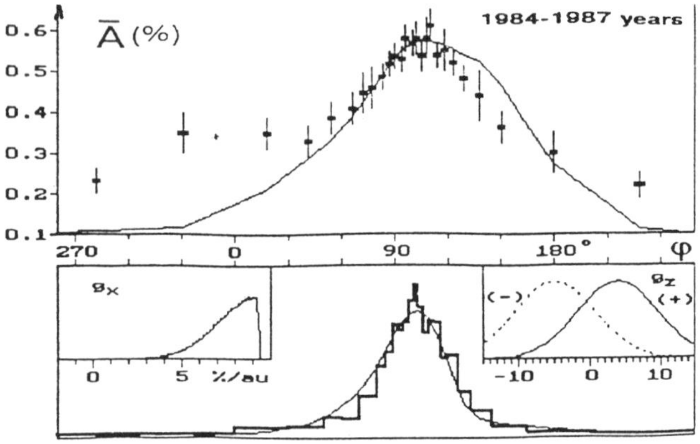

Fig . 7. The same as on Fig. 6, but for 1984 - 1987.

2.4. POSSIBLE STRUCTURE OF THE HELIOSPHERE AND EXPECTED NONLINEAR EFFECTS

A possible structure of the Heliosphere according to Dorman (1991) is shown on Fig. 8.

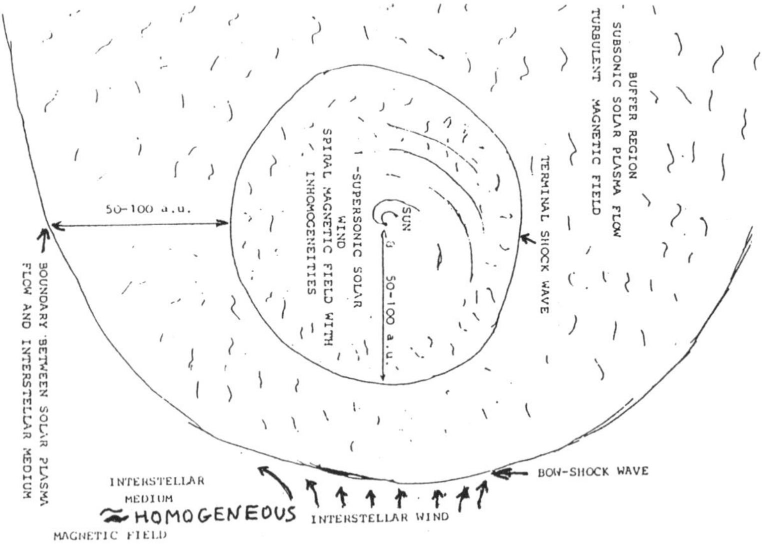

Fig. 8. Expected structure of the Heliosphere.

In the region of inner planets the dynamic pressure of solar wind is much larger than the cosmic ray pressure but at larger distances these pressures became of the same order and the nonlinear effects became important. The problem is what is the size of the Heliosphere (Dorman , 1991). About 15 -30 years ago many scientists came to conclusion that the radius of the Heliosphere is not more than 10 - 15 AU, but from investigations of cosmic ray hysteresis phenomena we decided that this size must be not smaller than the size of effective modulation region for small energy particles, i. e. not smaller about 100 AU (see Sect.2.1). If the size of the Heliosphere is as big as shown in Fig.8, then the dynamical pressure of solar wind in the outer part of the heliosphere becomes comparable to the cosmic ray pressure and it is necessary to take into account the influence of galactic cosmic ray pressure on solar wind propagation. This was first done by Axford and Newman (1965), and then by Dorman and Dorman (1968, 1969), Babayan and Dorman (1977, 1979a,b, 1981, 1990). It was shown that solar wind radial deceleration by pressure of galactic cosmic rays becomes important in the outer heliosphere. The cosmic ray modulation in the interplanetary space isn't spherical-symmetrical, the modulation is expected to be stronger in the low--heliolatitude region. Therefore one expects that cosmic ray pressure in high--heliolatitude region will be higher than in low--heliolatitude region. If it is so, we will expect the transverse compression of solar wind streams by cosmic ray pressure caused by the transverse cosmic ray density gradient (Dorman and Dorman, 1969; Babayan and Dorman, 1977). The effects of cosmic ray stream instability was first considered for cosmic ray propagation in interstellar medium by Ginzburg (1965), and for the processes in the outer heliosphere - by Babayan et al. (1987, Dorman et al. (1990), Zirakashvili et al. (1991).

3. Cosmic Ray Pressure Effects in the Heliosphere

3.1. EFFECTS IN RADIAL DIRECTION.

The set of hydrodynamics equations that determines the solar wind propagation, taking into account the cosmic ray pressure P_{CR} and exchange between solar wind protons and atoms of interstellar neutral hydrogen is given below:

$$\frac{d}{dr}\left(r^2\rho\, u\right)=0\,;\quad \rho\, u\,\frac{d\,u}{d\,r}=-\frac{d\,P_{SW}}{d\,r}-\frac{d\,P_{CR}}{d\,r}-\rho\, u;$$
$$r^{-2}\frac{d}{dr}\left(r^2\rho\left(\frac{\rho\, u^2}{2}+\frac{\gamma\, P_{SW}}{\gamma-1}\right)\right)=-u\left(\frac{d\,P_{CR}}{d\,r}+Su\right), \qquad (3.1)$$

where ρ, u, P_{SW} are solar wind density, velocity and gas - dynamically pressure, $\gamma = 5/3$,

$$S=Q_{EX}\,\rho\, uN_H\,(r);\; P_{CR}=\frac{1}{3}\int_0^\infty N(T,r)pVdT\;. \qquad (3.2)$$

Here Q_{EX} is the cross - section for exchange processes, $N_H\,(r)$ is the density distribution of neutral hydrogen in the interplanetary space , $N(T,r)$ is the cosmic ray density distribution as a function of the particle kinetic energy T . The modulation of galactic cosmic rays in interplanetary space in the convection - diffusion approximation can be described by equation :

$$r^{-2}\frac{\partial}{\partial\, r}\left(r^2\left(uN-D\frac{\partial\, N}{\partial\, r}\right)\right)-\frac{1}{3}\frac{\partial}{\partial\, T}\left(a(T)T\, r^{-2}\frac{\partial}{\partial\, r}\left(r^2uN\right)\right)=0\,, \qquad (3.3)$$

where $a(T)=\left(T+2m_0\, c^2\right)/\left(T+m_0\, c^2\right)$. From (3.1) - (3.3) we obtain the integral - differential equation for determining the behavior of solar wind velocity :

$$\frac{du}{dr} = -\frac{r^2}{3\,m}\int_0^\infty \frac{\partial\, N(T,r)}{\partial\, r} pVdT - Q_{EX}\, u\, N_H\,(r)\ . \tag{3.4}$$

Here $m = \rho_1\, u_1\, r_1^2$, where index 1 denotes that these values are taken at $r = r_1 = 1$ AU. Some examples of expected dependence of solar wind velocity on the radial distance for the isotropic diffusion and spherical - symmetric problem are shown on Fig. 9 and Fig. 10.

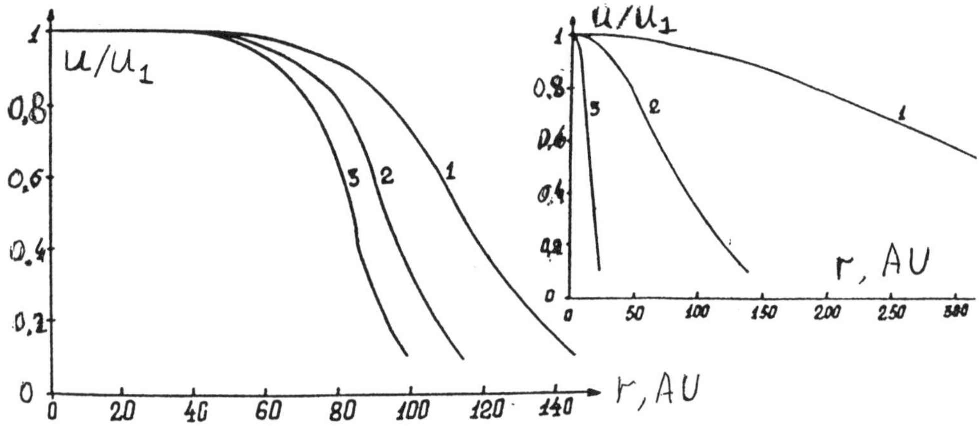

Fig. 9. Ratio u/u_1 versus r/r_1 for $u_1 = 3.10^7$ cm/s, $N_0 \propto R^{-2.5}$. Curves 1, 2, 3 - for $R_{min} = 0.1,\ 0.01$ and 0.001 GV. $D \propto r^s,\ s = 0.$

Fig. 10. The same as on Fig. 9, but for $s = 1$, $N_0 \propto E_K^{-2.5}$, curves 1, 2, 3 - for $E_{K\ min} = 0.1$, 0.01 and 0.001 GeV.

3.2. EFFECTS IN TRANSVERSE DIRECTION.

As we mentioned above, the cosmic ray distribution in the Heliosphere is not spherical - symmetric and transverse gradients of cosmic ray energy density (and corresponding transverse cosmic ray pressure gradients) are expected. If the solar wind plasma initially moved in radial direction, from transverse cosmic ray pressure gradients we expect some plasma moving in directions perpendicular to radial direction. To estimate approximately whether this effect is important or not (see Fig. 11), let us consider a model of solar wind concentrated near equator in the heliolatitudes $\pm 25^\circ$ and $\pm 30^\circ$ and that at the boundary the cosmic ray density is equal to interstellar cosmic ray density. The results of Fig.11 show that for definite conditions, the considered transverse cosmic ray gradients can be important for solar wind dynamics.

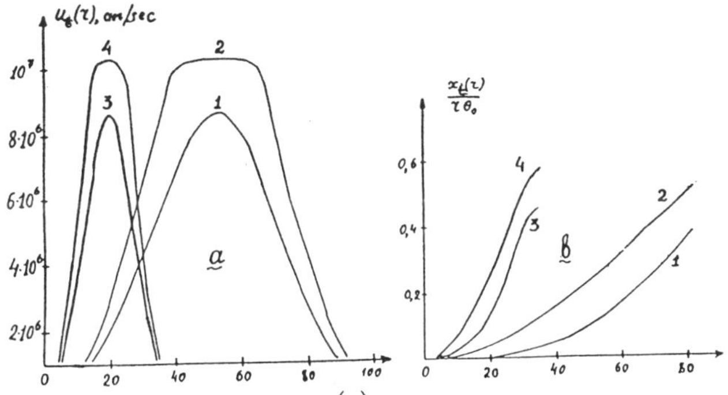

Fig. 11. The expected transverce velocity $u_{\perp}(r)$ and relative transverse displacement $x_{\perp}(r)/r\theta_0$ for the same conditions as on Fig. 9. Curves: $1-E_{K\min}=0.1\ GeV,\ \theta_0=30^0$; $2-0.1\ GeV,\ 25^0$; $3-0.01\ GeV,\ 30^0$; $4-0.01\ GeV,\ 25^0$.

4. Expected Mach Number Change

If we introduce the Mach number $M = u/V_S$, where $V_S = (\gamma P_{SW}/\rho)^{1/2}$ is the sound velocity, then on the basis of Eq. (3.4), taking into account the analytical solution of (3.3) for isotropic diffusion, we obtain :

$$M^{-2}\frac{dM^2}{dr} = \left[8M^2 r^3\left(\frac{dP_{CR}}{u\,dr} + Q_{EX}\,\rho\,uN_H(r)\right) - 12m\left(1+\frac{M^2}{3}\right)\right] \Big/ \left[3m\,r\left(1-M^2\right)\right]. \tag{4.1}$$

The behavior of $M(r)$ for several cases is shown in Fig. 12.

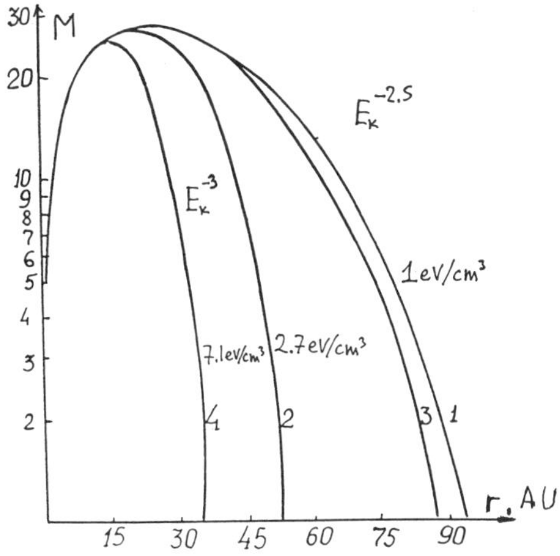

Fig.12. Expected change of M versus radial distance r for $N_0 \propto E_K^{-2.5}$ (curves 1 and 3) and for $N_0 \propto E_K^{-3}$ (curves 2 and 4) for different cosmic ray energy density.

5. On the Type of Transition Layer from Supersonic to Subsonic Fluid of Solar Wind.

The especially important problem is the behavior of $M(r)$ and $u(r)$ near critical point $M^2 = 1$: what type of transition layer from supersonic regime to subsonic regime is realized - of gradual type or of shock wave type? To solve this problem Babayan and Dorman (1990) considered the behavior of the Eq. (4.1) near $M^2 = 1$. Let us suppose that near the critical point $r = r_C$

$$d P_{CR} / u\, dr \approx K r^j \ , \tag{5.1}$$

where K and j are some constants. Then from (4.1) we obtain :

$$M^{-2}\frac{dM^2}{dr} = \left[\frac{8}{3} M^2 r^3 \left(K r^j + m\ {}^{-2} L_{EX}^{-1} - m r^{-3} L\, L_{EX}^{-1}\right) - 4m\left(1 + M^2/3\right)\right] \Big/ \left[mr\left(1 - M^2\right)\right] \tag{5.2}$$

where m is determined above in Section 3.1 , $L_{EX} = Q_{EX} N_{HO}$ and L is the distance, for which $N_H(r)$ decreases by a factor of e. The behavior of integral curves of Eq. (5.2) near critical point $r = r_C$ will be determined by the characteristic equation

$$X^2 - BX + C = 0 \ , \tag{5.3}$$

where

$$B = (16/3)K r_C^{3+j} + m r_C L_{EX}^{-1} - mL\, L_{EX}^{-1} - 20m/3 \ ; \ C = (8 m r_C/3)\left[K r_C^{2+j}(3+j) + m L_{EX}^{-1}\right] . \tag{5.4}$$

The position of the critical point $r = r_C$ is determined by the equation which follows from (5.2):

$$K r_C^{3+j} + m r_C L_{EX}^{-1} - m L L_{EX}^{-1} - 2m = 0 \ . \tag{5.5}$$

According to (5.3), Eq. (5.2) has critical point $r = r_C$ (which is determined by (5.5)) of the type knot (gradual transition) if $B^2 - 4C \geq 0$ and of type focus (shock wave transition) if $B^2 - 4C < 0$. It means that if $j \leq j_C$ we obtain near the critical point $r = r_C$ gradual transition from supersonic flow to subsonic flow, but if $j > j_C$ we expect near the critical point $r = r_C$ shock wave transition from supersonic flow to subsonic flow. The value of j_C is determined from the condition $B^2 - 4C = 0$:

$$j_C = -2 - \left(1 + 2LL_{EX}^{-1}\right)/2\left(2 - r_C L_{EX}^{-1} + LL_{EX}^{-1}\right) . \tag{5.6}$$

According to Semar (1970) , Mc Donough and Bracc (1971) , probable value of $L \approx 3 - 5AU \approx (4.5 - 7.5)*10^{13}\,cm$ and $L_{EX} \approx 5*10^{15}\,cm(\text{at } N_{H0} = 0.1\ cm^{-3})$, so $2LL_{EX}^{-1} \approx$ $\approx (1.8 - 3.0)*10^{-2}$. We expect that $r_C \approx 100AU$, so $r_C L_{EX}^{-1} \approx 0.3$ and $j_C = -(2.29 - 2.30)$. The value j_C depends very weakly on parameters L, L_{EX} and r_C. For example, if $r_C = 150AU$, then $j_C = -(2.322 - 2.328)$. From data on radial cosmic ray gradient obtained for the inner Heliosphere on space probes Pioneer , Voyager and others the value $j > j_C$ (therefore we expect shock wave transition), but what will be the situation near the critical point $r = r_C$? It needs a special investigation, including consideration of kinetic stream instability in the outer Heliosphere, which can change the diffusion coefficient and therefore the dependence (5.1).

6. Effects of Cosmic Ray Kinetic Stream Instability in the Heliosphere

First we will consider (Dorman et al., 1990) only the effects of cosmic ray kinetic stream instability in the outer Heliosphere without taking into account effects of cosmic ray pressure on the movement of the solar wind. Therefore we assume that the velocity of solar wind $u = 400\,km/s = const$, the radius of the Heliosphere $r_0 = 60AU$, the spiral interplanetary magnetic field has components

$$H_R = (r/r_S)^2 H_S\ ;\ H_\phi = (r_S^2\,\Omega_S/ur)H_S \sin\theta\ , \tag{6.1}$$

where $r_S = 7.10^{10}\,cm$ is the radius of the Sun, $H_S \approx 2\ Oe$ is the strength of Sun's general magnetic field on the surface, $\Omega_S = 2.7*10^{-6}\ s^{-1}$ is the angular velocity of solar rotation, θ is the polar angle ($\theta = \pi/2$ is the solar equator). At large distances from the Sun the full spiral interplanetary magnetic field H and the angle α between magnetic field and radial direction will be

$$H \approx H_\phi = H_1(r_1/r)\sin\theta\ ;\ \cos\alpha \approx r_1(r\sin\theta)^{-1}\ , \tag{6.2}$$

where $r_1 = 1AU$, $H_1 = H(r_1) = 5*10^{-5}\ Oe$ .The galactic cosmic ray anisotropy generated in the interplanetary space is determined by the spiral magnetic field, particle scattering and cosmic ray gradients and convection. Near the Earth's orbit the average anisotropy has an amplitude about 0.5 % and is perpendicular to the radial direction. In the first approximation the amplitude of average anisotropy must be $\propto(\cos\alpha)^{-1} \propto r\sin\theta$ and at large distances from the Sun we expect a large amplitude of cosmic ray anisotropy. Therefore we expect that the effects of stream instability in the outer Heliosphere must be very important.

According to the Sect. 3 of our Paper I we consider generation by cosmic ray stream instability of MHD waves propagated along the magnetic field (axis Z). Growth rate $\Gamma(k_Z)$ determines the evolution of spectral energy density of MHD waves with accidental phases $W(k_Z)$:

$$dW(k_Z)/dt = 2\Gamma(k_Z)W(k_Z)\ ;\ \langle(\delta H)^2\rangle = 4\pi\int dk_Z\,W(k_Z)\ . \tag{6.3}$$

Resonance interaction of particles with momentum p and cosines of pitch angle $\mu = (\mathbf{pH})/(pH)$ will be with waves $k_Z = \pm ZeH(pc\mu)^{-1}$. We consider here only particles with velocity $V \approx c \gg V_A = H/(4\pi\rho)^{0.5} \approx 5*10^6\ cm/s$ for conditions in the interplanetary space. On the basis of Eq. (3.1) we obtain :

$$\Gamma(k_Z) = \frac{\pi^2 Z^2 e^2 V_A^2}{2c^2}\ *$$

$$*\int d^3p(1-\mu^2)\left[\frac{\partial f(p,\mu)}{\partial p} + \left(\frac{k_Z V}{\omega_A(k_Z)} - \mu\right)\frac{\partial f(p,\mu)}{p\,\partial\mu}\right]\left[\left(k_Z\mu - \frac{ZeH}{pc}\right) + \left(k_Z\mu + \frac{ZeH}{pc}\right)\right] \tag{6.4}$$

Here $\omega_A = k_Z V_A$. The connection between MHD waves and cosmic ray propagation is determined by the effective frequency of particle scattering on waves :

$$\nu^{\pm} = 2\pi^2\omega_H k_{res} W^{\pm}(k_{res})H^{-2}\ ,\ \text{where}\ k_{res} = ZeH(pc|\mu|)^{-1}\ ,\ \omega_H = ZeHV(pc)^{-1} \tag{6.5}$$

Let us consider the isotropic part of cosmic ray distribution function $f_0 = \int f(p,\mu)d\Omega/4\pi$.

For the times $\Delta t \gg \nu^{-1}$ and distances $\Delta r \gg V\nu^{-1}$ the transport equation in diffusion approximation will be

$$\frac{\partial f_0}{\partial t} - \nabla_i D_{ij} \nabla_j f_0 + u_i \nabla_i f_0 - (\nabla_i u_i) \frac{p \partial f_0}{3 \partial p} = 0 , \tag{6.6}$$

where

$$D_{ij} = D_{||} h_i h_j , \quad D_{||} = \frac{V^2}{2} \int_0^1 \frac{(1-\mu^2) d\mu}{\nu^+ + \nu^-} , \quad h_i = H_i / H . \tag{6.7}$$

Cosmic ray particles penetrating into Heliosphere generate MHD waves along spiral magnetic field in direction to the Sun $(k_Z < 0)$, therefore we will use sign "$-$" , i.e. ν^-, W^- . Further we don't use the index "z" and the sign "$-$" . Let us introduce the function $E(k) = kW(k)$. Then we obtain:

$$-r^{-2} \frac{\partial}{\partial r} \frac{r_1^4 cpVH_1^2}{4\pi^2 Zer^2} \int_0^1 d\mu \, (1-\mu^2) \left(E(ZeH/pc|\mu|) \right)^{-1} \frac{\partial f_0}{\partial r} + u \frac{\partial f_0}{\partial r} - \frac{2up}{3r} \frac{\partial f_0}{\partial p} = 0 , \tag{6.8}$$

$$u \frac{\partial E}{\partial r} + \frac{3uE}{r} - \frac{2\pi V_A r_1^2 ZeH_1}{ckr^2} \int_{ZeH/ck}^{\infty} dp \, p^2 c \left(1 - \frac{ZeH}{pck} \right)^2 \frac{\partial f_0}{\partial r} = 0 . \tag{6.9}$$

For approximate calculations we exchange in the integral of Eq. (6.8) $E(ZeH/pc|\mu|)$ by $E(ZeH/pc)$ and introduce functions $E(p) = E(k = ZeH/pc)$ and $\Gamma(p) = \Gamma(k = ZeH/pc)$. Then from Eq. (6.8) and (6.9) we obtain

$$-r^{-2} \frac{\partial}{\partial r} \frac{r_1^3 cpVH_1}{6\pi^2 ZerE(p) sin\theta} \frac{\partial f_0}{\partial r} + u \frac{\partial f_0}{\partial r} - \frac{2up}{3r} \frac{\partial f_0}{\partial p} = 0 , \tag{6.10}$$

$$u \frac{\partial E(p)}{\partial r} + \frac{3u}{r} E(p) - \frac{4\pi}{3} \frac{V_A \, V p^4 r_1}{r \, sin\theta} \frac{\partial f_0}{\partial r} = 0 . \tag{6.11}$$

For the boundary condition $f_0(r = r_0, p) = f_E(p)$, where $f_E(p)$ is the cosmic ray distribution function out of the Heliosphere, the approximate solution of equations (6.10) and (6.11) will be

$$f_0(r,p) = \frac{f_E(p)}{1 + B(p) f_E(p) (r_0^3 - r^3)} \; ; \; E(p) = \frac{4\pi V_A}{3u} V p^4 f_0(r,p) \frac{r_1}{r \, sin\theta} , \tag{6.12}$$

where

$$B(p) = \frac{8\pi^3 ZeV_A \, p^3}{3 c r_1^2 H_1} . \tag{6.13}$$

Expected anisotropy $A(r,p)$ along interplanetary spiral magnetic field and radial diffusion coefficient $D_{RR}(r,p)$ are as follows :

$$A(r,p) = -\frac{3ursin\theta}{Vr_1} \; ; \; D_{RR}(r,p) = D_{||} \cos^2\alpha = \frac{ucH_1 r_1^2 \left(1 + B(p) f_E(p) (r_0^3 - r^3)\right)}{ZeV_A \, p^3 f_E(p) \, r_0^2} \left(\frac{r_0}{r} \right)^2 . \tag{6.14}$$

For $r = r_0 = 60AU$, $V = c$, $\theta = \pi/2$ we obtain $A \approx 20\%$. The expected dependence of D_{RR} from kinetic energy of particles is shown in Fig. 13, the depth of modulation in the outer Heliosphere f_0/f_E in the interval $40-60\,AU$ - in Fig. 14.

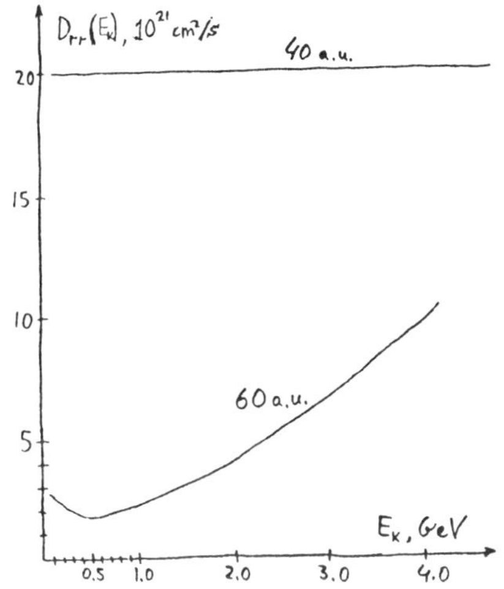

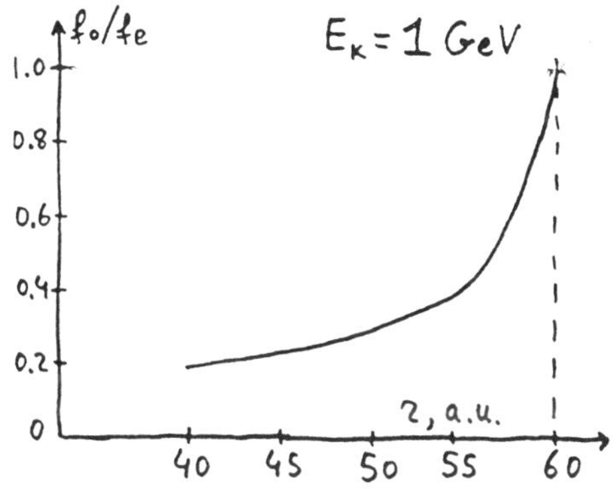

Fig. 13. Cosmic ray radial diffusion coefficient $D_{RR}\left(E_K\right)$ in units $10^{21}\ cm^2/s$ at heliocentric distances 40 AU (according to space -probe measurements) and 60 AU (as calculated in Dorman et al.,1990).

Fig. 14. Cosmic ray modulation in outer heliosphere at distances 40 - 60 AU for particles with kinetic energy 1 GeV, expected according to Dorman et al. (1990).

7. Effects of Cosmic Ray Pressure and Kinetic Stream Instability in the Heliosphere (Self-Consistent Problem)

In Sect. 6 it was supposed that the solar wind velocity is constant up to the boundary of the Heliosphere and we don't take into account the influence of cosmic ray pressure on solar wind moving , although it was shown in Sect. 5 that this influence is very important in the Outer Heliosphere. To take into account both effects (stream instability and cosmic pressure) Zirakashvili et al. (1991) considered self - consistent problem on the basis of the following set of equations in the hydrodynamic approximation (Drury and Volk, 1981; McKenzie and Webb, 1984):

$$\frac{\partial}{\partial r}\left(r^2\rho\, u\right)=0\ ;\ \rho\, u\frac{\partial u}{\partial r}=-\frac{\partial}{\partial r}\left(P_G+P_{CR}+P_W+P_M\right)\ ;\ \frac{\partial}{\partial r}r^2\left[\rho\, u\left(\frac{u^2}{2}+\frac{\gamma_G}{\gamma_G-1}\frac{P_G}{\rho}\right)+\right.$$

$$\left.+\frac{1}{\gamma_{CR}-1}\left(\gamma_{CR}\, u\, P_{CR}-D\frac{\partial P_{CR}}{\partial r}\right)+3uP_W+2uP_M\right]=0\ ;\ u\,\frac{\partial P_{CR}}{\partial r}= \qquad (7.1)$$

$$=\frac{1}{r^2}\frac{\partial}{\partial r}\left[\frac{r^2}{\gamma_{CR}-1}\left(\gamma_{CR}\, u\, P_{CR}-D\frac{\partial P_{CR}}{\partial r}\right)\right];\ \frac{1}{r^2}\frac{\partial}{\partial r}\left(3ur^2P_W\right)=u\,\frac{\partial P_W}{\partial r}\pm V_A\,\frac{\partial P_{CR}}{\partial r}\cos\alpha\,.$$

Here the new values $P_W=(1/2)\int W(\omega)\,d\omega$ is the pressure of MHD turbulence, P_M is the magnetic pressure. We will consider an inner boundary problem at $r=r_0>>1AU$ where $P_W=0$ and the boundary of Heliosphere (the transition from supersonic to subsonic flow) will be determined on the basis of solution of self - consistent problem (7.1). To obtain some rough analytical solution of the problem (7.1) we suppose that the dimension of layer between r_0 and the transition layer is small relative to the radial distance r so that we can consider as one-dimensional problem. In this case, in the set of equations (7.1) we can consider r^2 as a slow function and take it out from differentiation $\partial/\partial r$ and the set (7.1) will transform to :

$$\frac{\partial}{\partial r}(\rho u)=0\,;\ \rho u\frac{\partial u}{\partial r}=-\frac{\partial}{\partial r}(P_G+P_{CR}+P_W+P_M)\,;\ \frac{\partial}{\partial r}\left[\rho u\left(\frac{u^2}{2}+\frac{\gamma_G}{\gamma_G-1}\frac{P_G}{\rho}\right)+\right.$$
$$\left.+\frac{1}{\gamma_{CR}-1}\left(\gamma_{CR}uP_{CR}-D\frac{\partial P_{CR}}{\partial r}\right)+3uP_W+2uP_M\right]=0\,;\ u\frac{\partial P_{CR}}{\partial r}= \tag{7.2}$$
$$=\frac{\partial}{\partial r}\left[\frac{1}{\gamma_{CR}-1}\left(\gamma_{CR}uP_{CR}-D\frac{\partial P_{CR}}{\partial r}\right)\right];\ \frac{\partial}{\partial r}(3uP_W)=u\frac{\partial P_W}{\partial r}\pm V_A\frac{\partial P_{CR}}{\partial r}\cos\alpha\,.$$

Because the magnetic field is frozen in moving plasma, we obtain from the first equation of (7.2) $H=H_0(r_0u_0/ru)$, for magnetic pressure $P_M=H^2/8\pi\approx P_{M0}(u_0/u)^2$, for Alfven velocity $V_A=V_{A0}\sqrt{u_0/u}$, for $\cos\alpha=(u/u_0)\cos\alpha_0$ and on the basis of (7.2) we obtain

$$\rho u=\rho_0u_0\,;\ P_{CR}=P_{CR0}+\rho_0u_0(u_0-u)+P_{M0}\left[1-(u_0/u)^2\right]; \tag{7.3}$$

$$D\left[1-\frac{2P_{M0}}{\rho_0u_0^2}\left(\frac{u_0}{u}\right)^3\right]\frac{\partial u}{\partial r}=(u-u_0)(u-u_\infty)\left[\frac{\gamma_{CR}+1}{2}+\frac{P_{M0}(2-\gamma_{CR})}{\rho_0uu_\infty}\right]; \tag{7.4}$$

$$\frac{u_\infty}{u_0}=\frac{1}{2}\frac{\gamma_{CR}-1}{\gamma_{CR}+1}+\frac{\gamma_{CR}}{\gamma_{CR}+1}\frac{P_{CR0}+P_{M0}}{\rho_0u_0^2}+\sqrt{\left(\frac{1}{2}\frac{\gamma_{CR}-1}{\gamma_{CR}+1}+\frac{\gamma_{CR}}{\gamma_{CR}+1}\frac{P_{CR0}+P_{M0}}{\rho_0u_0^2}\right)^2+\frac{2P_{M0}}{\rho_0u_0^2}\frac{2-\gamma_{CR}}{\gamma_{CR}+1}} \tag{7.5}$$

We can find the pressure of Alfven turbulence P_W from the equation

$$2u\frac{\partial P_W}{\partial u}+3P_W=V_{A0}\rho_0u_0\cos\alpha_0\left(\frac{u}{u_0}\right)^2\left[\frac{2P_{M0}}{\rho_0u_0^2}\left(\frac{u_0}{u}\right)^3-1\right] \tag{7.6}$$

at the boundary condition $P_W(u=u_0)=0$:

$$P_W=V_{A0}\rho_0u_0\cos\alpha_0\left[\left(\frac{1}{4}+\frac{P_{M0}}{\rho_0u_0^2}\right)\left(\frac{u_0}{u}\right)^2-\frac{P_{M0}}{\rho_0u_0^2}\left(\frac{u_0}{u}\right)^3-\frac{1}{4}\right]\left(\frac{u}{u_0}\right)^2. \tag{7.7}$$

Because $P_{M0}/\rho_0u_0^2\approx 1/450$ and according to (7.5) $u_0/u<7$ for $\gamma_{CR}=4/3$, we obtain approximately

$$P_W\approx(1/4)V_{A0}\rho_0\cos\alpha_0\left[(u_0/u)^2-1\right](u/u_0)^{1/2}. \tag{7.8}$$

Cosmic ray diffusion coefficient is determined by

$$D(p)=D_{||}(p)\cos^2\alpha+D_\perp(p)\sin^2\alpha\,;\ D_{||}(p)=\frac{V^2}{3\nu(p)}\,;\ D_\perp(p)=\frac{V^2}{3\omega_H}\frac{\nu(p)}{\omega_H}, \tag{7.9}$$

where .

$$\omega_H=ZeHc/E(p)\,,\ \nu(p)=\pi\omega_H\omega_RW(\omega_R)/4P_M\,,\ \omega_R=(V_A/V)\omega_H\,. \tag{7.10}$$

Approximately $\omega_RW(\omega_R)\approx 2P_W$ and for $E(p)\approx 2\ GeV$ $(V\approx c)$, $r_0=60AU$ we find the expected dependence of D on r (see Fig. 15). In Fig 16 is shown the expected dependence of $P_{CR}/\rho_0 u_0^2$ and u/u_0 from r at $r \geq r_0 = 60AU$

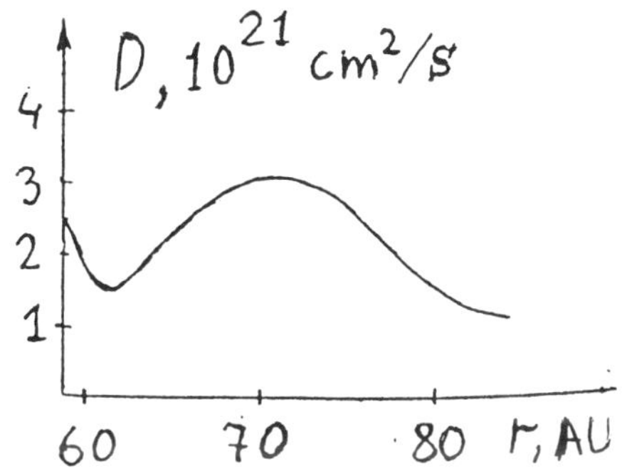

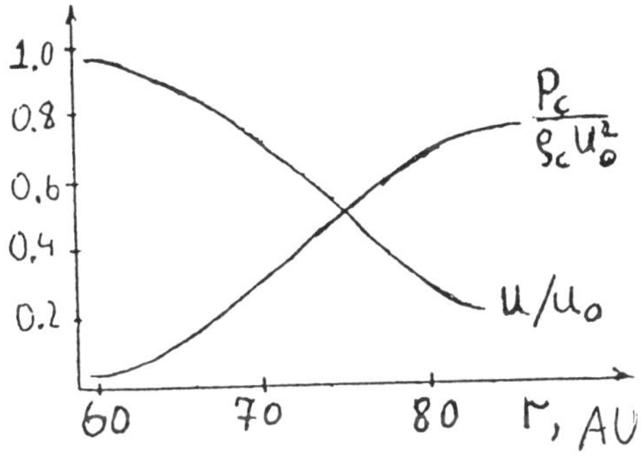

Fig. 15. Expected cosmic ray radial effective diffusion coefficient D in units 10^{21} cm^2/s as function of heliocentric distance r .

Fig. 16. Ratio of cosmic ray pressure to initial dynamic pressure of the solar wind and ratio of the solar wind speed u to initial speed u_0 = 400 km/s as functions of distance r .

It follows from (7.4) that the terminal transition will be of shock wave type only in the case when

$$1 - \frac{2P_{M\,0}}{\rho_0 u_0^2}\left(\frac{u_0}{u}\right)^3 = 0 \text{ at } u_\infty < u < u_0 \quad . \tag{7.11}$$

When the condition (7.11) can't be satisfied, the terminal transition must be of gradual type (without formation of terminal shock wave). For $P_{M\,0}/\rho_0 u_0^2 \approx 1/450$, (7.11) becomes equal to 0 at $u_0/u \approx 6.1$. Therefore if $u_0/u_\infty > 6.1$ we obtain shock wave transition. The value u_0/u_∞ is determined by Eq.(7.5) . If we consider cosmic rays as relativistic gas $(\gamma_{CR} = 4/3)$ then $u_0/u_\infty \approx 7$ and we expect shock wave transition. The problem is that also non relativistic cosmic rays contribute to cosmic ray pressure and stream instability and in this case the value u_0/u_∞ becomes smaller and for final solution we need some additional analyses.

8. Conclusion.

It is shown that cosmic ray nonlinear effects (pressure and stream instability) considered here play a vital role in the heliosphere: influence on solar wind propagation; a role in formation of terminal shock wave and a boundary between solar plasma and interstellar medium, in generation of MHD waves and formation of cosmic ray diffusion coefficient and anisotropy; significant influence on cosmic ray propagation and modulation in interplanetary space. It is expected that these nonlinear effects can play an important role in dynamics of other stellar winds.

ACKNOWLEDGMENTS. I would like to thank Prof-s M. Shapiro, J. Wefel and R. Silberberg for the possibility to attend the famous School of Cosmic Ray Astrophysics, for interesting discussions and comments, and the staff of Ettore Majorana Center for very kind hospitality.
I would like to thank also Prof. N. Iucci and Dr. G. Villoresi for their kind invitation to work in Rome and stimulating discussions. This work was partly supported by 3-rd Rome University.

9. References

Axford I.W and Newman R.C., 1965. Proc. 9-th ICRC, London, Vol.1, p. 173.
Babayan V.Kh. and Dorman L.I., 1977. Proc. 15-th ICRC, Plovdiv, Vol. 3, p. 107, p. 113.
Babayan V.Kh. and Dorman L.I., 1979a. Proc. 16-th ICRC, Kyoto, Vol. 3,p. 117; p. 123.
Babayan V.Kh. and Dorman L.I., 1979b. Izvestia Ac. of Sc. of USSR, ser.. phys., Vol. 43, p. 793.
Babayan V.Kh. and Dorman L.I., 1981. Proc. 17-th ICRC, Paris, Vol. 3, p. 373.
Babayan V.Kh. and Dorman L.I., 1990. In Physics of the Outer Heliosphere, ed. S.Grzedzielski and D.E.Page, Pergamon Press, p. 204.
Babayan V.Kh., Dorman L.I., Ptuskin V.S., 1987. Proc Intern. Conf. on Plasma Physics, Kiev, Vol. 4, p. 228.
Belov A.V., Dorman L.I., Eroshenko E.A., Oleneva V.A., 1988. Geomagn. and Aeronomy (Moscow), Vol. 28, N 4, p. 550.
Belov A.V., Dorman L.I., Eroshenko E.A., Oleneva V.A., 1990. Proc. 21-th ICRC, Adelaida, Vol. 6, p. 52.
Belov A.V., Dorman L.I., Eroshenko E.A., Rudnev Yu.F., 1991. Proc. 22-th ICRC, Dublin, Vol. 3, p.457
Belov A.V., Dorman L.I., Eroshenko E.A., Rudnev Yu.F., 1993. Proc. 23-th ICRC, Calgary, Vol.3, p.641.
Berezinsky V.S., Bulanov S.V., Ginzburg V.L., Dogiel V.A., Ptuskin V. S., 1990. *Cosmic Ray Astrophysics* (Ed. V.L.Ginzburg). Fizmatgiz, Moscow.
Dorman I.V. and Dorman L.I., 1967. J. Geophys. Res., Vol. 72, p.1513.
Dorman I.V. and Dorman L.I., 1968. Geomagn. and Aeronomy, Vol. 8, p. 817; p. 1008.
Dorman I.V. and Dorman L.I., 1969. Izv. Ac. of Sci. of USSR, ser. Phys., Vol. 33, p. 1908.
Dorman L.I., 1974. *Cosmic Rays: Variations and Space Exploration.* North - Holland, Amsterdam.
Dorman L.I., 1975a. *Variations of Galactic Cosmic Rays.* Moscow Univ. Press, Moscow.
Dorman L.I., 1975b. *Experimental and Theoretical Principles of Cosmic Ray Astrophysics,* Fizmatgiz, Moscow.
Dorman L.I., 1978. *Cosmic Rays of Solar Origin.* VINITI, Moscow.
Dorman L.I., 1991. Nuclear Physics B (Proc. Suppl.), Vol. 22B, p. 21.
Dorman L.I., Iucci N., Villoresi G., 1994. Proc. 14-th ECRS, Balatonfured (in press).
Dorman L.I. and Miroshnichenko L.I., 1968. *Solar Cosmic Rays.* Fizmatgiz, Moscow. English translation: 1976, NASA, Washington, DC.
Dorman L.I., Ptuskin V.S. and Zirakashvili V.N., 1990. In Physics of the Outer Heliosphere, ed. S.Grzedzielski and D.E.Page, Pergamon Press, p. 205.
Dorman L.I. and Venkatesan D., 1993. Solar cosmic rays. Space Sci. Rev., Vol. 64, p. 183.
Drury L.O'C. and Volk H.J., 1981. Astrophys. J., Vol. 248, p. 344.
Ginzburg V.L., 1965. Astron. J. (Moscow), Vol. 42, p. 1129.
McDonough T.R., Bracc N.M., 1971. Icarus, Vol. 15, p. 505.
McKenzie J.F. and Webb G.H., 1984. J. Plasma Physics, Vol. 31, p.375.
Moraal H., 1993. Nuclear Physics B (Proc. Suppl.), Vol. 33A,B, p. 161.
Semar C.L., 1970. J. Geophys. Res., Vol. 75, p. 6892.
Simpson J.A., 1963. Proc. 8-th ICRC, Bombay, Vol.2, p. 152.
Zirakashvili V.N., Dorman L.I., Ptuskin V.S., 1991. Proc. 22-th ICRC, Dublin, Vol. 3, p. 585.

PULSARS: RECENT RESULTS ON THEIR PHYSICS AND THEIR DYNAMICS

F. Graham Smith
Jodrell Bank
University of Manchester
United Kingdom

1. INTRODUCTION

A neutron star is most simply described as a very condensed, inert, and cold star, with mass between 1 and 2 solar masses and a radius of 10 km. There is an outer crust, which is a strong crystal lattice of heavy nuclei, and the interior is a superfluid consisting of neutrons with a small proportion of electrons and protons. The composition in no way reflects the composition of the star which was the progenitor of the neutron star, and gives no clue as to the details of the process of collapse in which it was formed. There are however two important legacies from the progenitor: spin and magnetic field. Angular momentum is certainly conserved in the collapse. Magnetic flux is probably also conserved: the outcome is a star with a polar magnetic field of $10^{8\text{-}12}$ gauss and a rotational period between 1.5 milliseconds and 4 seconds. Without rapid spin and high magnetic field, a neutron star would be almost unobservable; only the nearest might be detectable in X-ray telescopes by their thermal radiation.

The pulsed radiation which we observe at radio wavelengths, and now also in optical, X-ray, and gamma-ray radiation, is generated in beams which originate over the magnetic poles. The strong electric field generated by the rotation extracts charged particles from the surface and creates a magnetosphere which extends from the surface out to the velocity of light cylinder, where corotation would bring the velocity up to c. Two separate locations are involved in the radiation processes; the first part of this paper reviews the evidence for placing the radio source at a radial distance about one tenth of the radius R_L of the velocity of light cylinder, while the high energy source is at 0.8 or 0.9 R_L.

Timing the arrival of pulses from some 300 of the known pulsars is now a routine observation. The rotation rate is easily observed to decrease according to a simple power law, which relates to the large scale configuration of the magnetic field and the flux of particles. There are also steps in the rotation rate, known as the glitches; these give an extraordinary insight into the physics of the interior of the pulsars. Recent work here is concerned with the youngest known pulsar, the Crab, and with some of the older pulsars which have now been observed long enough to obtain a statistical picture of their glitches.

A major survey of the southern sky for pulsars is in progress at the Parkes radio telescope in Australia. The survey uses rapid sampling, which allows observation of millisecond pulsars; there is evidently a large population of these distributed throughout the galaxy. The total catalogue of pulsars now exceeds 600. An increasing proportion are found to be in binary systems; one third of the millisecond pulsars in globular clusters are binary.

M. M. Shapiro et al. (eds.), Currents in High-Energy Astrophysics, 209–224.

Finally, the velocities of pulsars are now found to be about twice as large as previously reported. The origin of such large velocities is a mystery since there is no obvious velocity asymmetry in supernova remnants. A considerable proportion of pulsars will leave the galaxy, or at least form an extended halo. Is this related to the unknown origin of the gamma-ray bursts?

2. THE LOCATION OF THE RADIO EMISSION

The magnetosphere of a typical pulsar is sketched in Fig. 1. The magnetic field is not much distorted by the plasma, and it is shown as a pure dipole. Field lines which originate near the equator close within the light cylinder and charged particles within this region cannot escape from the magnetosphere. Field lines from the polar region cross the light cylinder, and there is a flow of particles across the cylinder. It is this flow which provides the energy for the observed optical and X-ray radiation from the Crab Nebula.

The shaded area in Fig. 1 is the location of the radio emission. For a pulsar with a 1 second rotation period, this is at a height of about 100 km, i.e. at about 10 stellar radii. The angular width of the polar region corresponds well with the observed overall beam widths; lower radio frequencies originate farther out, giving wider beams. The emission

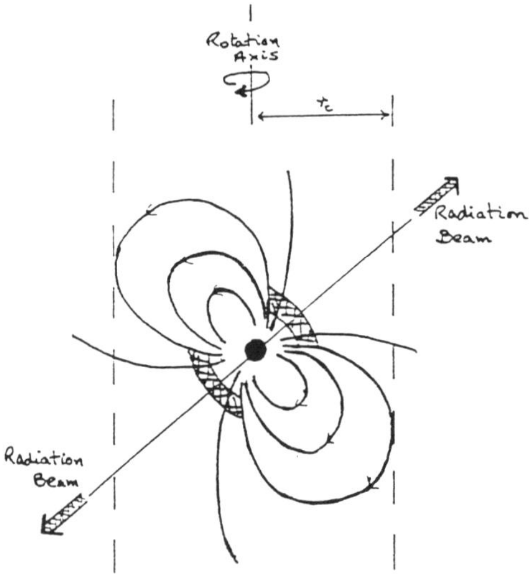

Figure 1. The Pulsar Magnetosphere

is not uniform over this polar cap region, and we shall later refer to the distribution of sources within the overall beam. There is remarkably little dependence of beam width or any other characteristic on rotation rate; millisecond pulsars have wider beams, but spectral index, power, and polarization are very similar over a range of 1000 to 1 in rotation rate. The radiation mechanism is curvature radiation. This has similar characteristics to synchrotron radiation, in which electrons are gyrating round a magnetic field line. In curvature radiation the field strength is so large that electrons can only stream along the field lines, all orthogonal motion being lost by radiation damping. The characteristic pattern of polarization is seen in Fig. 2a (Gil et al., 1992), as observed in a pulsar where only a single isolated region of the polar cap is radiating. A computed pattern for an ideal source of curvature radiation is shown in Fig. 2b. The reversing pattern of circular polarization is particularly significant in identifying the radiation mechanism.

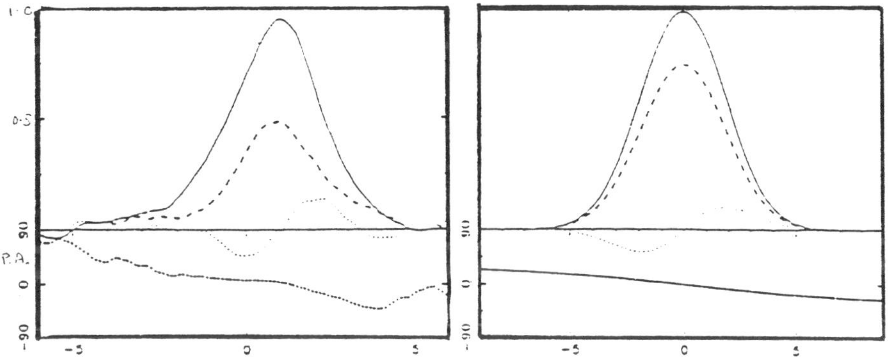

Figure 2. Pulse profile, showing intensity (full line), linear polarization (dashed) and circular polarization (dots). Position angle is shown in the lower panel. (a) Observed, (b) ideal curvature radiation.

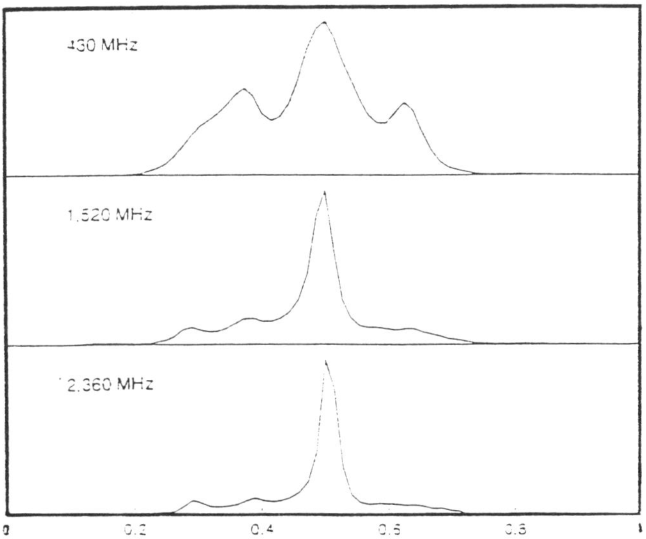

Figure 3. A typical pulse profile containing several discrete components.

A typical pulse profile, obtained by integrating some hundreds of individual pulses, is shown in Fig. 3. Incidentally, this pulsar is the recently discovered binary pulsar PSR J0437-4715, with a rotation period of 5.7 milliseconds and an orbital period of 5.7 days (Johnston et al., 1993). There are four components in this profile, which is a cut across the radiated beam. Is there any pattern in the arrangement of such components in the beam? There have been many attempts to classify patterns, especially according to their symmetry, and a distinction has been drawn between central components known as cores and outer components which are regarded as parts of cones. The most recent analysis confirms the view of Lyne & Manchester (1988) that the distribution of components across the beam is random and that no distinction should be made between core and cone components. There is, however, a gradient of spectral index and component width from the centre to the periphery.

3. THE SOURCE OF THE OPTICAL, X-RAY, and GAMMA-RAY EMISSION

Throughout most of the magnetosphere the very high electric fields induced by the rotation are shorted out by a charge separated plasma. There are two regions where there is a vacuum gap in the magnetosphere, where charged particles are accelerated to very high energies. Immediately over the magnetic poles such a gap covers the whole of the polar cap; within it there is a cascade process in which energetic electrons radiate gamma-rays, which then create electron-positron pairs which are themselves accelerated. Charged particles stream out along the field lines, and are responsible for the radio emission. The second region is located at the edge of the region of closed field lines, as shown in Fig. 4. A sketch of the cascade process in this gap in Fig. 5 shows a gamma-ray creating an electron-positron pair, as in the polar cap. In this case however the cascade tends to drift sideways, preserving the vacuum in the gap.

Why do we consider the radio and high energy regions to be distinct? The spectrum of emission from the Crab Pulsar (Fig. 6) shows a very clear gap between the steep spectrum of the radio and the flat spectrum in the optical and X-ray regions. Most importantly the intensity of the radio is so high that it must be coherent radiation in which bunches of particles move and radiate together. The high energy radiation is not

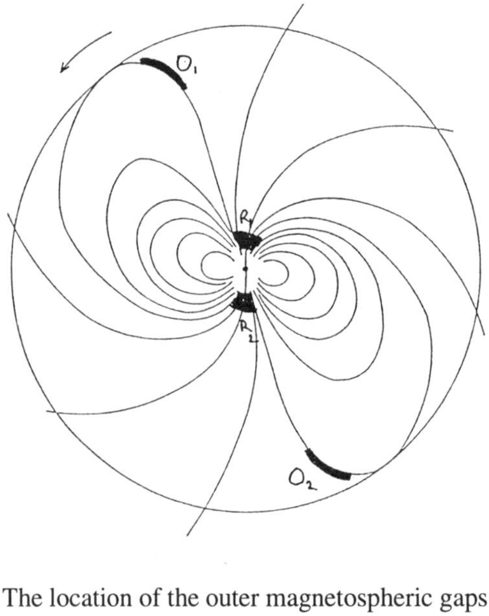

Figure 4. The location of the outer magnetospheric gaps O_1, O_2.

coherent, and it shows none of the structure of the radio, such as the fluctuations from pulse to pulse and the separate beam components. The beam shapes of the Vela pulsar, shown in Fig. 7 for radio and for optical and gamma-rays, are also very different for the two regimes.

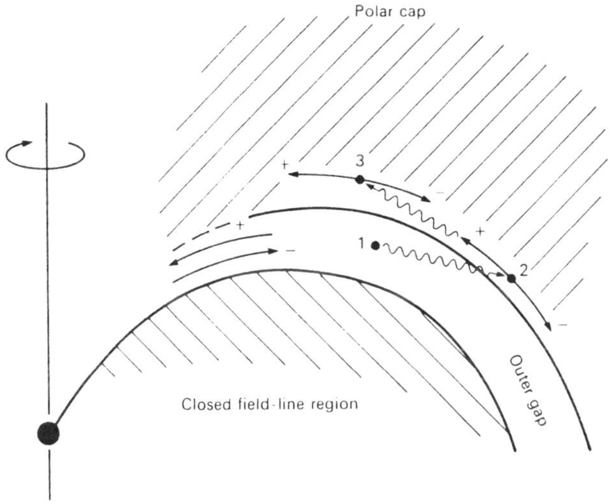

Figure 5. Acceleration and pair creation in the outer gap.

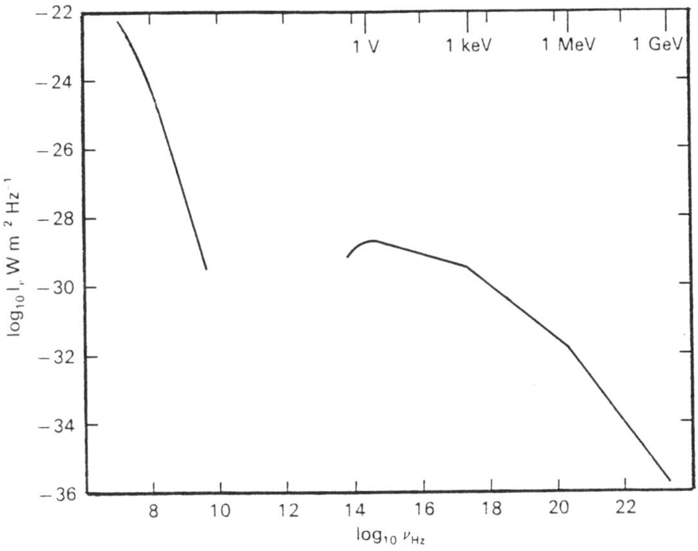

Figure 6. Spectrum of radiation from the Crab Pulsar.

An important test of the location and nature of the high energy emission region is the polarization of the optical emission, which has been measured in detail for the Crab Pulsar (Smith et al 1988). Detailed modeling based on this measurement places the emission region in the outer gap at a radial distance of 0.8 to 0.9 RL.

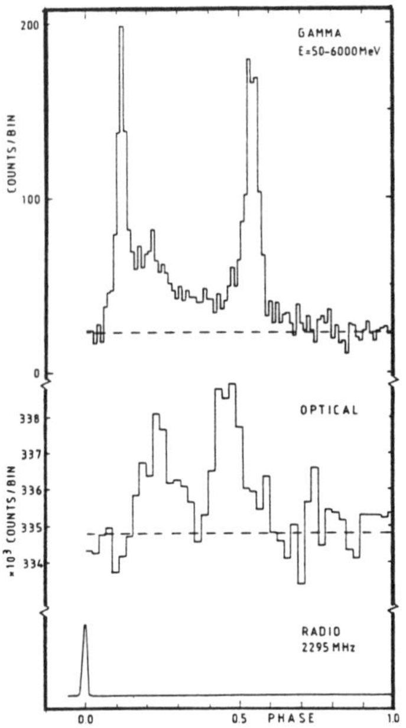

Figure 7. Pulse profiles of the Vela Pulsar

4. ROTATION SLOWDOWN

A rotating magnetic dipole radiates an electromagnetic wave at the rotation frequency ω, losing energy and slowing at a rate given by

$$I\,\omega\mathring{\omega} \propto -M^2\,\omega^4$$

where I is the moment of inertia and M is the component of the dipole perpendicular to the rotation axis. We therefore expect a slowdown following a power law where

$$\mathring{\omega} \propto \omega^n$$

the index n is expected to be 3. For three well-observed pulsars including the Crab, the indices are 2.04. 2.50 and 2.84. The difference from 3 must be related to the additional loss of energy in a particle stream; at least in the Crab Pulsar this is known to provide the energy for the optical and X-ray emission from the surrounding nebula.

The smoothest slowdown is observed in the millisecond pulsars. We would expect an isolated spinning body to behave according to a perfect power law; there are however irregularities in the form of low level noise and the steps known as glitches. The noise can be as low as that in the best terrestrial clocks. We may soon be able to establish a pulsar time scale smoother than 1 in 10^{14}.

Glitches are minor perturbations; even the largest steps in rotation rate in pulsars such as the Vela Pulsar are only at the level of 1 in 10^6. We now have data on glitching in some dozens of pulsars and can see an interesting pattern developing from the youngest to some of the oldest (but not yet in the millisecond pulsars which appear to be

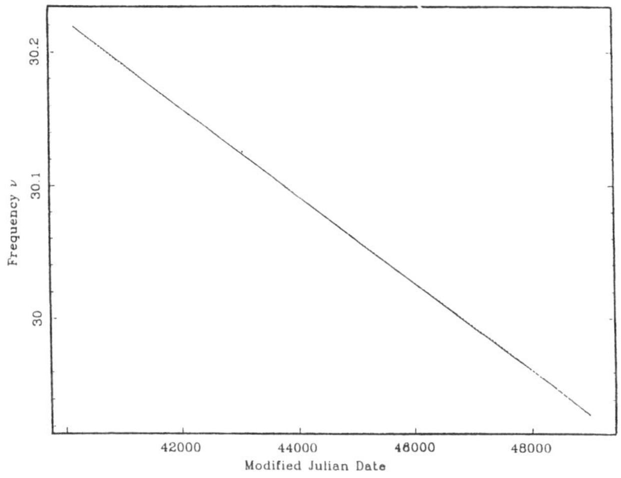

Figure 8. The rotation rate of the Crab Pulsar.

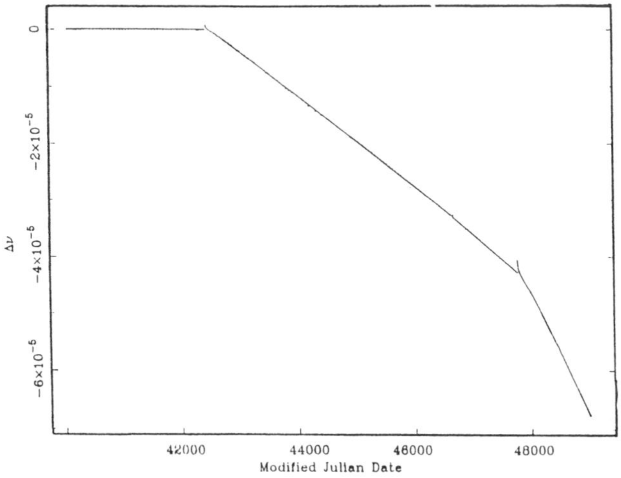

Figure 9. As Fig. 8 with magnified vertical scale and initial slope removed.

very stable). The youngest is the Crab Pulsar. Its slowdown and glitching behavior is shown in the plots of rotation rate in Figs. 8, 9 and 10 (Lyne, Smith & Pritchard, 1993).

Fig. 8 shows the slowdown over a period of 23 years. On this scale the slowdown appears to be perfectly linear. We now concentrate on differences from a linear slowdown by fitting a Taylor series to the first few years of slowdown and plot the differences obtained in Fig. 9. The scale is now expanded by a factor of 4000 and the two large glitches are now seen as changes in the slope. For both glitches the slowdown rate increases. Expanding further in Fig. 10, by removing the slope between glitches, shows that there are transient effects at and immediately after the glitches; contrary to our

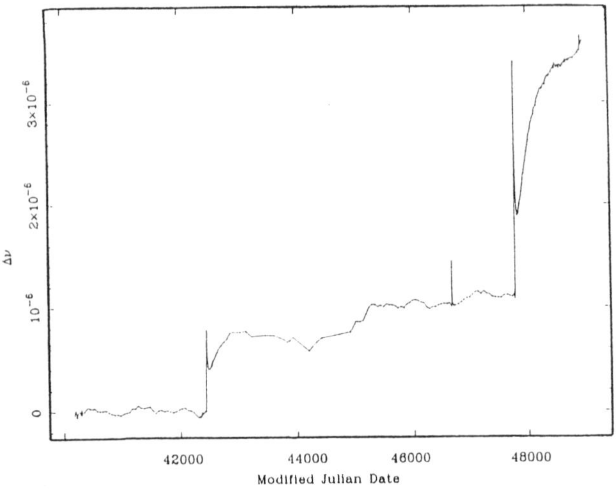

Figure 10. As Fig. 9, again magnified, with slopes removed.

original impressions, these are not the main effect of a glitch which is the step increase in slowdown rate. There are at least three transients involved, with time constants from 1 to 100 days.

In older pulsars the behavior is quite different. Fig. 11 shows the series of glitches observed in PSR 1737-30. Here the plot shows the residuals in pulse arrival times, i.e. the phase of the rotator compared with a model with uniform slowdown. These glitches are steps in rotation rate and phase, but not in slowdown rate.

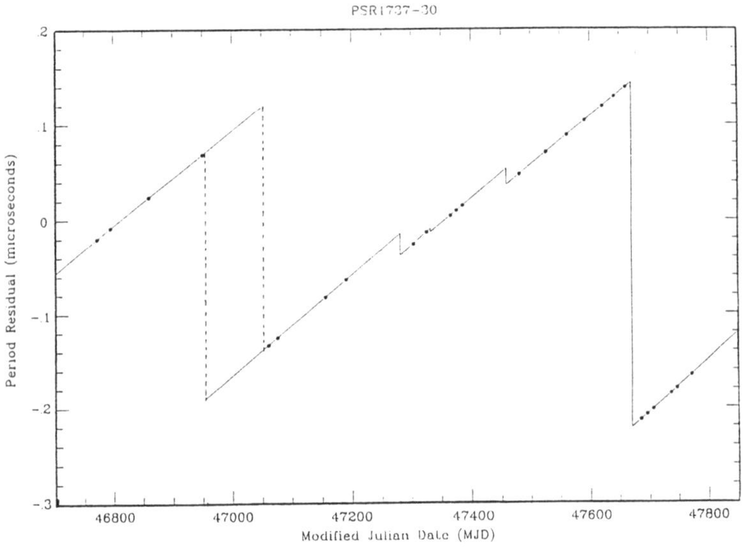

Figure 11. Glitches in PSR 1737-30.

The behavior of the Vela Pulsar, in which the glitches are steps in rotation rate of 1 in 10^6, is shown in Fig.12. The glitches are so large and frequent that they can only be described in terms of steps in rotation rate, with partial recovery which is incomplete at the next glitch; it is impossible to see any changes in slowdown rate.

The glitching behavior of pulsars can only be interpreted in terms of the rotation of the superfluid interior. The main bulk of the neutron superfluid is firmly attached to the solid crust through the strong magnetic field, since the neutron fluid contains an equilibrium proportion of electrons and protons, and is therefore highly conducting.

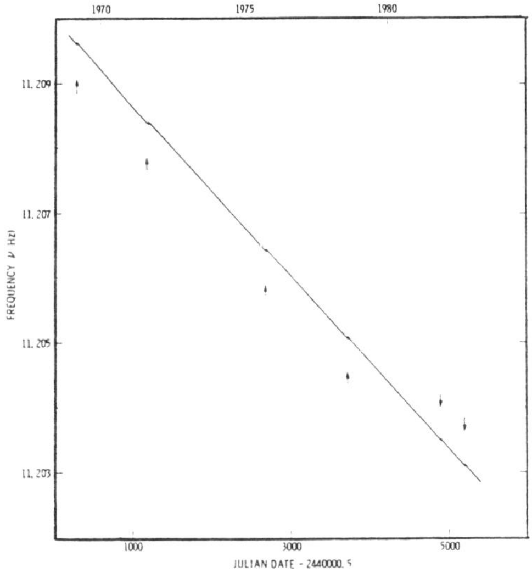

Figure 12. Glitches in the Vela Pulsar (after Cordes et al., 1988).

There is however a part of the neutron fluid which is a $^1S_0$ superfluid, which is not a conductor. This permeates the inner part of the crust and can move independently of the rest of the star. It contributes about 2% of the total moment of inertia. As the rotation of the star slows, this portion of the superfluid can continue to rotate faster until its coupling with the crystal lattice is suddenly increased at the time of a glitch. The rotation of the fluid is quantized, and the angular momentum appears as vortices each carrying a quantum of angular momentum; the area density of vortices gives the angular velocity of the superfluid. The vortices can move steadily outwards, corresponding to a steady slowdown, or they can pin onto the lattice; in this latter case they cannot move outwards, their density is constant, and the angular velocity of the superfluid is constant.

The sequence of events now seems to be as follows. In the Crab Pulsar, which at an age of 940 years is the youngest pulsar, the $^1S_0$ superfluid is becoming progressively pinned at each glitch. This component of the moment of inertia is therefore becoming progressively detached, and the rate of slowdown of the whole star increases at each step. This will proceed for only a limited time; during our observations over 25 years, the effective total moment of inertia has already decreased by 0.05%, so that after less than 1000 years the whole of this superfluid component, i.e. 2% of the star s moment of inertia, would be pinned. Note however that the rotation of this superfluid is fixed at a high rate, and the differential between the rotation of the superfluid and the lattice is increasing. Eventually the vortices tear away from the lattice in a catastrophic event and move outwards until the equilibrium vortex density is again reached. In doing so, they transfer angular momentum to the star whose rotation increases in a major glitch. This is the situation in an older pulsar, the Vela Pulsar, which is about 10,000 years old.

In the Vela pulsar, and to a lesser extent in the Crab, we see a relaxation following the glitch. This indicates that part of the vortex array in the superfluid is normally creeping outwards against the tendency to pin to the lattice; at the glitch the whole becomes pinned and the relaxation is the re-establishment of the creep. The observation of a long relaxation time constant, typically some tens of days, is strong support of the basic assumption of superfluid behavior. Finally, the older pulsars are showing only a series of steps in rotation rate, with little or no steps in rotation rate and no exponential recovery.

This outlined sequence is already complicated, but there will be more to explain as more observations are made. For example, it already appears that PSR1509-58, which is only 2000 years old, is not glitching, at least on the scale of either the Crab or the Vela pulsars. Can this be regarded as having completed the Crab phase and not yet having reached the Vela phase?

5. THE GALACTIC POPULATION

The pulsar catalogue (Taylor, Manchester, and Lyne, 1993) is updated at frequent intervals, and is available by e-mail. There is about 600 entries; the number is increasing mainly as the Parkes survey of the southern sky is analyzed. Fig. 13 shows the distribution over

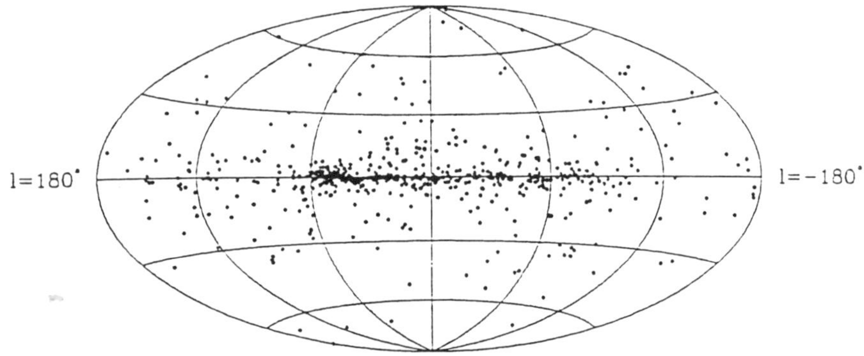

Figure 13. The galactic coordinates of known pulsars.

the sky of 558 pulsars. There are some obvious selection effects; for example, there is a concentration on the galactic plane near longitude 50°, which is where the plane goes overhead at Arecibo. There is a real concentration to the plane, and especially to the galactic centre, but the difficulty of observing through the ionized interstellar medium discriminates against short period pulsars in this general direction. The Parkes survey is arranged to improve sensitivity for millisecond pulsars, and is extending to high galactic latitudes. This survey has so far discovered 78 pulsars.

The millisecond pulsars form an almost entirely separate population. Fig.14 shows the catalogued pulsars plotted according to period P and its first differential $\dot{P}$. Young pulsars, like the Crab and Vela, are found in the top left of this plot; as they age, the normal pulsars move down and to the right. Their lifetime is of order 10^7 years; after that, their radio signal fades or cuts off, and they are no longer observable. The pulsars at the bottom left of the plot, those with short periods and with the smallest slowdown rates, are the millisecond pulsars. These are generally supposed to be derived from the normal population by a process of spin-up in a binary system; indeed many of them still have binary partners.

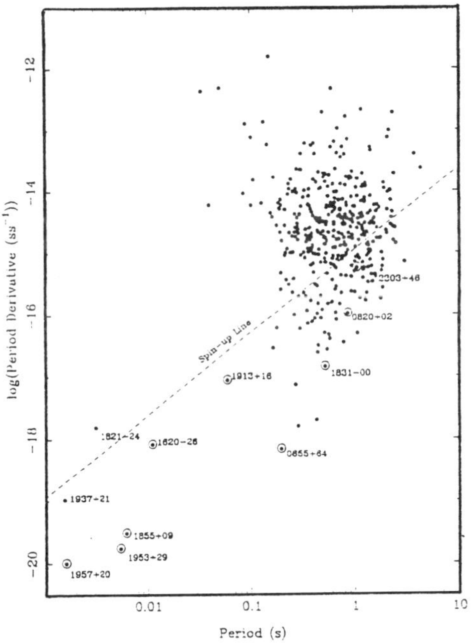

Figure 14. Pulsars plotted by period P and slowdown rate $\dot{P}$.

The concentration of X-ray binaries in the globular clusters, and the hypothesis that X-ray binaries represented the process of spin-up, led to the discovery of a large population of millisecond pulsars in the globular clusters. There are in fact 12 such X-ray sources, but no fewer than 33 pulsars known in a total of 13 globular clusters. Of these, one third are in binary systems.

The total population of pulsars is of order 100,000. It is more difficult to find the population of millisecond pulsars in the whole of the galaxy. The Parkes survey has now found a total of 25, and a new estimate leads to a total population of around 30,000. It is too early to establish whether they all originated in supernova explosions or in binary spin-ups, but there are suspicions that this is not the whole story.

6. BINARY PULSARS

Almost all of the main population of pulsars are solitary. The millisecond pulsars, on the other hand, are often in binary systems, and it is possible that all millisecond pulsars originated in binaries, some of which have subsequently been disrupted.

The signature of a binary pulsar system is easily recognized. Fig. 15 shows the variations of the measured period of PSR J0437-4715 over an interval of 15 days. (This is the same recently discovered pulsar whose pulse profile we saw in Fig. 2; it is worth a reminder, as it is one of the closest known pulsars, and a good candidate for detection in gamma rays.) It has period of 5.7 milliseconds, which varies sinusoidally with a period

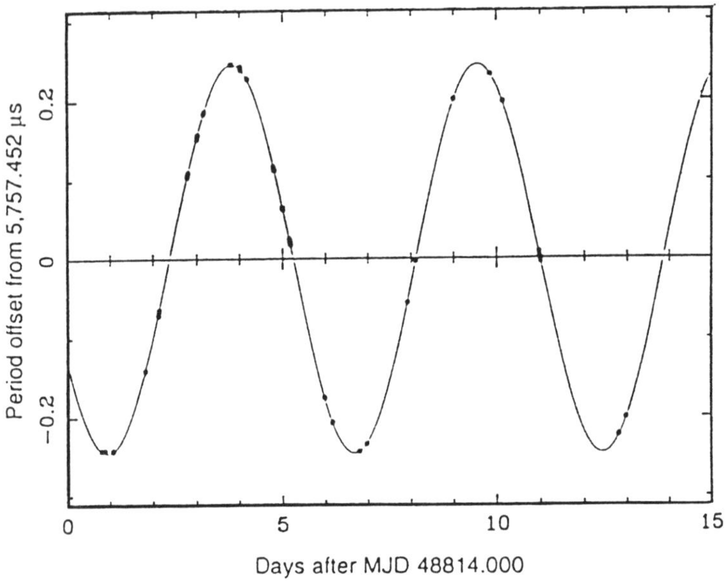

Figure 15. A binary system with low eccentricity.

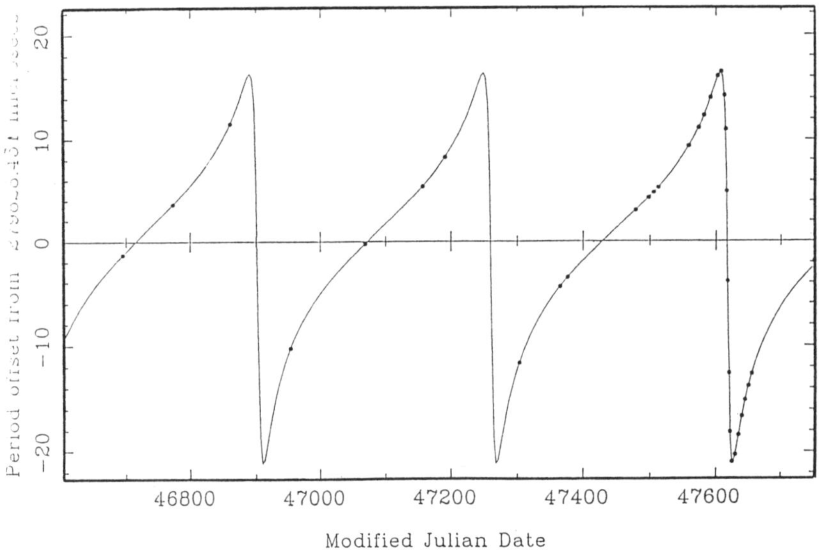

Figure 16. A binary system with high eccentricity.

of 5.7 days and an amplitude of 0.2 microseconds. This is an example of a pulsar in a circular orbit. The orbit is a remarkably good circle, with an eccentricity less than 10^{-6}.

There are also highly elliptical orbits as in PSR 1820-11 seen in Fig. 16. Eccentricities up to 0.967 have been measured. Binary pulsar orbits include some of the most eccentric as well as some the most circular orbits of any astronomical bodies.

The masses of several neutron stars have been measured, notably the two neutron stars in the relativistic binary PSR 1913+16. In all cases the mass is close to 1.4 solar masses, and we use this value in obtaining the mass of a binary companion from the period modulation. The orbit inclination, i, is unknown and we have available only the mass function

$$f(m_p, m_c, i) = m_c^3 (m_p + m_c)^{-2} \sin^3 i$$

Despite this limitation, it is evident that there are two distinct classes of companion with low and high mass. Furthermore these two classes have distinct ranges of eccentricity. The two classes are:

Low Mass, with circular orbits;
High Mass, with elliptical orbits.

There are now several optical identifications of pulsar companions, and these confirm the division. The low mass binary companions are all condensed stars, mainly white dwarfs, while the high mass companions are young bright stars, typically of type Be (where e indicates emission lines). The Be stars are believed to be the forerunners of X-ray binaries.

As might be expected, nearly all the globular cluster binaries have low mass companions. There is an exception in the recently discovered PSR 1718-19 which has an orbital period of 1 second and a high mass companion. The occurrence of a high mass star in a globular cluster is a surprise; the cluster stars are all supposed to be old and highly evolved. Another surprise is the discovery of a low mass binary with an elliptical orbit, again in a globular cluster; this is PSR 1802-07 with $e = 0.21$.

Most binary orbits bring pulsars very close to their companions; the closest orbits would be contained within the radius of an ordinary main sequence star. In several cases such as PSR 1744-24A the radio pulses are attenuated or lost in a stellar wind from the companion; in the Be stars this is the normal stellar wind, but for the white dwarfs, the wind must be generated by the incidence of radiation from the pulsar on the white dwarf; the pulsar is in the process of evaporating its companion. This may be the process which ends up with an isolated millisecond pulsar.

7. SUPERNOVA REMNANTS

There are now nine identifications of young pulsars with supernova remnants (SNRs). Although the main population of pulsars is believed to have originated in supernova explosions, we do not expect many identifications with SNRs, since the remnants fade from sight after some tens of thousands of years. It has nevertheless been surprisingly difficult to identify young pulsars with viable remnants; the reason is now emerging. There has recently been an identification of PSR 1758-24 with the remnant G5.4-1.2 (Frail & Kulkarni, 1991). This is an extensive remnant possibly extending to 30' diameter, with a bright knot at one side. The pulsar is located at the extreme edge of this knot. If this is a correct identification, the pulsar has been moving very rapidly from its original location in the centre of the SNR. Another example is shown in Fig. 17, where PSR B1338-62 is located in the shell of the SNR G308.8-0.1 (Kaspi et al., 1992), about 10' from the centre.

These identifications suggest that pulsars leave the site of their birth with very large velocities. There are two ways of measuring pulsar velocities using either their proper motion or their scintillation characteristics. Both methods have until recently underestimated pulsar velocities by a factor of about two. Proper motion has been

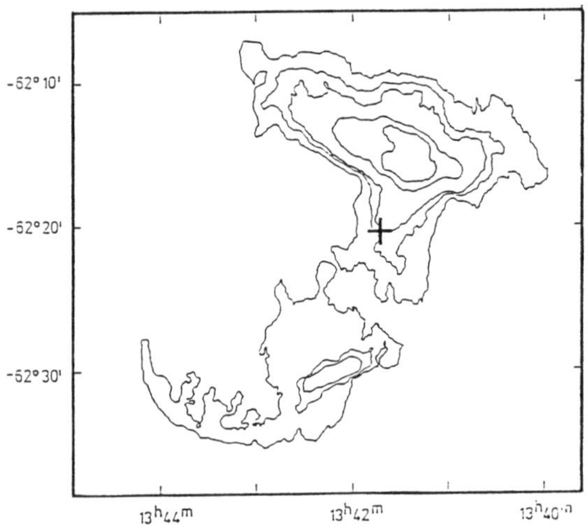

Figure 17. Location of PSR B1338-62 in a Supernova remnant.

measured using interferometry for some dozens of pulsars, for example by Harrison, Lyne and Anderson (1993). The problem here has not been the actual angular measurements but the distances of the pulsars. A revision of the distance scale particularly for the closest pulsars has resulted from a reassessment of the interstellar electron density which gives distances from the observed dispersion measures.

The revision of the scintillation distances also depends on the distribution of interstellar electrons. Scintillation is due to the irregularities in the interstellar gas rather than the mean density. There is now good evidence that the irregularities are concentrated to the galactic plane and are therefore closer to the Sun than previously estimated. The effect is to reduce the observed scintillation velocities by a leverage factor.

The present estimates of pulsar velocities from proper motion and scintillation measurements give a mean space velocity of 450 km s^{-1}. This is now high enough to explain the large displacements in the SNR associations which require a mean velocity of 690 km s^{-1} (but not with high statistical accuracy!).

At such high velocities we expect a large proportion of pulsars to leave the galaxy. The observed directions of their proper motions confirm that this is happening; Fig. 18 shows the predominant motion away from the plane in measurements by Harrison, Lyne, and Anderson (1993). Old neutron stars must exist in a large galactic halo; perhaps related to the problem of identifying the gamma-ray bursters.

If all pulsars have high velocities it is difficult to account for the existence of pulsars in the globular clusters. Here the escape velocity is only about 50 km s^{-1} and the present population should have evaporated long ago.

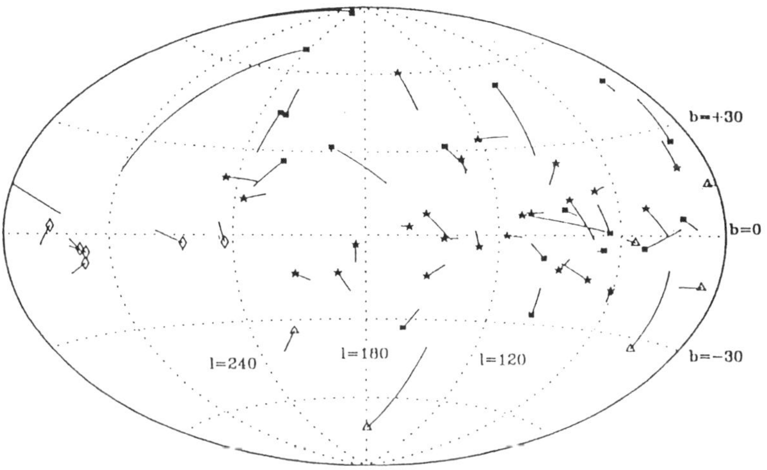

Figure 18. Proper motions of pulsars, showing a predominant motion away from the galactic plane.

These high velocities are hard to account for in the standard theory of supernova explosions. Naturally the theory starts with a symmetrical collapse, and it may be that a more sophisticated approach would reveal an instability that favored a growing asymmetry. There would then remain an observational problem: if a pulsar with 1.4 solar mass is ejected, we would expect to see a mass carrying a similar inertia in the opposite direction. This is not so far observed, either as a rapidly moving small component or as a smaller general velocity of the whole remnant.

8. PLANETS ROUND PULSARS

There has recently been a clear confirmation of the existence of planetary bodies round PSR 1257 + 12 (Wolszczan & Frail, 1992). The complicated pattern of Doppler shifts in period for this pulsar (Fig. 19) shows that there are three planets in orbit, with periods nearly, but not quite, in the ratios of 4:2:1. These orbits are expected to interact gravitationally, and it is the observation of the evolution of the orbits that has provided the confirmation of the planetary system. Another such system is known to exist, in the young PSRB 1828-11; although details of the orbits are not yet as certain, it also has three planets in the same ratio of periods.

These planetary systems provide another challenge to the standard theory of pulsar birth in supernovae. It seems unreasonable to suggest that such a system of planets could survive a Type II supernova explosion which ejects the major part of a 10 solar mass star with very high velocity. The quieter collapse of a white dwarf as a result of accretion may provide a more reasonable scenario, but it still remains to account either for the survival of the planets or their assembly in the turmoil immediately following the collapse.

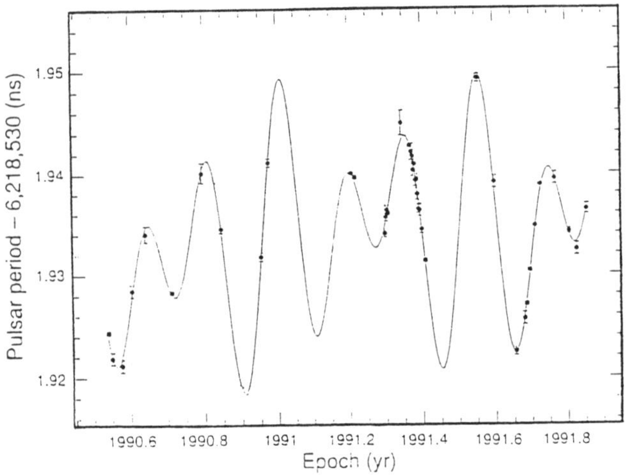

Figure 19. Doppler shifts due to the planetary system round PSR 1257+12.

References

Cordes J.M., Downs G.S. & Kraus-Polstorff, J., *Astrophys. J.* **330,** 847 (1988).
Frail D A. & Kulkarni S.R., *Nature* **352**, 785 (1991).
Gil J.A., Lyne A.G., Rankin J.M., Snakowski J.K., & Stinebring D.R., Astron. Astrophys. **255** 181 (1992).
Lyne A.G. & Anderson B.A., *MNRAS* **261**, 113 (1993).
Johnston et al., *Nature* **361,** 613 (1993).
Kaspi V.M., Manchester R.N., Johnston S., Lyne A.G., & D'Amico N., *Astrophys. J.* **399,** L155 (1992).
Lyne A.G. & Manchester R.N., *MNRAS* **234**, 477 (1988).
Lyne A.G., Pritchard R.S., & Smith F. Graham, *MNRAS* **265**, 1003 (1993) .
Smith F. Graham, Jones D.H.P., Dick J.B. & Pike C.D., *MNRAS* **233**, 305 (1988).
Taylor J.H., Manchester R.N., & Lyne A.G., *Astrophys. J. Suppl. Ser.* **88**, 529 (1993).
Wolszczan A., & Frail D. A., *Nature* **355**, 145 (1992).

SYNCHROTRON AND INVERSE COMPTON RADIATION IN SUPERNOVA REMNANTS

OCKER C. DE JAGER
Space Research Unit
Potchefstroom University for CHE
Potchefstroom 2520, South Africa

ABSTRACT. This lecture serves as an introduction to synchrotron and inverse Compton processes in pulsar-driven Supernova Remnants. TeV to PeV electrons are injected by a central pulsar into the remnant, which is observed as an X-ray synchrotron nebula. However, the same electrons also scatter soft photons to gamma-ray energies, and the discussion below allows the reader to calculate the expected gamma-ray flux, which can be compared with space-borne (MeV to GeV) and ground-based (TeV to PeV) observations. Such observations allow us to estimate the magnetic field strength of the nebula.

1. Introduction

Textbooks on SNR generally deal with the synchrotron features as seen in radio (from all SNR - young and old), but also in optical and X-rays as seen from the young pulsar-driven remnants such as the Crab Nebula. However, the same electrons spiralling around a magnetic field line (resulting in synchrotron radiation) can also scatter the soft synchrotron photons (radio to optical) (Gould 1965), the 2.7K background photons, and thermal photons from heated dust grains near SNR to gamma-ray energies.

The synchrotron spectrum depends on the magnetic field strength B and the electron spectrum, whereas the inverse Compton spectrum depends on the same electron spectrum, as well as the spectrum of the soft photon field. The latter is usually known, so that we obtain a measure for B if both the synchrotron and inverse Compton spectra are observed. For the Crab it was shown that B obtained in this way is consistent with B obtained from other considerations (De Jager & Harding 1992 (DH)).

A good review on synchrotron and inverse Compton radiation in general can be found in the textbook of Rybicki & Lightman (1979). A summary of the relevant aspects will be given in this lecture. The last part of this discussion is concerned with the sources of soft photons, which can act as targets for IC scattering. In some cases we may be confronted with the possibility of IC losses dominating synchrotron losses, resulting in an *inverse Compton nebula*, as opposed to the well-known con-

M. M. Shapiro et al. (eds.), Currents in High-Energy Astrophysics, 225–234.

cept of a synchrotron nebula, since most of the electron energy is then channeled into γ-rays.

2. Synchrotron Radiation

Any charge that accelerates is known to emit electromagnetic radiation. The special case where the electron spirals around a magnetic field line in a helix motion gives rise to synchrotron radiation. The frequency spectrum (or radiated power per unit frequency) is given by the Fourier transform of the Poynting flux, integrated over all directions:

$$P(\nu)d\nu = \frac{\sqrt{3}e^3}{mc^2} B_\perp \frac{\nu}{\nu_c} \int_{\nu/\nu_c}^{\infty} K_{5/3}(t)dt\, d\nu \quad (1)$$

Note that this expression has been given in c.g.s. units - the units mostly used in Astrophysics. The function K is the modified Bessel function, and $B_\perp = B\sin\theta$ is the perpendicular component of the magnetic field strength (measured in Gauss), with θ the pitch angle. Eqn (1) reaches a maximum near the *characteristic frequency*

$$\nu_c = \frac{3e}{4\pi mc} B_\perp \gamma^2 = 4.20 \times 10^6 B_\perp \gamma^2 = 6.26 \times 10^{18} B_\perp E^2 \text{ Hz} \quad (2)$$

where the electron energy is given by $E = \gamma mc^2$ (in units of ergs). Using this expression we can determine the typical electron energies that are responsible for synchrotron emission at a given frequency.

The total amount of energy radiated per second is obtained by integrating Eqn (1) over all frequencies giving

$$-(\frac{dE}{dt})_{\text{syn}} = \frac{2e^4}{3m^2c^3} B_\perp^2 \gamma^2 = 2.368 \times 10^{-3} B_\perp^2 E^2 \text{ ergs s}^{-1} = \frac{4}{3}\sigma_T c \gamma^2 \left[\frac{<B^2>}{8\pi}\right] \quad (3)$$

The last expression allows us to write dE/dt in terms of the magnetic energy density $<B^2>/8\pi$ for isotropic pitch angles, and $\sigma_T = (8\pi/3)r_0^2 = 6.65 \times 10^{-25}$ cm^2 (where $r_0 = e^2/mc^2$ is the classical electron radius) is the *Thomson cross section*. The *lifetime* of such an electron is given by the timescale

$$\tau_{\text{syn}} = \frac{E}{|dE/dt|} = \frac{422.3}{B_\perp^2 E} \text{ sec.} \quad (4)$$

Thus, a high energy electron with energy E will lose its energy relatively fast (τ_{syn} is relatively small). Initially, E, ν_c and $|dE/dt|$ will be large, but these three quantities will decrease with time as a result of losses. However, low energy electrons will take a much longer time to lose a significant amount of energy, and their corresponding characteristic frequencies ν_c (Eqn (2)) will also be much lower. We see this effect in pulsar-driven SNR where relativistic electrons are injected by a central pulsar: the radio nebula is large (as a result of relatively low values of E), but the nebular scale size shrinks with increasing ν_c. We see this effect clearly in the Crab Nebula when comparing the radio, optical and X-ray images. DH have shown that this phenomenon should continue into the γ-ray range, until the nebular size for high energy γ-rays is comparable to the size of the electron accelerating region.

3. Inverse Compton Scattering

To fully understand the fundamentals of inverse Compton scattering, we have to depart at Quantum Electrodynamics. This is however beyond the scope of this lecture, and we will just give the basic results as summarised by Blumenthal & Gould (1970) for the isotropic scattering case. See also Rybicki & Lightman (1979).

Consider a highly relativistic electron with energy $E = \gamma mc^2$ moving through a soft isotropic photon gas with number density $dn(\epsilon) = n(\epsilon)d\epsilon$. In fact, the same electrons that radiate synchrotron photons will also be able to scatter some of these ambient soft photons to γ-ray energies $\epsilon_1 = \gamma mc^2 \hat{\epsilon}_1 \gg \epsilon$. Here $\hat{\epsilon}_1 \leq 1$ is a convenient dimensionless quantity that determines the energy of the scattered photon in terms of E.

Jones (1968) has calculated the general production rate of γ-rays (for isotropic scattering) with energy ϵ_1 per electron with $\gamma \gg 1$ giving

$$\frac{dN_{\gamma,\epsilon}}{dt d\epsilon_1} = \frac{2\pi r_0^2 c}{\gamma^2}\frac{n(\epsilon)d\epsilon}{\epsilon}\left[2q\ln q + (1+2q)(1-q) + \frac{1}{2}\frac{(\Gamma_\epsilon q)^2}{1+\Gamma_\epsilon q}(1-q)\right] \quad (5)$$

where the dimensionless parameter $\Gamma_\epsilon = 4\epsilon\gamma/mc^2$ and $q = \hat{\epsilon}_1/[\Gamma_\epsilon(1-\hat{\epsilon}_1)]$. The parameter Γ_ϵ determines the domain of the scattering and for $\Gamma_\epsilon \ll 1$ we are in the *Thomson limit* whereas the *Klein-Nishina limit* corresponds to $\Gamma_\epsilon \gg 1$. If we look at the incident photon from the rest frame of the electron, then we see that the Thomson limit corresponds to $\lambda' \gg h/mc$, whereas the Klein-Nishina limit corresponds to $\lambda' \ll h/mc$, where λ' is the wavelength of the incident photon in the electron rest frame, and h/mc is the Compton wavelength.

The range of values of $\hat{\epsilon}_1$ follows from the kinematics of the problem $1 \gg \epsilon/E \leq \hat{\epsilon}_1 \leq \Gamma_\epsilon/(1+\Gamma_\epsilon)$ and the corresponding range for q is $1 \gg 1/4\gamma^2 \leq q \leq 1$. We can also express the scattered γ-ray energy in terms of its maximum value by introducing the parameter z (between 0 and 1) to give

$$\hat{\epsilon}_1 = \frac{\epsilon_1}{E} = \left[\frac{4\epsilon\gamma/mc^2}{1+4\epsilon\gamma/mc^2}\right] z \quad (6)$$

The parameter z is a stochastic variable with distribution given by Eqn (5) after the proper normalization. Apart from calculating inverse Compton spectra from Eqn (5), the latter equation also allows us to calculate the mean energy of a scattered photon: Fig. 1a shows the spectral distribution (the area under the expression in brackets in Eqn (5) has been normalized to unity) of the scattered photons in terms of z (from Eqn (6)) for the *Thomson limit* where $\Gamma_\epsilon \ll 1$ was taken, so that the average scattered energy for an isotropic photon gas is given by $<\epsilon_1> = \frac{4}{3}\gamma^2 <\epsilon>$. Fig. 1a also shows the spectral distribution for the *Klein-Nishina* limit (with $\Gamma_\epsilon \gg 1$). We see that this distribution is concentrated towards $z = 1$, so that the scattered

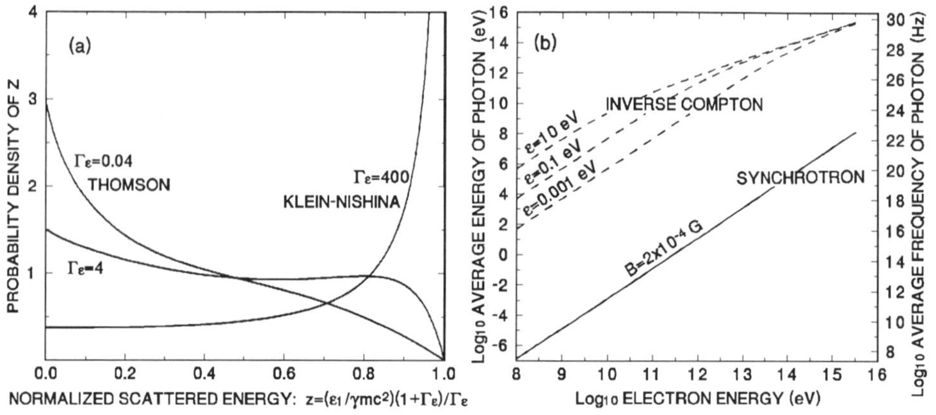

Fig. 1 (a) The probability density function (Eqn (5)) of the energy of an inverse Compton scattered photon in terms of z as given by Eqn (6) for small Γ_ϵ (Thomson limit), and large Γ_ϵ (Klein-Nishina limit). (b) The average energy of an inverse Compton scattered photon (short dashed lines). The soft photon energy ϵ is indicated above each curve, and the corrresponding electron energy is indicated by the horizontal scale. The solid line shows the average synchrotron energy (from Eqn (2)) of the same electron when moving in a magnetic field with strength as indicated above the solid line.

γ-ray receives nearly all the energy of the electron with $<\epsilon_1>\sim E=\gamma mc^2$, whereas the electrons lose only tiny amounts of energy per scattering in the Thomson limit. However, in the Klein-Nishina limit the production rate is suppressed relative to the Thomson limit, so that relatively fewer γ-rays are produced.

Fig. 1b shows the average scattered energy $<\epsilon_1>$ (calculated from the first moment of Eqn (5)) as a function of the electron energy E for three incident soft photon energies $\epsilon = 0.001$ eV, 0.1 eV and 10 eV as indicated above each corresponding dashed line in Fig. 1b. The range for E selected here corresponds to the those energies expected in a young Crab-like SNR, and the three ϵ values selected represent the range for which scatterings are mostly expected to take place in such SNR: at low electron energies we are in the Thomson limit where $\epsilon_1 \propto E^2$, but the slope flattens by one power towards higher energies resulting in the $\epsilon_1 \propto E$ relation in the Klein-Nishina limit. Also shown in Fig. 1b is the characteristic energy of synchrotron radiation from Eqn (2) as a function of E for $B_\perp = 2\times 10^{-4}$ G. We see here that synchrotron radiation produces much lower energy photons than obtained from inverse Compton scattering in the same environment.

The energy loss rate of an electron due to inverse Compton losses in the presence of a soft isotropic photon spectrum $dn_\epsilon = n(\epsilon)d\epsilon$ (or energy density defined by $U_\epsilon = \int \epsilon dn_\epsilon$) in the Thomson limit is then given by

$$-(\frac{dE}{dt})_{\rm IC} = \frac{4}{3}\sigma_T c\gamma^2 U_\epsilon \left[1 - \frac{63}{10}\frac{\gamma <\epsilon^2>}{mc^2 <\epsilon>}\right] \tag{7}$$

Note that Eqn (7) is similar to the loss rate given for synchrotron radiation in

Eqn (3), except that the magnetic energy density has been replaced with the photon energy density. The last factor in square brackets is the first-order correction to the Thomson limit, which is necessary if the condition $\gamma\epsilon \ll mc^2$ is violated.

4. Pulsar-driven Supernova Remnants

The *Plerions* or *filled-center* remnants may or may not show a shell interacting with the Interstellar Medium, but the distinctive feature here is the presence of a central source (a *pulsar*) which injects electrons with large energies into the remnant. This is manifested by a centrally peaked radio to X-ray nebula (Seward 1989) with the X-ray size smaller than the radio size. This topic is of importance in this lecture, since these electrons are then also capable of scattering soft photons to γ-ray energies. In fact, these remnants are capable of producing γ-rays through inverse Compton scattering with energies above 10^{12} eV (TeV energies), which are observable with ground-based air Cherenkov telescopes (Weekes 1988).

4.1 THE INJECTION OF RELATIVISTIC ELECTRONS INTO THE REMNANT

A significant fraction of the pulsar spindown power L_m is usually spent in the acceleration of electrons, which are in turn injected into the remnant (or nebula). The injection rate is usually a power law of the form $Q(\gamma)d\gamma = Q_o\gamma^{-\Gamma}d\gamma$. The details of the acceleration are uncertain, but for the Crab Nebula there is evidence of the existence of a *pulsar wind shock* at a radius $r_s \sim 0.05r_N$ from the pulsar, where r_N is the outer radius of the nebula. This shock acts as a medium which decelerates the relativistic outflow from the pulsar (Rees & Gunn 1974) from a velocity $V \sim c$, down to the outer expansion speed $V_N \ll c$ ($V_N \sim 1500$ km/s for the Crab nebular expansion). Electron acceleration is seen at this shock, and ROSAT pictures of this region clearly show an underluminous region at $r < r_s$, but a synchrotron bright region at $r > r_s$ as a result of accelerated electrons convected outwards into the nebula.

The maximum electron energy $E_{\rm max}$ is obtained by balancing the acceleration rate $t_{\rm acc}^{-1} = \alpha\nu_g$ (which is some fraction α of the gyrofrequency $\nu_g = eBc/2\pi E_{\rm max}$) with the fastest loss rate $t_{\rm cool}^{-1}$. If synchrotron losses dominate, then $t_{\rm cool} = \tau_{\rm syn}$ as given by Eqn (4) and we can rewrite the equality $t_{\rm acc} = \tau_{\rm syn}$ in terms of the maximum observable characteristic synchrotron energy (using Eqn (2)) $h\nu_{\rm max} \sim 32\alpha f_D$ MeV. The fastest acceleration timescale for any Fermi- or resonant-type acceleration in a plasma is the electron gyroperiod, so that $\alpha_{\rm max} = 1$, and $f_D \sim O(1)$ is the Doppler boosting factor due to the relativistic plasma flow relative to the observer. Thus, the observation of a spectral cutoff at $h\nu_{\rm max}$ would confirm the existence of acceleration in the nebula near maximum efficiency. Note that this limit does not hold for electric field acceleration in the pulsar magnetosphere (A.K. Harding 1994, personal communication).

4.2 THE STEADY STATE ELECTRON SPECTRUM IN A SNR

Reynolds & Chanan (1984) made a detailed study of the evolution of an injected spectrum in the remnant. The most obvious feature expected and observed is the modification of the injected spectrum by different loss processes. The dominant energy loss mechanisms for electrons are adiabatic expansion (with $-(d\gamma/dt)_{\rm ad} = \gamma V_r/r$, given the expansion of the plasma at a velocity V_r) and synchrotron losses with an energy loss rate given by Eqn (3). Inverse Compton losses are usually negligable compared to synchrotron losses, and we can calculate the steady state particle spectrum from

$$N_e(\gamma)d\gamma = \frac{Q_o}{(\Gamma-1)}\frac{\gamma^{-\Gamma+1}d\gamma}{[(-d\gamma/dt)_{\rm ad} + (-d\gamma/dt)_{\rm syn}]} \tag{8}$$

This simplistic approach neglects the time evolution of the injection process, but gives us a basic understanding of the observed synchrotron spectrum. At low energies we find that adiabatic losses dominate so that the spectral index Γ remains unaffected by losses. Thus, $N_e(\gamma) \propto \gamma^{-\Gamma}$ in this case. However, at high energies, synchrotron losses (because of the γ^2 dependency in the loss term) dominate and the electron spectrum steepens by one power, so that $N_e(\gamma) \propto \gamma^{-\Gamma-1}$. The energy where the two loss rates are approximately equal is obtained by equating the synchrotron energy loss rate with the adiabatic energy loss rate. For adiabatic losses, the expansion timescale r/V_r is taken as the age T of the remnant. This approximation is at least reasonable for the Crab, since the expansion rate of 0".23 per year for the Crab Nebula is roughly consistent with the known age of T = 940 years. The electron energy at the break can now be written in terms of its characteristic synchrotron frequency ν_b through Eqn (2), resulting in (Marsden et al. 1984)

$$\nu_b \sim 1.3 \times 10^{13}(\frac{3\times 10^{-4}\ {\rm G}}{B})^3(\frac{T}{10^3\ {\rm yr}})^{-2}\ {\rm Hz} \tag{9}$$

4.3 THE ELECTRON SPECTRUM FROM THE SYNCHROTRON SPECTRUM

Suppose we know the synchrotron spectrum $F_{\rm syn}(\nu)$ from broad-band observations. We can then obtain the electron spectrum $N_e(\gamma)d\gamma$ by inverting the expression (given B and using Eqn (1)

$$4\pi d^2 F_{\rm syn}(\nu) = \int P(\nu)N_e(\gamma)d\gamma \propto K_\gamma B_\perp^{(p+1)/2}\nu^{-(p-1)/2} \tag{10}$$

The inversion process can be quite complicated (DH) if the steady state spectrum has not yet been reached (e.g. for the Crab Nebula) and if the radial dependence has to be taken into account. However, the inversion is analytic under the assumption of a steady state spectrum (which may hold more accurately for older SNR) resulting in an expression of the form $N_e(\gamma)d\gamma = K_\gamma\gamma^{-p}d\gamma$ as indicated by the proportionality in Eqn (10) in the range where the observed energy flux $F_{\rm syn} \propto \nu^{-(p-1)/2}$.

Radio observations of SNRs show that the radio index $(p-1)/2 = (\Gamma-1)/2$ is between 0.3 and 0.7 (Green 1988), so that the injected spectra range between Γ =1.6 to 2.4. Above the break frequency ν_b, the synchrotron spectrum is expected

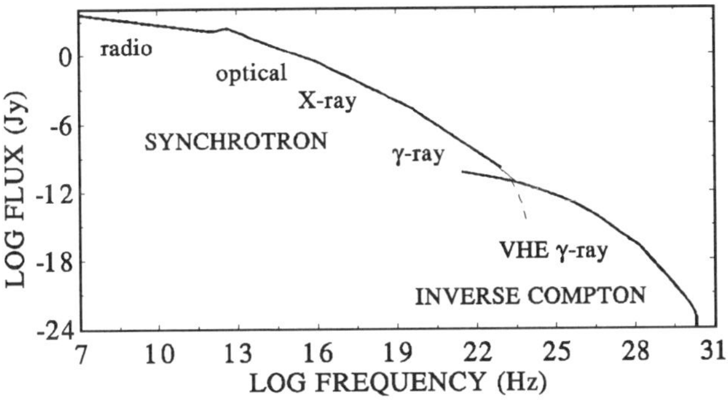

Fig. 2 The observed synchrotron and calculated inverse Compton spectrum of the Crab Nebula (from DH).

to steepen to an energy index of $(p-1)/2 = [(\Gamma+1)-1]/2 = 0.8$ to 1.2. We see this steepening when comparing the radio and higher frequency spectra of pulsar-driven remnants (see Asaoka & Koyama [1990] for the X-ray spectra).

4.4 THE INVERSE COMPTON SPECTRUM

From the inverted electron spectrum $N_e(\gamma)d\gamma$ (obtained from Eqn (10)) we calculate the expected inverse Compton spectrum at Earth using

$$4\pi d^2 F_{\rm IC}(\epsilon_1) = \epsilon_1 \int\int N_e(\gamma)d\gamma(\frac{dN_{\gamma,\epsilon}}{dtd\epsilon_1}) \tag{11}$$

Note that $n(\epsilon)$ (employed in Eqn (5)) should contain all sources of soft photons (see Section 5). Furthermore, $F_{\rm IC}(\epsilon_1)$ will also depend on B since the calculation of N_e from Eqn (10) also involves knowledge of B. However, if we observe both $F_{\rm syn}(\nu)$ and $F_{\rm IC}(\epsilon_1)$, then B can be calculated provided that we know all sources of soft photons.

5. The Crab Nebula Synchrotron-self Compton Spectrum

The Crab Nebula is probably the best observed SNR, which is driven by a 33 ms pulsar. The energy input into the nebula is $L_m \sim 5 \times 10^{38}$ erg s^{-1}. This is also the most energetic pulsar known to us. The synchrotron spectrum has been observed in radio, infrared, optical, X-ray, and γ-rays. A summary of the spectrum over all frequencies is given by DH, which has been reproduced in Fig. 2. We see the typical flat-spectrum in the radio range, but a steepening towards higher frequencies. The spectral break at 10^{13} Hz translates to an average magnetic field strength (from Eqn (13)) of $B_\perp \sim (2\ -\ 3) \times 10^{-4}$ G in the remnant, given the age of $T = 940$ years. However, we see more than one spectral break, and the reader is again referred to the work of Reynolds & Chanan (1984) for a detailed discussion of such phenomena.

The inverse Compton spectrum calculated from the scattering of synchrotron photons by relativistic electrons (after taking the proper radial dependence of the field distribution, the electron-, and photon distributions into account) is also shown in Fig. 2 (note that the assumption of a steady state electron spectrum was not made in this case). This spectrum dominates the synchrotron spectrum above a few hundred MeV, and is consistent with observations in the GeV to TeV range (Lewis et al. 1993). This consistency confirms the value of $B_{\perp} \sim (2 - 3) \times 10^{-4}$ G. The ultrahigh energy cutoff at 10^{30} Hz corresponds to the expected maximum electron energy of $E_{\rm max} = 3 \times 10^{15}$ eV (these electrons also radiate synchrotron photons at $h\nu_{\rm max}$ - see Section 4.1), which scatters a soft photon to $\epsilon_1 \sim E_{\rm max}$ in the Klein-Nishina limit.

6. Inverse Compton Scattering of Photons from External Sources

In the preceding Section we discussed the case where the target photon density $n(\epsilon)$ is specified by the density of synchrotron photons produced in the remnant. However, the surface brightness of synchrotron radiation, and hence $n(\epsilon)$ is too small to make any significant contribution to the inverse Compton component in older remnants which are more extended. Thus, older remnants, which still inject ultrahigh energy electrons into the remnant, are unable to function properly as synchrotron-self Compton sources. However, other sources of soft photons will contribute to $n(\epsilon)$ which would boost the inverse Compton component:

6.1 SCATTERING OF THE 2.7K MICROWAVE BACKGROUND.

The isotropic $T = 2.7$K microwave background has an energy density of $U_{2.7K} = aT^4 = 4.4 \times 10^{-13}$ erg cm^{-3}. The older, but larger Crab-like SNR (such as Vela and MSH15-52) have much smaller magnetic fields with typical $B \sim 10^{-5} B_{-5}$ G, as obtained from the assumption of equipartition between particles and fields (obtained from the observed synchrotron spectra). The importance of synchrotron losses relative to inverse Compton losses due to the scattering of microwave photons, is given by the ratio of Eqns (3) and (7) (in the Thomson limit):

$$\frac{(dE/dt)_{IC}}{(dE/dt)_{\rm syn}} \sim \frac{U_{2.7K}\left[1 - 24.15\gamma kT/mc^2\right]}{B^2/8\pi} = 0.1\left[1 - \frac{\gamma}{10^8}\right] B_{-5}^{-2} \tag{12}$$

provided that $\gamma < 10^8$, otherwise higher-order corrections, or, the full Klein-Nishina formulation must be used. Thus, we see that the electrons will lose about 10% of their energies to inverse Compton γ-rays of typical energy (in the Thomson limit)

$$\epsilon_1 \sim \gamma^2(2.7kT) = 3.8 \times 10^{12}(\frac{h\nu_c}{1\ \rm keV})B_{-5}^{-1}\ \ \rm eV \tag{13}$$

In the last expression we have rewritten the electron energy $E = \gamma mc^2$ in terms of its corresponding characteristic synchrotron energy through Eqn (2), and we see that TeV γ-rays can result from the scattering of 2.7K microwave photons by electrons radiating keV X-rays. Thus, the TeV energy flux will be roughly

10% of the synchrotron energy flux around a keV, and we illustrate this effect on the Crab-like SNR MSH15-52 for which $B_{-5} \sim 0.7$ (Seward et al. 1983), and $L_m \sim 2 \times 10^{37}$ ergs/s. The measured X-ray energy spectrum due to synchrotron radiation is (Kawai et al. 1993) $F_{\rm syn}(\epsilon) = 0.079\epsilon_{\rm keV}^{-1.15}$ keV cm^{-2} s^{-1} keV^{-1} for the interval 1 to 60 keV. We therefore expect a TeV flux (calculated from Eqn (2.65) of Blumenthal & Gould 1970) of the order

$$F_{\rm IC}(\epsilon_1) = 1.4 \times 10^{-11} \left[\frac{7 \times 10^{-6}\ {\rm G}}{B_\perp}\right]^{2.2} \left(\frac{\epsilon_1}{1\ {\rm TeV}}\right)^{-1.2} \ {\rm TeV\ \ cm^{-2}s^{-1}TeV^{-1}} \quad (14)$$

which is of the same order as the Crab flux in TeV γ-rays as given by Lewis et al. (1993). See Du Plessis et al (1994) for a more detailed discussion on the MSH15-52 spectrum.

Another example is the Vela SNR (driven by the Vela pulsar with $L_m \sim 7 \times 10^{36}$ ergs/s), for which the synchrotron energy index is also ~ 1.15 (Pravdo et al. 1978), but the X-ray flux is larger than seen from MSH15-52. The magnetic field strength is also of the order of 10^{-5} Gauss, so that this source (with a $\sim 1°$ extent, Willmore et al. 1992) is expected to be brighter in TeV γ-rays than MSH15-52. This source was recently reported (a preliminary report) to be a TeV source by the CANGAROO collaboration operating in Australia.

6.2 SCATTERING OF IR PHOTONS BY HEATED DUST.

Dwek & Arendt (1992) have discussed the importance of SNR in the study of grain heating, leading to infrared emission. The basic idea is that dust grains are swept up by the supernova explosion, and heated by either radiation (in the case of the Crab Nebula), or by gas-grain collisions in the shock transition in the SNR shell. The Crab-like remnants with identified IR emission are the Crab Nebula, MSH15-52, and G54.1+0.3. The dust temperatures are usually between 30K and 200K.

Another example is the dust shells blown by the stellar winds from young OB stellar associations. One clear detection of such a shell is the Vela IRAS shell (Sahu 1992) which most likely surrounds the Vela SNR. The distance to this shell is 450 pc (compared to the ~ 500 pc distance to the Vela SNR), and the radius of the shell is 65 pc. The temperature of the dust grains is 25K and the IR energy density is $U_{\rm IR} \sim 6 \times 10^{-13}$ erg cm^{-3}. This is larger than the 2.7K energy density, and would also contribute significantly to the TeV γ-ray flux from the Vela SNR.

6.3 SCATTERING OF GALACTIC BACKGROUND RADIATION.

The galactic IR and optical radiation field is considered as an important contributor to the energy losses of Cosmic Ray electrons in our galaxy. It is found that this radiation field is about 10^{-13} erg cm^{-3} in the solar neighbourhood, but increases to $\sim 6 \times 10^{-11}$ erg cm^{-3} near the galactic center. Thus, the photon energy density may exceed the magnetic energy density (if $B \sim 10^{-5}$ G) for such cases. I therefore suggest calling a SNR for which the photon energy density exceeds the magnetic energy density, as an *inverse Compton nebula*.

Acknowledgements

I am indebted to A Harding, A Mastichiadis and V. Ginzburg for useful discussions, and F. Aharonian for stressing the importance of the 2.7K and galactic background radiation.

References

Asaoka, I, & Koyama, K.: Pub. Astr. Soc. Japan **42**, 625 (1990).
Blumenthal, G. R., & Gould, R. J.: Rev. Mod. Phys. **42**, 237 (1970).
De Jager, O. C., & Harding, A. K.: Ap. J. **396**, 161 (1992 (DH)).
Du Plessis, I. et al.: submitted to Ap. J., (1994).
Dwek, E. & Arendt, R. G.: Ann. Rev. Astr. Astrophys. **30**, 11 (1992).
Gould, R. J.: Phys. Rev. Lett. **15**, 577 (1965).
Green, D. A.: Astrophys. Sp. Sci. **148**, 3 (1988).
Jones, F. C.: Phys. Rev. **167**, 1159 (1968).
Kawai, N. Okayasu, R., & Sekimoto, Y.: in Proc. Compton Symp., ed. M. Friedlander, N. Gehrels, & D. J. Macomb, p. 280 (AIP 280, NY, 1993).
Lewis, D. A. et al.: in Proc. 23rd ICRC **1**, 279 (1993).
Marsden, P. L., et al.: Ap. J. **278**, L29 (1984).
Pravdo, S. H. et al.: Ap. J. **223**, 537 (1978).
Rees, M. J., & Gunn, J. E.: MNRAS **167**, 1 (1974).
Rybicki, G. B. & Lightman, A. P.: in Radiative Processes in Astrophysics (John Wiley & Sons, NY, 1979).
Reynolds, S. P., & Chanan, G. A.: Ap. J. **281**, 673 (1984).
Sahu, M. S.: in A Study of the ISM in Puppis-Vela including the Gum Nebula, Ph.D. Thesis (Rijksuniversiteit Groningen, 1992).
Seward, F. D. et al.: Ap. J. **267**, 698 (1983).
Seward, F. D.: Sp. Sci. Rev. **49**, 385 (1989).
Weekes, T. C.: Phys. Rep. **160**, 1 (1988).
Willmore, A. P. et al.: MNRAS **254**, 139 (1992).

THE BIG BANG AND THE INFRARED SKY AS SEEN BY COBE

John C. Mather
Code 685, Laboratory for Astronomy and Solar Physics
NASA Goddard Space Flight Center, Greenbelt, MD 20771 USA

ABSTRACT. The Cosmic Background Explorer (COBE) was developed by NASA Goddard Space Flight Center to measure the diffuse infrared and microwave radiation from the early universe. It also measured emission from nearby sources such as the stars, dust, molecules, atoms, ions, and electrons in the Milky Way, and dust and comets in the Solar System. It was launched 18 November 1989 on a Delta rocket, carrying one microwave instrument and two cryogenically cooled infrared instruments. The Differential Microwave Radiometers (DMR) measured the fluctuations in the CMBR originating in the Big Bang, with a total amplitude of 11 parts per million on a 10° scale. These fluctuations are consistent with scale-invariant primordial fluctuations. The Far Infrared Absolute Spectrophotometer (FIRAS) mapped the sky at wavelengths from 0.01 to 1 cm, and compared the CMBR to a precise blackbody. The spectrum of the CMBR differs from a blackbody by less than 0.03%. The Diffuse Infrared Background Experiment (DIRBE) spanned the wavelength range from 1.2 to 240 μm and mapped the sky at a wide range of solar elongation angles to distinguish foreground sources from a possible extragalactic Cosmic Infrared Background Radiation (CIBR).

1. COBE Concept

The three COBE instruments cover the entire wavelength range from 1.2 μm to 1 cm. The FIRAS (Far Infrared Absolute Spectrophotometer) spans 105 μm to 1 cm in two bands with a 7° beam. The DMR (Differential Microwave Radiometer) covers 31.5, 53, and 90 GHz (9.5, 5.6, and 3 mm), also with a 7° beam. The DIRBE (Diffuse Infrared Background Experiment) spans 1.2 to 240 μm with 10 bands, and also measures polarization in its three shorter bands. Previous works have described the COBE and early results (Mather *et al.* 1990; Mather *et al.* 1991a,b; Janssen and Gulkis 1991; Wright 1990, 1991; Hauser 1991a,b; Smoot *et al.* 1991; Smoot 1991; Bennett 1991; Boggess 1991; Mather 1982; Gulkis *et al.* 1990; Boggess 1992), and the engineering of the spacecraft has also been published (Barney 1991; Milam 1991; Hopkins and Castles 1985; Hopkins and Payne 1987; Volz and Ryschkewitsch 1990; Volz *et al.* 1990, 1991; Volz and DiPirro 1992; Mosier 1991; Colodonato *et al.* 1990; Sampler 1990; Bromberg and Croft 1985).

A liquid helium cryostat cooled the FIRAS and DIRBE to about 1.5 K for 10 months, and after the evaporation of the helium the temperature slowly rose to about 60 K. The DMR radiometers are radiatively cooled, with the 31.5 GHz receivers at 290 K and the others at 140 K. A conical shield protects them from the Sun and Earth, which do not illuminate the instruments directly. The orbit is a sun-synchronous polar orbit, at 900

M. M. Shapiro et al. (eds.), Currents in High-Energy Astrophysics, 235–254.

km altitude and 99° inclination, with a 6 PM ascending node. The Earth's gravitational quadrupole precesses the orbit plane to maintain a 6 PM ascending node, so that the plane is always approximately perpendicular to the Sun. The spacecraft is oriented with its spin axis away from the Earth and about 94° to the Sun. It spins at 0.8 rpm and the DMR and DIRBE beams therefore scan the range from 64° to 124° from the Sun.

The COBE design evolved significantly since it was first proposed in 1974. The initial concept was proposed by a Goddard-MIT-Princeton team, while versions with only microwave radiometers were proposed by teams at JPL and Berkeley. In 1976 NASA formed a science team of 6 people, drawn from all three groups, and selected Goddard to support the mission. The science team eventually grew to include 20 members, and the Goddard effort required about 1500 civil servants and contractors. The initial concept called for launch on a Delta rocket, but in the late 1970's NASA decided to launch all payloads on the Shuttle. Following the Challenger explosion in 1986, the COBE was redesigned for launch on the Shuttle. To achieve this, its weight was cut in half, but no compromise in performance and few changes were needed in the scientific instruments. Only three years after the redesign was approved, the COBE was launched. Its liquid helium lasted for 10 months, but the DMR and the short wavelength channels of the DIRBE continued operation until January 1994. The COBE now serves for satellite communication tests.

2. FIRAS

2.1 FIRAS INSTRUMENT

The COBE FIRAS (Far Infrared Absolute Spectrophotometer) was designed to compare the CMBR spectrum to a blackbody with great precision. The most recent results show that the CMBR deviates from blackbody form by less than 0.03% of the peak intensity over the wavelength range from 0.5 to 5 mm (Mather *et al.* 1994). Our interpretation of these results has also been given (Wright *et al.* 1994a). The FIRAS also showed that the cosmic dipole has the expected shape of the derivative of the Planck function with respect to temperature (Fixsen *et al.* 1994a). The primary basis of the comparison is a full beam external blackbody calibrator that can be adjusted to match the temperature of the sky. This calibration is described in Fixsen *et al.* (1994b). The FIRAS is the first instrument to have this opportunity in the protected space environment. The instrument is designed with multiple modes of operation and multiple detectors to test for possible systematic errors.

The instrument is a polarizing Michelson interferometer used as a spectrometer (Mather, Shafer, and Fixsen 1994). It has a beamwidth of 7° , defined by a compound parabolic concentrator. It is symmetrical, with differential inputs and differential outputs. It is a two-beam interferometer, whose output signal (called an interferogram) depends on the path difference between the two beams. A monochromatic input produces an output proportional to the cosine of the phase difference between the two beams. A general input produces an interferogram which is the Fourier transform of the input spectrum. These interferograms are detected, amplified, filtered, digitized, filtered and averaged again, and transmitted to the computers on the ground. They are then Fourier transformed numerically and calibrated to determine the input spectra. The mirror mechanism can scan at two different scan speeds and over two different stroke lengths, to obtain low and high spectral resolution.

The FIRAS was used to map almost the entire sky. To derive cosmological results, two terms were subtracted from the spectrum of each pixel. A dipole term compensated for the Doppler shift of the spectrum, due to the motion of the Earth. The mean spectrum of the dust in the Galaxy was also derived from the map, and a fraction of this mean spectrum was fitted to each pixel spectrum and subtracted. The resulting sky map showed almost no trace of the strong Galactic emission, so that a single dust brightness map times a single characteristic spectrum is a very good representation of the far infrared Galaxy.

To quantify the distortion of the mean spectrum of the whole sky, it was fitted to a sum of additional terms. The first two terms are a Planck function for the mean temperature, and a temperature adjustment times the derivative of the Planck function with respect to temperature. The second term just allows for the possibility that the first may not be calculated for the correct mean temperature, since we do not know it a priori. The third term is a small number times the function $g(\nu)$ that represents the mean Galactic spectrum. This allows for the possibility that the initial subtraction done on each pixel might have been incomplete or biased. The fourth term represents the cosmic distortion. There are two interesting forms for this: a μ form and a y form (Zeldovich & Sunyaev, 1969, 1970). The μ form is a chemical potential distortion, that might arise between the redshifts of about 3×10^6 and 10^5. The second form is a Comptonization form that could arise after $z = 10^5$.

Figure 1 (Mather *et al.* 1994) shows the results of these least squares fits. The residuals are for the case where $y = \mu = 0$ and show that the CMBR spectrum is the same as a blackbody within 0.03% of the peak intensity. The weighted rms residual is only 0.01% of the peak brightness. Nevertheless, these residuals are about twice as large as those expected from detector noise alone. We do not presently know the cause of this but have arbitrarily increased the size of the error bars throughout. The y and μ curves show the shapes of distortion that would be produced if these coefficients have the 95% confidence limit values of $|y| = 2.5 \times 10^{-5}$ and $|\mu| = 3.3 \times 10^{-4}$. These limits allow for the increased uncertainty noted above, and for variations of the derived y and μ as the boundary of the selected data set moves farther from the Galactic plane.

Other possible distortions can also be explored. As an example, if the CMBR spectrum is a graybody, then we can limit its emissivity to 1 ± 0.00041 (95% CL). If the true cosmic signal contains a distortion with the same shape as the Galactic signal, then we cannot determine it with this method alone. Similarly, a cosmic energy release that simply changed the CMBR temperature could not be recognized.

The temperature of the CMBR is also important, although few cosmological calculations require a more precise number. The measured temperature is primarily needed for comparisons with different experiments, such as ground-based measurements and interstellar CN measurements. The FIRAS has two ways to determine this number. First, the thermometers in the external calibrator are the primary scale. When the calibrator is set to match the sky brightness, its thermometers measure the sky temperature. This method gives 2.730 K. The second method uses the wavelength scale of the FIRAS to determine color temperatures. In effect, we measure the frequency of the peak brightness of the CMBR and use fundamental constants to compute a temperature. The wavelength scale was determined from FIRAS observations of interstellar [C II] and confirmed with other lines of

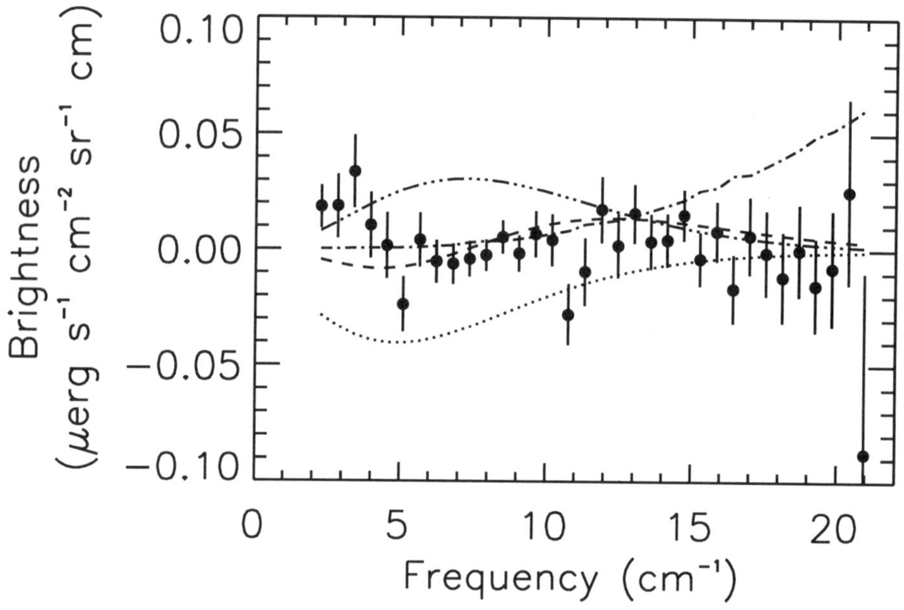

Figure 1: FIRAS measured CMBR residuals, $I_0 - B_\nu(T_0) - \Delta T(\partial B_\nu/\partial T) - G_0 g(\nu)$. Spectrum model components: the maximum allowed distortions (95% CL) $y = 2.5\times10^{-5}$ (— — —) and $|\mu| < 3.3\times10^{-4}$ (· · ·); the Galaxy spectrum $g(\nu)$ scaled to one fourth the flux at the Galactic pole (· –), and the effects of a 200 μK temperature shift in T_0, $0.0002(\partial B_\nu(T)/\partial T)$, (· · · —). (Mather *et al.* 1994)

[N II], [C I], and CO. The color temperature method yields a CMBR temperature of 2.722 K. These numbers differ by much more than the random errors, which are only a few μK. We have no obvious explanation for the difference and simply take the mean, obtaining T = 2.726 ± 0.010 K (95% confidence). Clearly this is not a statistical determination of the error bar, which is entirely systematic error.

Fortunately, there are other confirmations that this number is approximately correct. A sounding rocket experiment launched only weeks after the COBE by Gush, Halpern, and Wishnow (1990) with a similar instrument and greatly superior detectors obtained the result 2.736 ± 0.017 K. Interstellar CN results are also in agreement with these numbers (Roth, Meyer, & Hawkins 1993, and Kaiser & Wright 1990). We can also determine the temperature from the spectrum of the dipole anisotropy. This can be done from both DMR and FIRAS data. The FIRAS result is determined only from the color temperature, since the absolute value of the Earth's velocity is not known. The FIRAS yields 2.714 ± 0.022 K (95% CL). The DMR result (see below) is 2.76 ± 0.18 K, and is derived from the known variation of the Earth's velocity around the Sun. The spectrum of the FIRAS dipole was fit to an amplitude of 3.343 mK. The dipole residuals are very small and are close to the value expected from detector noise alone.

2.2 FIRAS INTERPRETATION

Our interpretation of these spectrum distortions is given by Wright *et al.* (1994a) and summarized here. Large CMBR spectrum distortions are very difficult to produce in plausible versions of the hot Big Bang universe. After the annihilation of positrons, the CMBR energy density far exceeded even the rest mass energy density of the baryonic matter until quite recently. Consequently, there are few processes involving the baryonic matter that can liberate much energy and change the CMBR spectrum significantly. It is even more difficult to produce enough energy to create the whole CMBR radiation from anything except a hot Big Bang.

Therefore, the most immediate conclusion of the FIRAS spectrum distortion measurements is that the hot Big Bang is the only natural explanation for a nearly perfect blackbody. Carefully tailored models are required if one desires to produce the whole energy content of a blackbody spectrum by adding up non-blackbody contributions at different redshifts, as required by alternatives to the hot Big Bang. If one imagines that dust in intergalactic space can thermalize the radiation, then that dust must have substantial optical depth over an interval of cosmic history. That moment cannot be recent, or we would not be able to see distant galaxies at far infrared wavelengths. The IRAS galaxy at $z = 2.286$ demonstrates that one can see very far, and if the millimeter wave optical depth were large we would not see even such a spectacular object.

The next conclusion is that rather little of the energy of the CMBR was added to it after the first year of the expansion. The fraction of the CMBR energy added is approximately 0.71μ in the redshift range $3 \times 10^6 > z > 10^5$. For later redshifts, the fraction is $4y$. A more precise calculation gives the results shown in Fig. 2.

There are many possible sources of such energy conversions, including decay of primeval turbulence, elementary particles, cosmic strings, or black holes. The growth of black holes, quasars, galaxies, clusters, and superclusters might also convert energy from other forms. The FIRAS data, together with the DMR anisotropy measurements, provide a limit on the spectral index of primordial density fluctuations. Wright *et al.* (1994a) found an upper limit of 1.9, while a preprint by Hu, Scott, and Silk (1994) claims an upper limit of about 1.55. It is interesting that these calculations give tighter limits than existing direct measurements, even though the FIRAS numbers are a non-detection. These results are dependent on assuming that a power law is the correct form for the fluctuations over 7 orders of magnitude of scale. There is little possibility of observational evidence to confirm this assumption over such a wide range, since small scale fluctuations have long since been replaced by nonlinear phenomena.

Wright *et al.* (1994a) also give limits on hydrogen burning following the decoupling. These results depend on using geometrical arguments (a csc $|b|$ fit) to estimate the maximum amount of extragalactic energy that could have a spectrum similar to that of our own Galactic dust. We found a limit that is a factor of about 3 smaller than the polar brightness of the Milky Way. A better understanding of the Galactic dust would help produce a tighter limit on these extragalactic signals.

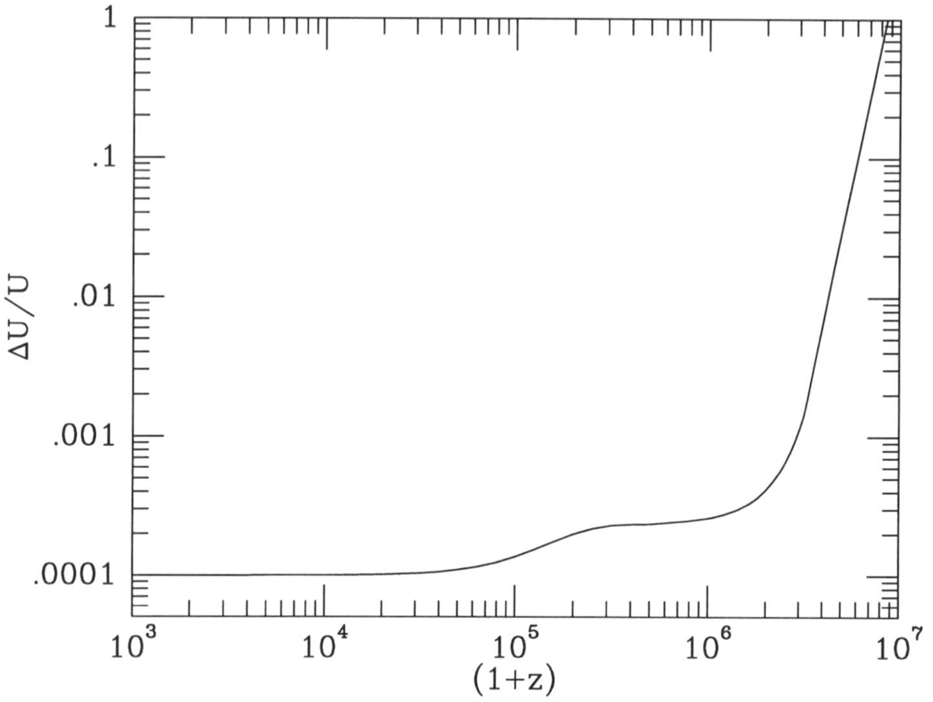

Figure 2: Limits on Energy Release Prior to Recombination (Wright *et al.* 1994a)

Consider first population III stars liberating energy that is converted by dust into far infrared light (using an optical depth of 0.02 per Hubble radius), and assume that $\Omega_b h^2 = 0.015$. In that case less than 0.6% of the hydrogen could have been burned after $z = 80$. As a second example, consider evolving infrared galaxies as observed by the IRAS. For reasonable assumptions, we found that less than 0.8% of the hydrogen could have been burned in evolving IR galaxies.

We also obtained limits on the heating and reionization of the intergalactic medium. It does not take very much energy to reionize the medium, relative to the CMBR energy, because there are so few baryons relative to CMBR photons. Even the strict FIRAS limits permit a single reionization event to occur as recently as $z = 5$. More detailed calculations by Durrer (1993) show that the energy required to keep the intergalactic medium ionized over long periods of time is much more substantial and quite strict limits can be obtained. If the FIRAS limits were about a factor of 5 more strict, then it would be possible to test the ionization state of the IGM all the way back to the decoupling.

If the IGM were hot and dense enough to emit the diffuse X-ray background light, it should distort the spectrum of the CMBR by inverse Compton scattering. This is a special case of the Comptonization process, with small optical depth and possibly relativistic particles. Calculations show that a smooth hot IGM could have produced less than 10^{-4} of the X-ray background, and that the electrons that do produce the X-ray background must also have a filling factor of less than 10^{-4}.

2.3 INFRARED ASTRONOMY WITH FIRAS

The FIRAS produced the first nearly all-sky far IR survey of Galactic emission at wavelengths greater than 120 μm (Wright *et al.* 1991). The general features seen in the maps are very similar to those found by the IRAS satellite at 60 and 100 μm . Dust is concentrated in the Galactic plane and is heated by starlight to a temperature ranging from 15 to 30 K, depending on the intensity of the starlight incident on the dust. This dust controls radiative transfer of energy through the Galaxy, and thus the heating and cooling of the interstellar gas. The dust has a mean spectrum which is approximated well by the form $\nu^{\alpha} B_{\nu}(T)$, where the index $\alpha = 1.65$ and $T = 23.3$ K. The mean dust spectrum of the Galaxy can also be well represented by the sum of two such terms with the index $\alpha = 2$, with cold dust with a temperature of about 4.8 K and warmer dust with a temperature of 20.4 K. In this representation, the optical depth of the cold dust is 6.7 times that of the warmer dust component. There are many possible explanations for this kind of dust spectrum. It is possible that there are multiple types of dust grains, so that some can have a very different equilibrium temperature than others. Dust grains could have unusual shapes, like needles (Wright 1982) or fractals (Wright 1987). Small grains with low heat capacities could have variable temperatures, as they are excited by discrete UV photons or cosmic rays. Some dust is protected within optically thick clouds and must reach much lower temperatures than that exposed to the full intensity of the interstellar radiation field, and other dust is very close to stars and must be much warmer than average. The total dust luminosity of the Galaxy is $(1.8 \pm 0.6) \times 10^{10} L_{\odot}$ (Wright *et al.* 1991).

The FIRAS also provided the first panoramic and unbiased spectral survey of the far IR line emissions from the interstellar medium, including CO, [C I], [C II], and [N II] lines. These results were reported by Wright *et al.* (1991) and by Bennett *et al.* (1994a). The FIRAS map of the emission of [C II] at 158 μm shows it to be extremely widespread, both in and out of the Galactic plane. This line is the dominant cooling mechanism for most of the interstellar gas, and is the brightest emission line of the Galaxy, accounting for 0.3% of the dust luminosity. The strength of the line is so great that it can be seen by eye in the raw interferograms as an undamped sinusoid, and with more analysis the Galactic rotation curve can be seen in the line center frequency. The [C II] line is very strongly correlated with the dust emission, as expected from the theory of photodissociation regions. The heating of the gas is provided mostly by photoelectrons emitted from dust grains, and must be in balance with the cooling provided mostly by the [C II] line. Deviations from this correlation are expected where the dust properties change, where the interstellar radiation field changes color, or where other heating and cooling mechanisms become important at extremes of temperature or density. In particular, the ratio of [C II] to dust decreases to nearly zero at the Galactic poles.

The FIRAS also detected the emission of ionized nitrogen [N II] at 205 μm for the first time and measured its frequency well enough for further astronomical (Petuchowski and

Bennett 1992) and laboratory (Brown *et al.* 1994) observations. This is the second brightest far IR line from the interstellar gas and probes a different phase of the interstellar medium because nitrogen ionizes less easily than hydrogen and carbon. The nitrogen and carbon line intensities are also strongly correlated but show the relation [N II] $\propto$ [C II]$^{3/2}$. The meaning of this correlation was explored by Petuchowski and Bennett (1993).

Carbon monoxide is the standard tracer for interstellar molecular gas. The FIRAS has mapped the brightness of the lines of CO from the 2-1 up to the 6-5 transitions, showing that the line ratios are highly dependent on location in the Galaxy. In particular, the CO lines are extremely bright relative to the others in the Galactic Center, where the gas temperature and density are very high and the UV field is shielded by dust. The two lines of neutral carbon [C I] are also seen and mapped by the FIRAS.

Weaker lines observed by the FIRAS include oxygen and water. The water line at 269 μm has never been observed before in astrophysical sources because of the extremely high optical depth of the terrestrial absorption line. The line is seen in absorption by FIRAS (Bennett *et al.* 1994a) against the continuum emission of the Galactic Center, but at a low level of significance relative to the detector noise and possible systematic errors. The observed absorption is comparable to that expected from the concentration of interstellar water vapor.

3. DMR

3.1 DMR INSTRUMENT

The Differential Microwave Radiometer (DMR) instrument has made the first detection of primeval anisotropy of the CMBR. It operates with two radiometers at each of three frequencies, 31.5, 53, and 90 GHz. These frequencies were chosen to minimize the effects of Galactic free-free, synchrotron, and dust emission on the cosmic measurements. The beamwidth is 7° , with two horns pointing 60° apart and 30° away from the spin axis of the spacecraft. The output of the DMR is a signal proportional to the difference of the brightnesses observed by the two antennas. As the spacecraft spins and orbits, and the orbit precesses through the year, all possible pairs of pixels 60° apart are compared by the DMR. The fundamental data analysis algorithm is a least squares fit of all the difference measurements to a map of the sky. The map has 6144 pixels, with 6 square faces of 32×32 pixels.

To achieve high sensitivity, the DMR uses mixer–preamplifier receivers with Gunn diode local oscillators. The input to each receiver is switched between the two horn antennas at 100 Hz by a ferrite switch. The output from each receiver is detected by a square law diode, and amplified again. The part of the output that is synchronous with the input switch is further amplified, integrated, and telemetered every half second.

The analysis of the data begins with inspection to ensure that no bad data are included in the final products. Corrections are made for known instrument errors, including drift of the DC output of the receiver, and sensitivity to the Earth's magnetic field. The instrument performance was monitored throughout the mission, and calibration with noise diodes was performed every two hours. The CMBR dipole was determined every day to verify the stability of the instrument. The Moon was used whenever it was visible to confirm the calibration stability and verify that the spacecraft orientation was correctly calculated. Detailed analysis of the Moon observations also yield an in-flight map of the antenna beam patterns.

3.2 DMR RESULTS

The initial results from the DMR were reported by Smoot *et al.* (1992), Bennett *et al.* (1992b), and Wright *et al.* (1992), based on the error analysis of Kogut *et al.* (1992) and the calibration reported by Bennett *et al.* (1992a). Recent papers by Bennett *et al.* (1994b), Wright *et al.* (1994b), Smoot *et al.* (1994), Kogut *et al.* (1994) and Gorski *et al.* (1994) have reported the results from analysis of two years of flight data.

The results are also consistent with a far IR balloon survey of the CMBR anisotropy, as reported by Ganga *et al.* (1993). They showed that there is a good statistical correlation between their maps and the one-year DMR maps. As we reported previously, the angular correlation function of the maps is also consistent with the scale-invariant $n = 1$ Peebles-Harrison-Zeldovich power law spectrum of primordial density fluctuations.

The second year results (Bennett *et al.* 1994b) are consistent with the first year, and averaging them together improves the precision and accuracy. The two-year 53 GHz data show RMS temperature fluctuations of $(\Delta T)_{rms}(7°) = 44 \pm 7$ μK and $(\Delta T)_{rms}(10°) = 30.5 \pm 2.7$ μK at 7° and 10° angular resolution respectively. The 53×90 GHz cross-correlation amplitude at zero lag is $C(0)^{1/2} = 36 \pm 5$ μK (68% CL) for the unsmoothed (7° resolution) DMR data. A likelihood analysis of the cross correlation function, including the quadrupole anisotropy, gives a most likely quadrupole-normalized amplitude, Q_{rms-PS}, of $12.4^{+5.2}_{-3.3}$ μK (68% CL) and a spectral index $n = 1.59^{+0.49}_{-0.55}$ (68% CL) for a power law model of initial density fluctuations, $P(k) \propto k^n$. With n fixed to 1.0 the most likely amplitude is 17.4 ± 1.5 μK (68% CL). Excluding the quadrupole anisotropy we find $Q_{rms-PS} = 16.0^{+7.5}_{-5.2}$ μK (68% CL), $n = 1.21^{+0.60}_{-0.55}$ (68% CL), and, with n fixed to 1.0 the most likely amplitude is 18.2 ± 1.6 μK (68% CL). Monte Carlo simulations indicate that these derived estimates of n may be biased by $\sim +0.3$ (with the observed low value of the quadrupole included in the analysis) and $\sim +0.1$ (with the quadrupole excluded). Thus the most likely bias-corrected estimate of n is between 1.1 and 1.3. The best estimate of the dipole from the two-year DMR data is 3.363 ± 0.024 mK towards Galactic coordinates $(\ell, b) = (264.4° \pm 0.2°, +48.1° \pm 0.4°)$, and the best estimate of the RMS quadrupole amplitude in our sky is 6 ± 3 μK (68% CL).

The statistical properties of the map can be presented in several forms. The angular correlation function given by Bennett *et al.* (1994b) shows the results when the portion of the sky at Galactic latitude greater than 20° are analyzed. A range of simulated skies was drawn from a distribution of primordial fluctuations with a spectral index of 1.6 and an amplitude (normalized to the quadrupole) of 12.6 μK. There is very good agreement between the model with these parameters and the observations. As noted above, the technique is subject to bias and this estimate of the spectral index must be lowered by about 0.3.

It is also possible to analyze the maps into spherical harmonics. There is considerable subtlety and difficulty in this since the removal of the Galactic plane causes the harmonics to be non-orthogonal, and produces strong correlations among the fitted amplitudes.

Nevertheless, it is informative to show Fig. 3, updated after Wright *et al.* (1994b). They use the Hauser-Peebles (1973) method to obtain the harmonic amplitudes. The figure also shows the results of other instruments. A scale-invariant power spectrum would produce a horizontal line on this plot.

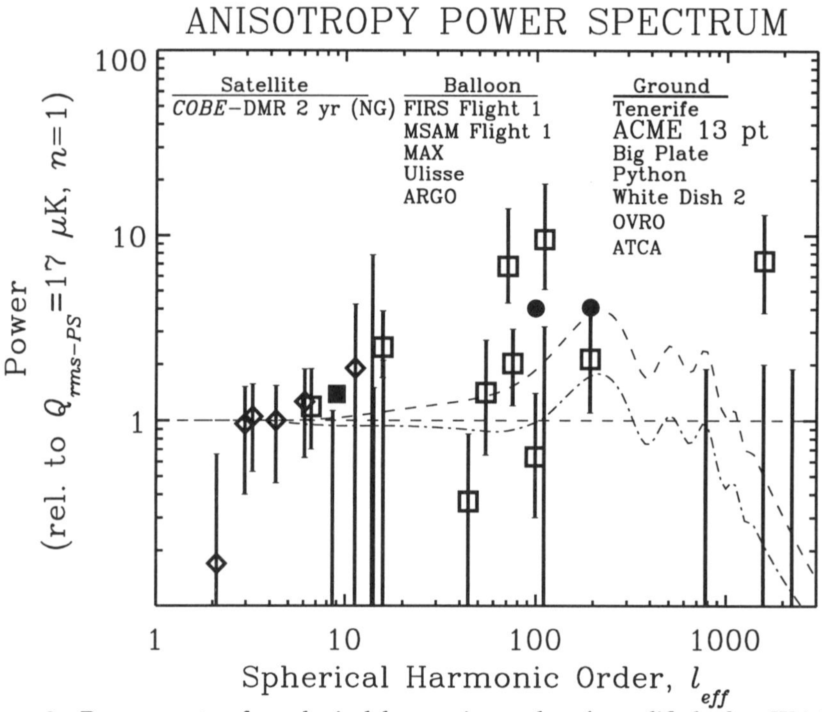

Figure 3: Power spectra for spherical harmonic number ℓ, modified after Wright *et al.* 1994b, *q.v.* for references. The unmodified $n = 1$ prediction would be a horizontal line. COBE data points are from the 2 year no galaxy DMR maps. The diamond points on the left are Galaxy-removed *COBE* DMR results from Wright et al. (1994b), and the small box at $\ell = 9$ is the normalization derived from the use of the overall *COBE* power spectrum by Górski et al. (1994). The two filled circles show the effects of not removing sources from the MSAM data. The two dashed curves are indicative of power spectra expected in dark matter model

Gorski (1994) has developed a new method for spherical harmonic analysis of the maps that accounts explicitly for the correlations induced by cutting out the Galactic plane. He uses a spherical harmonic basis set that has been orthogonalized on the cut sphere using a Gram-Schmidt algorithm. With this new basis set the correlations of the receiver noise and the true sky fluctuations can be given analytic forms, although large matrices are required. The method is unbiased and requires no Monte Carlo simulations, although they were useful to verify the accuracy of the method. The application of the method is described in Gorski *et al.* (1994). They find that the maximum likelihood Bayesian estimates of the spectral index are $n = 1.22$ (1.02) and $Q_{rms-PS} = 17$ (20) μK including (excluding) the quadrupole. The marginal likelihood function for the spectral index gives $n = 1.10 \pm 0.32$ (0.87 ± 0.36).

There is extensive theoretical activity regarding the interpretation of the CMBR fluctuations. Even within the context of the inflationary theory, there are more free parameters than there are observable quantities about the CMBR. The total density of the universe Ω, Einstein's Λ term, the total density of baryons Ω_b, and the Hubble constant H are all required. In addition, many theories have shown that the shape of the inflation potential function need not be simple, and the details affect the spectral index of the fluctuations. There can even be more than one episode of inflation. There can be more than one kind of dark matter, and of course the properties of dark matter are unknown and perhaps unmeasurable.

Even within the context of observations, there are inevitable limits to precision. The fluctuations are a random process, and a finite sample cannot characterize it perfectly. There are only a few measurements of each angular scale of the random field, $2\ell + 1$ for each spherical harmonic index ℓ. Smaller angular scales (higher ℓ) are less limited by this problem but require larger instruments because of diffraction. At scales less than a few degrees, the primordial fluctuations can no longer be observed because they have been modified by processes at and after decoupling.

It is also important to determine whether the primordial fluctuations are really Gaussian. This is still difficult because even with two years of DMR data the signal to noise ratio is only about 1.4 on an average pixel. In other words, only 2/3 of the variance is due to the sky, and the other 1/3 is from the instrument. There are a number of tests. The raw distribution of the temperature residuals should be close to Gaussian if the sky variance is Gaussian and the receiver noise is Gaussian. The receiver noise varies somewhat from pixel to pixel because the observation times are not all the same, but when this is taken into account there is extremely good agreement between these histograms (Smoot *et al.* 1994). There is no evidence that there is an excess of large deviations, as would be expected if there is an unknown population of point sources. On the other hand, considering the large beam of the instrument and the variance of both cosmic signals and receiver noise, it would be quite possible for interesting signals to be hidden in the data. Cosmic strings and so forth could well produce maps that are similar to those we observed. Strings might be much more observable with higher spatial resolution. Kogut *et al.* (1994) report a search for point sources in the two-year maps, and the result is negative.

It is also interesting to test the genus statistic of the maps. These results are also consistent with Gaussian statistics. The genus statistic was explored by Gott *et al.* (1990) and the COBE DMR calculations in Smoot *et al.* (1994) were based on their work. Smoot *et al.* found no evidence for deviations from Gaussian statistics, but most popular alternative theories of cosmic structure now produce similar statistics. A three point correlation function also tests the statistical properties of the maps. These results are also consistent with Gaussian statistics.

A more important question is whether the fluctuations are sufficient to produce the present large scale structure of the galaxy correlations. This is a difficult question. It seems desirable to introduce dark matter of both cold and hot types in the same model. Our initial interpretations were given by Wright *et al.* (1992), who used the tabulations of many cases by Holtzman (1989). However, there are many computations between concept and realization. I expect much further progress in this area before conclusions are firm.

4. Diffuse Infrared Background Experiment

4.1 DIRBE INTRODUCTION

The DIRBE was designed to search for the Cosmic Infrared Background Radiation (CIBR), recognizable as an isotropic flux after all foreground contributions have been removed. The predicted CIBR is faint relative to the foregrounds, so this is an exceptionally difficult task. The DIRBE accomplishes this by making absolutely calibrated maps of the full sky in 10 photometric bands at 1.25, 2.2, 3.5, 4.9, 12, 25, 60, 100, 140, and 240 μm. To characterize the light of the interplanetary dust, the DIRBE also measures linear polarization at 1.25, 2.2, and 3.5 μm.

The interplanetary dust is the brightest of these foregrounds at most of the wavelengths observed by DIRBE. Its brightness is a function of many variables: the position of the Earth in the dust cloud, the direction of observation, the wavelength, and the polarization of the light. To first order, the dust cloud is symmetrical above and below a plane which is close to the ecliptic plane, and rotationally symmetrical about an axis nearly perpendicular to that ecliptic plane. The dust grains gradually spiral in toward the Sun under the Poynting-Robertson drag, which produces a number density roughly proportional to $1/r$, where r is the radius of the orbit. However, the perturbations of Jupiter and the other planets have systematic effects that spoil the simple symmetry of the cloud. Moreover, there are significant discrete sources of dust, such as individual comets and asteroid families. These produce bands of dust which are not fully symmetrized nor stable in time. Finally, there are two roughly equal sources of dust particles: asteroid collisions and comets, which have quite different orbit parameters.

To observe this dust as well as possible, the DIRBE line of sight is offset from the COBE spacecraft spin axis by 30° . The observing pattern is a spherical cycloid, sweeping the DIRBE beam across half the sky in a few orbits. Over the course of a year, the DIRBE observes each direction many times at all accessible angles from the Sun, with a maximum range from 64° to 124° . The DIRBE achieves true color measurements of each pixel since all the detectors view the same field of view. The design sensitivity for each 0.7° square field of view is $\nu I_\nu = 1$ nW m^{-2} sr^{-1} (1 σ, 1 year), but this is not achieved at 140 and 240 μm . This is much less than the sky brightness.

The DIRBE instrument contains the only telescope on the COBE, and is optimized for radiometry of a diffuse source. The telescope is in an off-axis Gregorian configuration for strong rejection of stray light (Magner 1987, Evans 1983). It utilizes a 32 Hz tuning fork chopper to modulate the sky signal and compare it continuously with the cold interior of the instrument. The DIRBE is cooled to less than 2 K by the COBE cryostat. Its photometric zero point is calibrated by a cold shutter at the prime focus, and its responsivity is calibrated against known celestial point sources. The calibration is verified frequently by an internal reference source, and is reproducible within $\sim$ 1% or better. The DIRBE instrument has been described in more detail by Silverberg *et al.* (1993).

The initial results show the expected features on the maps, published by Hauser (1993) in four all-sky false color images. The Galactic plane is bright at all wavelengths, and discrete stars are numerous at wavelengths from 1.2 to 4.9 μm . At longer wavelengths, diffuse emission from cirrus dust clouds and dust in star forming regions is bright. The interplanetary dust is brightest at 12 and 25 μm and clearly detectable at all wavelengths. By coincidence, the ecliptic plane passes near the Galactic Center.

4.2 THE COSMIC INFRARED BACKGROUND RADIATION

The long-sought CIBR has not yet been measured by the DIRBE because the foreground sources are complex and difficult to model and subtract. Table 1 (Hauser 1994) lists the darkest sky brightness observed at each wavelength, prior to subtraction of the foregrounds. These are generally at high ecliptic or Galactic latitude, according to which source is dominant at each wavelength. The absolute calibration error listed is 20%. The FIRAS and DIRBE overlap at 140 and 240 μm and the calibrations are consistent within this accuracy.

Table 1: UPPER LIMITS ON THE COSMIC INFRARED BACKGROUND

λ μm	νI_ν nW m^{-2} sr^{-1}	Reference
1.25	480±96	DIRBE dark sky
2.2	190±38	"
3.5	83±17	"
4.9	260±52	"
12	2800±560	"
25	2400±480	"
60	300±60	"
100	90±18	"
140	140±28	"
240	28±6	"
500-5000	$680/\lambda(\mu m)$	Mather *et al.* (1994)

The DIRBE results are in general agreement with prior measurements of the same wavelengths, lines of sight, and time of year where these have been compared (Hauser *et al.* 1991ab; Noda *et al.* 1992). Prior estimates of isotropic backgrounds have also been made (Matsumoto *et al.* 1988; Matsumoto 1990; Noda *et al.* 1992). These estimates are difficult because of the limited sky coverage and observing time available before COBE, and because of possible systematic errors. Further comparisons await the DIRBE sky modeling and subtraction.

Infrared extragalactic sources have been observed from the ground and with the IRAS satellite. Extrapolation of the source counts with various models for galaxy formation and evolution yield estimates of the expected CIBR from known classes of sources. Cowie *et al.* (1990) estimated the CIBR at 2.2 μm to be $\nu I_\nu = 5$ nW m^{-2} sr^{-1} on the basis of deep exposures with a large telescope. Hacking & Soifer (1991) used the IRAS galaxy counts to estimate a CIBR (integrated to $z = 3$) at 25, 60, and 100 μm of 1, 2, and 4 nW m^{-2} sr^{-1} respectively. Beichman & Helou (1991) also used extrapolations of IRAS data. At 300 μm, their minimum estimated brightness (integrated to $z = 3$) is 2 nW m^{-2} sr^{-1}. The actual CIBR could be much brighter than these estimates, depending on the still-unknown process of galaxy formation and clustering, the evolution of galaxy luminosities, and the amount of dust in each.

A new line of reasoning suggests that the CIBR may be much larger than that estimated from the discrete sources. Stecker *et al.* (1992) showed that gamma rays from distant quasars could be used to infer the CIBR, since gamma rays are attenuated by pair production reactions with the CIBR. These data are obtained with great difficulty, from space instruments and high energy air shower array detectors. The analysis has been carried out by De Jager, Stecker and Salamon (1994) and by Dwek and Slavin (1994), with good agreement. The data yield a possible detection of a CIBR in the thermal dust emission range, and an upper limit for shorter wavelengths. These numbers are much larger than the extrapolations of discrete sources. Unfortunately, the analysis is entirely dependent on the unknown spectrum of the unattenuated source Mrk421, and reasonable but unprovable assumptions must be made.

The DIRBE results may be compared with the extensive predictions by Bond, Carr, and Hogan (1986). Most of the possible sources are not yet ruled out by the DIRBE data.

4.3 DIRBE INFRARED ASTRONOMY

The DIRBE has revealed new features of the Galaxy and the Solar System. Since infrared penetrates the dust toward the Galactic Center, a fresh view of the inner regions is quite revealing. The disk is composed largely of K and M giants. The dust obscuring this region produces a reddening curve that is consistent with other observations (Arendt *et al.* 1993, 1994), and is detectable in absorption out to 3.5 μm wavelength. The same dust is correlated with the polarized light observed by DIRBE near the Galactic Center (Berriman *et al.* 1993). The polarization is less than about 2%, and is produced by preferential extinction by dust grains oriented in a magnetic field. After correction for extinction and emission from the disk, the central stellar bulge is found to be asymmetrical (Weiland *et al.* 1993, 1994), consistent with a bar as described by Blitz and Spergel (1991) and suggested by Liszt and Burton (1980) and Sinha (1979). The DIRBE data allow detailed modeling of the shape and orientation of this bar, and estimates of its luminosity and mass (Dwek *et al.* 1994).

The DIRBE also confirms the IRAS observation that "cirrus" dust clouds are prevalent throughout the Galaxy. These clouds are prominent at 12 μm wavelength, even though the equilibrium temperature of dust grains in clouds is far too low to allow emission at 12 μm . The emission is explained by transient heating of small dust grains by energetic UV photons or collisions with cosmic rays. The grains apparently reach temperatures of the order of 300 K and even higher. There is little doubt of the reality of this phenomenon, although there are many possible versions of the theory. The grain materials and optical properties are known to differ in different sources. Grains observed in particular IR and UV sources include graphite, hydrogenated sheets of hexagonal carbon rings (polycyclic aromatic hydrocarbons), silicates, carbonates, sulfides. The structure of the grains is also unknown. It is likely that some grains are needle-shaped (Wright 1982, Hawkins and Wright 1988), others are fractal fluff balls (Wright 1987), and others have dense cores with ice mantles. Given the wide variety of materials, shapes, and size distributions, it is impossible to derive detailed descriptions of the dust from observations that average over these properties.

The Galactic plane is also seen to be warped in both its stellar and interstellar dust components, particularly in the outer galaxy (Freudenreich *et al.* 1993, 1994). The dust warp is similar to that observed in interstellar gas. Modeling is needed to determine whether the physical displacement of the stars is similar to that of the gas.

The DIRBE far infrared maps of the Galaxy are strikingly similar to maps of the gamma ray sky obtained by the EGRET. A dominant source of the gamma rays is inverse compton scattering, in which collisions with relativistic electrons convert far IR photons to gamma rays. The detailed comparison of the maps will be of great interest in understanding the origin of the gamma rays and mapping the population of relativistic electrons.

Closer to home, the DIRBE has observed 4 comets. The analysis was reported by Lisse (1993) and Lisse *et al.* (1994). Comet Austin was observed to be extremely dark (very low albedo), so that the infrared emission is far brighter than the visible scattered sunlight. A favorable orbit geometry enabled the mapping of the time history of the dust emission and the determination of the particle size spectrum. Large dust grains ejected by the comet stay near the comet nucleus, while small ones are blown rapidly away from the Sun by radiation pressure. An individual spot in the map of the comet corresponds to dust emitted at a particular time and with a particular ratio of solar radiation pressure to gravitational force. This ratio is primarily controlled by the particle size.

The detailed modeling of the zodiacal light emission is still in progress. The bands observed by IRAS are strongly confirmed in the DIRBE data, and are seen at more wavelengths.

5. Data Delivery

A set of initial data products from the Cosmic Background Explorer Satellite is now available for analysis, and further data releases are planned for summer of 1994. A COBE Guest Investigator Facility, which supports proposers funded under NASA's Astrophysics Data Program, has been established at the Goddard Space Flight Center. The data are distributed from the National Space Science Data Center (NSSDC). All of the initial data products are available via anonymous ftp and accessible on the World Wide Web; most of the datasets larger than about 10 MB will be distributed on a set of CD-ROMs.

The data now available, and the associated documentation, may be obtained by anonymous ftp to nssdca.gsfc.nasa.gov (equivalent IP address 128.183.36.23). Please provide your e-mail address as the password. Change to directory (cd) anon_dir:[000000.cobe] (" cd cobe" should work) and download the file called AAREADME.DOC for further instructions. Alternatively, the data and documentation may be obtained on tape by request to the Coordinated Request and User Support Office (CRUSO), NASA Goddard Space Flight Center, Code 633.4, Greenbelt, MD 20771. They may be requested by phone at 301-286-6695, or by e-mail from request@nssdca.gsfc.nasa.gov. For assistance in the use of

COBE data, contact David Leisawitz by phone (301-286-0807), FAX (301-286-1771), e-mail (leisawitz@stars.gsfc.nasa.gov), or surface mail at the address given above. Please address requests for preprints or reprints of COBE Team publications to Susan Adams as follows: Ms. Susan Adams, NASA Goddard Space Flight Center, Code 685, Greenbelt, MD 20771, phone: 301-286-4257, or FAX: 301-286-1617, or e-mail: adams@stars.gsfc.nasa.gov. The COBE data are accessible on the World Wide Web at the following Uniform Resource Locator (URL) address:

`http://www.gsfc.nasa.gov/astro/cobe_home.html`

6. Summary and Conclusions

The COBE was designed to answer three questions: what is the spectrum of the CMBR, what are its anisotropies, and what is the brightness of the CIBR? The first two have been answered. The CMBR spectrum is very close to a blackbody of T = 2.726 K, and the weighted rms deviation from the blackbody form is less than 0.01%. The CMBR has anisotropies on all size scales larger than 7° , with a size spectrum consistent with the predicted scale-invariant primordial fluctuations. They have an amplitude of 11 parts per million on a 10° scale. The CIBR has not yet been found, but the data are in hand and analysis is in progress. All three instruments have given new views of the Galaxy, showing large scale features due to the interstellar medium and the general distribution of material. The Galactic bulge at the center is revealed at wavelengths where it is less obscured by dust absorption, and the Galaxy appears to have a central bar. Line emission from neutral and ionized carbon, ionized nitrogen, and carbon monoxide are widespread, and water may have been seen in absorption against the Galactic Center. The interplanetary dust (zodiacal light) has been observed more extensively than ever before, at a wide range of elongations from the Sun and throughout the year, enabling a new and more complete understanding of the dynamics and sources of the cloud.

7. Acknowledgements

The National Aeronautics and Space Administration/Goddard Space Flight Center (NASA /GSFC) is responsible for the design, development, and operation of the Cosmic Background Explorer (COBE). GSFC is also responsible for the software development and the final processing of the mission data. The COBE SWG is responsible for the definition, integrity, and delivery of the public data products, and includes the following members:

NAME	AFFILIATION	SPECIAL ROLE
J. C. Mather	GSFC	Project Scientist and FIRAS Principal Investigator
M. G. Hauser	GSFC	DIRBE Principal Investigator
G. F. Smoot	UC Berkeley	DMR Principal Investigator

C. L. Bennett	GSFC	DMR Deputy PI
N. W. Boggess	GSFC–ret	Deputy Proj. Scientist
E. S. Cheng	GSFC	Deputy Proj. Scientist
E. Dwek	GSFC	
S. Gulkis	JPL	
M. A. Janssen	JPL	
T. Kelsall	GSFC	DIRBE Deputy PI
P. M. Lubin	UCSB	
S. S. Meyer	MIT	
S. H. Moseley	GSFC	
T. L. Murdock	Gen.Res.Corp.	
R. A. Shafer	GSFC	FIRAS Deputy PI
R. E. Silverberg	GSFC	
J. Vrtilek	NASA HQ	Program Scientist
R. Weiss	MIT	Chairman of SWG
D. T. Wilkinson	Princeton	
E. L. Wright	UCLA	Data Team Leader

Many people have made essential contributions to the success of COBE in all its stages, from conception and approval through hardware and software development, launch, and flight operations. To all these people, in government agencies, universities, and industry, the science team and I extend my thanks and gratitude. In particular, I thank the large number of people at the GSFC who brought this challenging in-house project to fruition.

8. Bibliography

Arendt, R. G., *et al.* , *Back to the Galaxy*, eds. S. S. Holt & F. Verter, AIP Conf. Proc 278, 222, New York (1993)
Arendt, R.G., *et al.* , *ApJL* 425, L85-L88 (1994)
Barney, R. D., *Illuminating Eng. Soc. J.*, 34, 34 (1991)
Beichman, C. and Helou, G., *ApJ* 370, L1 (1991)
Bennett, C. L. *Highlights Astron*, 335 (1991)
Bennett, C. L. *et al.* , *ApJ*, 391, 466 (1992)
Bennett, C. L. *et al.* , *ApJ*, 396, L7 (1992)
Bennett, C. L., *et al.* , *ApJ*, accepted (1994a)
Bennett, C. L., *et al.* , *ApJ*, accepted (1994b)
Berriman, G. B., *et al. Back to the Galaxy*, eds. S. S. Holt & F. Verter, AIP Conf. Proc, 278, 214, New York (1993)
Berriman, G.B., *et al.* , submitted to *ApJL* (1994); COBE Preprint 94-09
Blitz, L., and Spergel, D., *ApJ*, 379, 631-638 (1991)
Boggess, N. W. *Highlights Astron*, 273 (1991)
Boggess, N. *et al.* , *ApJ*, 397, 420 (1992)
Bond, J. R., Carr, B. J., & Hogan, C. J., *ApJ*, 306, 428 (1986)

Bromberg, B. W. & Croft, J. *Adv. Astron. Sci.*, 57, 217 (1985)
Brown, J.M., *et al.* , *ApJL*, 428, L37 (1994)
Coladonato, R. J. *et al.* , *Proc. Third Air Force/NASA Symp. on Recent Advances in Multidisciplinary Analysis and Optimization*, 370, Anamet, Hayward, CA (1990)
Cowie, L. L., Gardner, J.P., Lilly, S.J., and McLean, I, *ApJL*, 360, L1-L5 (1990)
De Jager, O.C., Stecker, F.W., and Salamon, M.H., preprint submitted to *Nature* (1994)
Durrer, R., "Early Reionization in Cosmology", *Infrared Phys. Technol.*, 35, 83-94 (1993)
Dwek, E., *et al.* , *ApJ* submitted (1994)
Dwek, E., and Slavin, J., *ApJ* submitted (1994)
Evans, D. C., *SPIE Proc.*, 384, 82 (1983)
Evans, D. C.& Breault, R. P., *SPIE Proc.*, 384, 90 (1983)
Fixsen, D. J. *et al.* , *ApJ* 420, 445-449 (1993)
Fixsen, D. J. *et al.* , *ApJ* 420, 457-473 (1994)
Freudenreich, H. T., *et al.* , *Back to the Galaxy*, eds. S. S. Holt & F. Verter, AIP Conf. Proc, New York, 278, 485 (1993)
Freudenreich, H.T., *et al.* , COBE Preprint 94-10, submitted to *ApJL* (1994)
Ganga, K., Cheng, E., Meyer, S., and Page, L., *ApJ*, 410, L57-L60 (1993)
Gorski, K., *ApJL*, submitted (1994)
Gorski, K. *et al.* , *ApJL*, submitted (1994); COBE Preprint 94-08
Gott, R. *et al.* , *ApJ*, 352, 1 (1990)
Gulkis, S., Lubin, P. M., Meyer, S. S., & Silverberg, R. F., *Sci. Amer.*, 262, 132 (1990)
Gush, H.P., Halpern, M., & Wishnow, E.H., Phys. Rev. Lett. 65, 537 (1990)
Hacking, P., and Soifer, B. T., *ApJL*, 367, L49-L53 (1991)
Hauser, M. G. & Peebles, P. J. E. 1973, *ApJ* 185, 757 (1973)
Hauser, M. G. *et al.* ,*After the First Three Minutes*, 222, 161, eds. S. S. Holt, C. L. Bennett, & V. Trimble, AIP Conf. Proc, New York (1991)
Hauser, M. G. *Highlights Astron.*, 291 (1991a)
Hauser, M. G., *Proc. Conf. Infrared Astronomy and ISO, Les Houches*, 479 (1991b)
Hauser, M. G., *Back to the Galaxy*, eds. S. S. Holt & F. Verter, AIP Conf. Proc, New York, 278, 201 (1993)
Hauser, M.G., in *Extragalactic Background Radiation*, eds. D. Calzetti, M. Fall, M. Livio, and P. Madau, Cambridge University Press, to be published (1994)
Hawkins, I., and Wright, E., *ApJ*, 324, 46-59 (1988)
Holtzman, J. A., *ApJS*, 71, 1 (1989)
Hopkins, R. A., & Castles, S. H., *Proc. SPIE*, 509, 207 (1985)
Hopkins, R. A., & Payne, D. A. *Adv. Cryogenic Engineering*, 33, 925 (1987)
Hu, W., Scott, D., and Silk, J. 1994, "Power Spectrum Constraints from Spectral Distortions in the Cosmic Microwave Background", preprint CfPA-THJ-94-12, submitted to *ApJL* (1994)
Janssen, M. A. & Gulkis, S., *Proc. The Infrared and Submillimetre Sky After COBE, Les Houches*, 391, ed. M. Signore & C. Dupraz, Kluwer, Dordrecht (1991)
Kaiser, M.E., and Wright, E.L., ApJ, 356, L1 (1990)
Kogut, A., *et al.* , *ApJ*, 401, 1-18 (1992)
Kogut, A. *et al.* , *ApJ*, 419, 1-6 (1993)
Kogut, A., *et al.* , *ApJ*, submitted (1994); COBE Preprint 94-06

Lisse, C. M., PhD thesis, University of Maryland, Astronomy (1993)
Lisse, C. M., *et al.* , *ApJL*, submitted (1994); COBE Preprint 94-11
Liszt, H.S., and Burton, W.B., *ApJ*, 236, 779-797 (1980)
Low, F. J. *et al.* , *ApJ*, 278, L19 (1984)
Magner, T. J., *Opt.Eng.*, 26, 264 (1987)
Mather, J. C., *Opt. Eng.*, 21, 769 (1982)
Mather, J. C. *et al.* , *IAU Colloq. 123, Observatories in Earth Orbit and Beyond, Proc.*, ed. Y. Kondo, 9, Kluwer, Boston (1990)
Mather, J. C. *et al.* , *AIP Conf. Proc. After the First Three Minutes*, 222, 43, ed. S. S. Holt, C L. Bennett, & V. Trimble,AIP, New York (1991a)
Mather, J. C., *Highlights Astron.*, 275 (1991b)
Mather, J.C., Shafer, R.A., and Fixsen, D. J., Proc. SPIE, 2019, 146-157, 1993.
Mather, J. C. *et al.* , *ApJ* 420, 439-444 (1994)
Matsumoto, T., Akiba, M. & Murakami, H., *ApJ*, 332, 575-595 (1988)
Matsumoto, T., in *The Galactic and Extragalactic Background Radiation*, IAU Symposium 139, ed. S. Bowyer and C. Leinert (Dordrecht: Kluwer), 317 (1990)
Milam, L. J., *Illuminating Eng. Soc. J.*, 34, 27 (1991)
Mosier, C. L. *AIAA*, 91-361 (1991)
Noda, M., *et al.* , *ApJ* 391, 456-465 (1992)
Petuchowski, S. J. & Bennett, C. L., *ApJ*, 391, 137-140 (1992)
Petuchowski, S.J., and Bennett, C.L., *ApJ*, 405, 595-598 (1993)
Roth, K.C., Meyer, D., & Hawkins, I., *ApJL*, 413, L67-L71 (1993)
Sampler, H. P. *Proc. SPIE*, 1340, 417 (1990)
Silverberg, R. F. *et al.* , *Proc. SPIE*, 2019, 180-189 (1993)
Sinha, R.P., Kinematics of H I near the Galactic Center, PhD Thesis, University of Maryland (1979)
Smoot, G. F. et al., *ApJ*, 360, 685 (1990)
Smoot, G. F. et al., *AIP Conf. Proc., After the First Three Minutes*, 222, 95, ed. S. S. Holt, C. L. Bennett, & V. Trimble, AIP, New York, 95 (1991)
Smoot, G. F. *Highlights Astron*, 281 (1991)
Smoot, G. F. *et al.* , *ApJ*, 371, L1 (1991)
Smoot, G. F. *et al.* , *ApJ*, 396, L1 (1992)
Smoot, G. F., *et al.* , *ApJ*, submitted (1994); COBE Preprint 94-03.
Sodroski, T. J., *et al.* , *Back to the Galaxy*, eds. S. S. Holt & F. Verter, AIP Conf. Proc, New York, 278, 218 (1993)
Stecker, F. W, De Jager, O.C., and Salamon, M.H., *ApJL*, 390, L49-L52 (1992)
Torres, S., *et al.* , *Data Analysis in Astronomy* eds. V. di gesu, L. Scarsi, & M. C. Maccarone, Erice, June 20-27, Plenum Press (1989)
Volz, S. M. & DiPirro, M. J., *Cryogenics*, 32, 77 (1992)
Volz, S. M. & Ryschkewitsch, M. G., *Superfluid Helium Heat Transfer, HTD*, 134, 23, ed. J. P. Kelly & W. J. Schneider AME, New York (1990)
Volz, S. M., Dipirro, M. J., Castles, S. H., Rhee, M. S., Ryschkewitsch, M. G., & Hopkins, R., *Proc. Internat. Symp. Optical and Opto-electronic Applied Sci. and Eng.*, 268, SPIE, San Diego (1990)

Volz, S. M., Dipirro, M. J., Castles, S. H., Ryschkewitsch, M. G., & Hopkins, R., *Adv. Cryogenic Engineering*, 37A, 1183 (1991)

Weiland, J. L., *et al.* , *Back to the Galaxy*, eds. S. S. Holt & F. Verter, AIP Conf. Proc, New York, 278, 137 (1993)

Weiland, J.L, *et al.* , *ApJL* 425, L81-L84 (1994)

Wright, E.L., *ApJ*, 255, 401-407 (1982)

Wright, E. L,*Proc. The Infrared and Submillimetre Sky After COBE, Les Houches*, 231, ed M. Signore & C. Dupraz, Kluwer, Dordrecht (1991)

Wright, E. L., *ApJ*, 320, 818-824 (1987)

Wright, E. L. *Ann. NY Acad. Sci., Proc. Texas-ESO-CERN Sym*, 647, 190 (1990)

Wright, E. L. *et al.* , *ApJ*, 381, 200 (1991)

Wright, E. L. *et al.* , *ApJ*, 396, L13 (1992)

Wright, E.L., et al., *ApJ*, 420, 450-456 (1994a)

Wright, E. L., *ApJ*, accepted (1994b)

Zeldovich, Ya. B., and Sunyaev, R.A., Ap&SS, 4, 301 (1969)

Zeldovich, Ya. B., and Sunyaev, R.A., Ap&SS, 7, 20 (1970)

INFLATION, THE TOP AND ALL THAT

YUVAL NE'EMAN
Wolfson Distinguished Chair of Theoretical Physics
Raymond and Beverly Sackler Faculty of Exact Sciences
Tel-Aviv University, Tel-Aviv, Israel 69978

and

Center for Particle Theory, Physics Department
University of Texas, Austin, Texas 78712

Introduction and Outline

Inflationary Cosmogony was suggested by Alan Guth in 1981 [1], as a solution to an apparent clash between two "Standard Models" – the 1974 SM of the Physics of Particles and Fields and Big Bang Cosmogony (BBC), as the latter theory had crystalized, after the 1965 discovery of the 3K Cosmological Background Radiation (CBR). As a matter of fact, there were already several paradoxes present within BBC itself, difficulties which were known as the Causality (or Homogeneity) and Fine-Tuning (or Flatness) issues. After 1975, the application of the SM new physics added a new paradox, that of the missing Domains and the related issue of the absence of Magnetic Monopoles. In lecture I, we review the basic cosmological formalism; we then discuss the emergence of causal "Horizons" (leading to the Causality Paradox). This is followed by an exposition of the Fine-Tuning issue. Lecture II describes the essentials of the SM of Particle Physics (including a diversion related to the recent discovery of the Top quark), emphasizing the mechanism of Spontaneous Symmetry Breakdown (SSB), which plays an important role, both in causing the apparent clash with cosmological observations and in the subsequent resolution of the difficulty, through the mechanism of Inflation. We discuss the mystery of the missing Domains and Monopoles. Lecture III introduces the Inflationary Model, explains how it resolves the various paradoxes and presents the "bonuses", namely (1) a mechanism for the Big Bang and (2) a process which generates all the matter present in the universe. Two related issues are then discussed: (a) the "Graceful Exit", namely the transition to the present observed linear expansion, through a mechanism known as "Extended Inflation", a relatively recent, important improvement of the theory, due to La and Steinhardt [2]; (b) "Eternity", originally due to Linde [3], a return to a Steady-State-like eternal universe,

M. M. Shapiro et al. (eds.), Currents in High-Energy Astrophysics, 255–280.

except for the much grander scale and a discrete, rather than continuous, creation of matter. However, eternity also involves a new application of Copernican levelling humility, putting "our" universe "in its place", as yet another "ordinary sample", out of an infinity of universes, making up "reality". That reality, however, also turns out to require a new conceptual extension, a philosophical reappraisal, which I have described as "Cosmological Surrealism" [4].

1. Lecture I: Friedmann Cosmology; The Horizon and Fine-Tuning Paradoxes

A. THE COSMOLOGICAL EQUATIONS.

Einstein's 1917 corrected equations (including the cosmological constant λ) for the gravitational field read ($R, R_{\mu\nu}, g_{\mu\nu}, T_{\mu\nu}$ are respectively the scalar curvature and the Ricci, metric and energy-momentum density tensors, c is the velocity of light and G is Newton's constant)

$$R_{\mu\nu} - \frac{1}{2} R g_{\mu\nu} + \frac{8\pi G}{c^4} \lambda g_{\mu\nu} = \frac{8\pi G}{c^4} T_{\mu\nu} \tag{1.1}$$

yielding for a Friedmann/Robertson-Walker spherically symmetric metric ($k = 1, 0, -1$ respectively for an elliptic, flat or hyperbolic geometry; $S(t)$ is the scale function, so that r is dimensionless)

$$ds^2 = c^2 dt^2 - S^2(t)\left[\frac{dr^2}{1 - kr^2} + r^2(d\theta^2 + sin^2\theta d\phi^2)\right] \tag{1.2}$$

the 1924 Friedmann equations ($\rho(r,t)$ is the energy density),

$$\left(\frac{\dot{S}}{S}\right)^2 + \frac{kc^2}{S^2} = \frac{8\pi G}{3c^2} (\rho(r,t) + \lambda) \tag{1.3a}$$

$$2\frac{\ddot{S}}{S} + \left(\frac{\dot{S}}{S}\right)^2 + \frac{kc^2}{S^2} = \frac{8\pi G}{c^2} (-p(r,t) + \lambda) \tag{1.3b}$$

It is instructive to multiply (1.3a) by S^2 and take $k = 0$; the l.h.s. is then seen to represent the kinetic energy, the r.h.s. is proportional to the potential energy GM/S, with $M = (4/3)\pi(\rho+\lambda)S^3$. We see that λ is the vacuum energy. In quantum physics, it is a legitimate component which can have a non-vanishing value regionally, in a localized gravitational equation, rather than the unique-value overall cosmological attribute of Einstein's 1917 equation. In (1.3b) λ appears as a negative pressure density. Replacing the Hubble factor (whose inverse H^{-1} is the "age" of the universe)

$$H(t) = \frac{\dot{S(t)}}{S(t)} \tag{1.3c}$$

in (1.3b) by its value from (1.3a), still with $k = 0$, we have

$$\ddot{S} = -\frac{4\pi G}{3c^2} (3p + \rho - 2\lambda)S \tag{1.3d}$$

where the role of λ as the only repulsive force in gravity is obvious: the only way in which the acceleration on the l.h.s. can be positive is to have $\lambda > (3p+\rho)/2$. This is indeed what happens in the de Sitter model (also invented in 1917): take $k = 0, \rho = 0$ in (1.3a), getting

$$H^2 = \frac{8\pi G}{3c^2}\lambda, \quad H = \left(\frac{8\pi G\lambda}{3c^2}\right)^{1/2}, \; S = exp(Ht) \tag{1.4}$$

i.e. a constant Hubble factor (proportional to the square root of λ)

and an exponential ("inflational") expansion. For $\rho \neq 0$, still with $k = 0$, we may define the corresponding "critical density" for a given time,

$$\tilde{\rho}_{critical}(t) = \frac{3c^2}{8\pi G}\left(H(t)\right)^2 \tag{1.5}$$

where $\tilde{\rho}$ may include a non-time-dependent piece (i.e. a λ).

B. THE SCHWARZSCHILD METRIC, HORIZONS AND TIME-ARREST.

The first solution to Einstein's 1915 equations of General Relativity already involved a *horizon*. This was the Schwarzschild solution (for the Kepler problem). The (spherically symmetric) metric obtained,

$$ds^2 = \left(1 - \frac{r_s}{r}\right)c^2dt^2 - \left(1 - \frac{r_s}{r}\right)^{-1}dr^2 - r^2(d\theta^2 + sin^2\theta d\phi^2) \tag{1.6a}$$

where

$$r_s = \frac{2GM}{c^2} \tag{1.6b}$$

is the *Schwarzschild radius*. For the sun, this is about 2km. Note that when $r \to r_s$, for a point-like infalling system, $ds^2 = 0$ and time is arrested, similarly to what happens in a photon's system in Special Relativity. Frederick Pohl's novel "Beyond the Blue Horizon" has a wonderful illustration of that state of affairs. The hero has lived in a permanent state of depression for many years, ever since he and his beloved had been exploring a black hole, when he happened inadvertantly to push her in (which thereby also made him recoil and saved his life). He especially suffers because he knows that throughout all the years since, in her time-arrested frame, she is still "all the time" (i.e. all of his time) living through just that one same second of the push and wondering why he was doing it to her... There are other physical issues, relating to horizons, in which these time-arrest considerations might well play an important role. One such case is a controversy relating to the violation of 'global' conservation laws (i.e. of charges which do not induce a dynamical gauge field), such as baryon conservation, in the process in which a black hole is formed - and evaporates (by quantum tunneling, in the context of Hawking radiation). Coleman and Hughes [5] have claimed that as these charges are anyhow violated in GUT (gauge-unification theories) but at a very low rate (e.g. the proton lifetime, which is longer than 10^{32} years), in a strong gravitational field this will be accelerated and they will decay long before falling into the black hole. Thus, there is no mysterious disappearance of charge, only its 'natural' decay. Note that quantum tunneling causes the black hole to evaporate, with an ever-shrinking Schwarzschild radius-as its mass keeps diminishing - and with the consequent release from full time-arrest. Note also in (1.6) that a small space-like body (i.e. small r) would spread out and cover the entire horizon; this point is important in the recent rebuttals [6] of the claims that quantum correlations are lost when matter goes down a black hole's (BH) horizon.

Figure 1 represents a Schwarzschild solution, using Kruskal-Szekeres coordinates [7], in which it is easier to visualize the horizon's effects:

$$u = A(r)cosh\,\frac{t}{2r_s}, \quad v = A(r)sinh\,\frac{t}{2r_s}, \quad A(r) = [\frac{r-r_s}{r}]^{1/2}\,e^{r/2r_s}$$
$$ds^2 = 4(r_s)^3 e^{-r/r_s}(-dv^2 + du^2) + r^2(d\theta^2 + sin^2\theta d\phi^2) \tag{1.7}$$
$$t = 2r_s tanh^{-1}\,\frac{v}{u}, \quad \left|\frac{v}{u}\right| < 1 \qquad t = 2r_s tanh^{-1}\,\frac{u}{v}, \quad \left|\frac{v}{u}\right| > 1$$

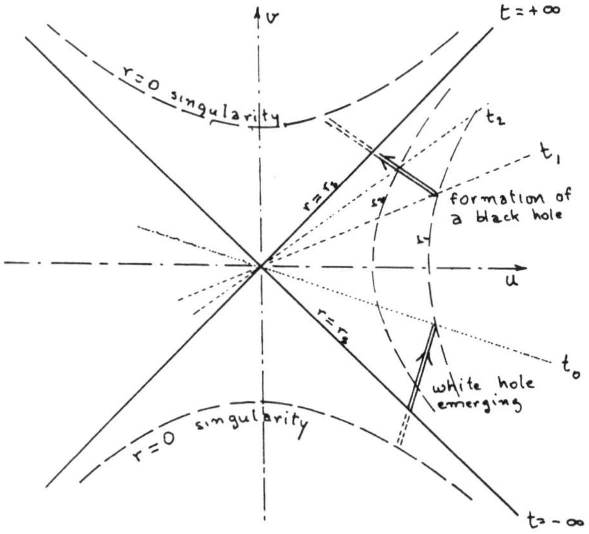

Figure 1: Horizons in the Schwarzschild Metric
Emergence of a White Hole; Collapse Leading to a Future Black Hole

Observe that, setting aside the angles θ, ϕ which play no role (having assumed spherical symmetry), we regain in v, u a set behaving as a 2-dimensional Minkowski spacetime. In the diagram, time appears as a rotating clock-hand (counter-clockwise), going in the right-hand quarter from $t = -\infty$ (at 04.30) to $t = +\infty$ (at 01.30). This quarter is the physical (external) Schwarzschild solution. A collapsing mass will forever, in this system, approach the Schwarzschild radius, but never cross it (or cross at $t = \infty$). However, for a particle inside the collapse, the free fall is "normal"; it will cross the Schwarzschild horizon after a few hours (in its time scale) and somewhat later hit the singularity. We shall return to these considerations in Lecture III.

Let us return to the discussion of the horizon's *time-arrest* feature, making the infalling particles appear like photons to the external observer. The analogy with a system moving at the velocity of light is not an accident. The Schwarzschild radius can be obtained, using Newtonian mechanics and equating the gravitational force $GMm/r^2 = mv^2/r$ to the centrifugal (M, m, v are respectively the BH mass, a test mass and the escape velocity) as that radius which would make $v = c$ (within a factor 2, as often happens in the Newtonian+Special Relativity approximations to GR, including the 1919 eclipse effect). This is how Laplace could conceive the idea of a BHs two hundred years ago – *as a body so heavy that even light would just get into orbit around it but would not be able to escape its pull.*

C. HORIZONS, GRAVITATIONAL BINDING ENERGY AND THE CREATION OF MATTER.

Another important aspect relates to the *gravitational binding energy.* Using again a combination of Newtonian physics with Special Relativity, we equate the gravitational binding energy ($\sim GM^2/r$) to Mc^2, or, in other words, ask *when does the (negative) gravitational binding energy "eat up" the mass-energy entirely?* Again, within a numerical factor of order 1, we get the Schwarzschild radius. In the de Sitter metric (see (1.4)) r_s is replaced by cH^{-1} (with precisely the same role of horizon). This explains why the Inflationay model (based on a de Sitter exponential expansion in an earlier phase of the universe's history) does provide that "free lunch": (1) the vacuum-energy *density* remains fixed throughout the exponential expansion in volume; (2) the *global* vacuum energy thus increases exponentially; (3) when this vacuum energy materializes, in a process known as *tunneling from the false to the true vacuum* (which we shall discuss in lecture III), all the matter in the observable universe, some 10^{90} particles, is created "out of nothing"; (4) the total [mass + binding energy] retains all the while its zero value, so that no conservation law is violated.

D. THE CAUSALITY PARADOX.

In a Friedmann universe (1.3), *the scale function* $S(t)$ represents the growth of distances in the expanding universe; $c(H(t))^{-1}$ is the *horizon.* The reason why we cease seeing things at a distance of 10^{10} light-years is that the galaxies on that horizon are receeding at the speed of light, and are thus in time-arrested frames. We observe them forever in the same frozen instant, they are infinitely red-shifted – and do not "have enough time" (our time, which has to be infinite for that purpose) to emit photons (they are in the same state as that poor girl in Pohl's novel). The ratio between the distance-scale and the horizon is thus

$$\frac{S(t)}{cH^{-1}} = \frac{S(t)}{cS(t)}\,\dot{S}(t) = \frac{\dot{S}}{c}, \tag{1.8}$$

which, in a linear expansion, is smaller or equal to 1. There is thus no possibility for matter outside the horizon to have been "put there" by the scale-growth at some previous stage, if the expansion was always linear. Morevover, the furthermost galaxies have just "recently" been reached by the horizon (which is growing faster than the distance-scale) and thus could not have gotten there "naturally". They thus represent regions which were separated from our own at the BB (e.g. they might have been at spatial intervals from "us" when the BB ocurred) until the time when our horizon reached them, perhaps a billion years ago. The *causality paradox* is the fact that there is no sharp difference between the galaxies at that distance and those in our neighborhood – except for the effects of age. For instance, why are there no sharp jumps in the densities?

We can anticipate and observe that *inserting a de Sitter phase [1] in the past history of the universe would provide a solution to the paradox.* In the de Sitter model (1.4), the distance-scale $S = e^{Ht}$ grows exponentially, whereas the horizon cH^{-1} is a constant! During such a de Sitter phase, all the matter outside our horizon could have been "laid there", after first "sharing our history" for a while. There would thus have been a period of causal contact – and there should thus be no surprise in observing a family resemblance!

E. THE FLATNESS AND FINE-TUNING PARADOXES

Using equation (1.5) for the critical energy-density, and inserting the value of the Hubble factor (from observations, i.e. 50-100 km/(sec Mpc)), we find (in mass units, i.e. dividing by c^2), $\tilde{\rho}_{critical} = .25 - 1.0 \times 10^{-29} g/cm^3$. The estimate for the actual value of the energy-density, as based upon observations of visible matter, detected neutrinos and electromagnetic radiation (including the 3K CBR) is about 10-100 times lower. This in itself is somewhat of a surprise: a priori, it could have taken any value, perhaps a billion times higher or lower. Considering the fact that there is evidence (from the dynamics of galactic clusters, etc) for "dark" (i.e. optically unobservable) matter, it could even be that the present density is precisely the critical! And since there is no reason why the present era should be special in any way, it is then possible that the density is always at the critical value! Remembering that this critical value corresponds to taking $k = 0$, i.e. a *flat* Friedmann universe, we understand why this has been called the *flatness* paradox.

There are additional considerations relating to this problem. Looking at equations (1.3a-b), denoting the ratio $\tilde{\rho}/\tilde{\rho}_{cr.}$ by Ω, we can rewrite (1.3a) as

$$(H(t))^2 + k\left(\frac{c}{S(t)}\right)^2 = \Omega (H(t))^2$$

or

$$1 + k\left(\frac{c}{\dot{S}(t)}\right)^2 = \Omega \tag{1.9}$$

This indicates the instability of the $k \neq 0$ case: in the roughly linear Friedmann expansion, having $k = 1$ or $k = -1$ implies $\Omega > 1$ or $\Omega < 1$, respectively (i.e. a larger than critical density closes the universe's geometry, a smaller than critical density opens that geometry). However, the difference increment in Ω falls off like S^{-2}, thus rapidly tending towards flatness. Moreover, inserting these results in (1.3b) and taking for instance the observed zero value for the pressure and a low value for λ, the second derivative $\ddot{S}$ is forced to be large and negative; calculations show that collapse (Wheeler's Big Crunch) would follow within a very brief time. Back in 1964/5, Novikov and I [8] had suggested that quasars (this was a brief time after their discovery) might represent lagging cores from the cosmological expansion ("white holes", see figure 1). In calculations with the late Gerald Tauber, our problem was always how to make them last billions of years - it did require tremendous fine-tuning - the same difficulties later encountered by Friedmann models as such. To preserve a semi-linear expansion over 10^{10} years with $k \neq 0$ requires fine-tuning of the parameters, an almost impossible task. Before 1981, this feature used to be explained by *anthropic* considerations. Personally, I am quite allergic to this notion. In Weinberg's recent book "Dreams of a Final Theory", he discusses Astrology and Superstition and gives "man-centeredness" as a criterion distinguishing non-science from science. I fully agree. The only sense in which I consider "anthropic" arguments to be scientific is in those cases in which the word really stands for *evolutionary* and has nothing to do with humanity. In the concrete example we are discussing here, one would first need to have a theory in which billions of universes are created, with one mutational fluke taking on the fine-tuned values. But the only consistent model with eternally recurring Big Bangs is precisely the Inflationary model, described in these lectures. This is anyhow a theory requiring neither evolutionary nor "anthropic" arguments: the de Sitter stage has $\Omega = 1$ (entirely "made of" vacuum energy) which is then also the initial value of Ω in the Friedmann stage after Inflation – which would make it very close to unity in the present.

2. Lecture II. The Standard Model (and the "Top" Quark): Spontaneous Symmetry Breakdown, Domains and Magnetic Monopoles

A. THE STANDARD MODEL AND THE ERLANGEN PROGRAM

Sometime in 1973-75, a new grand synthesis emerged in the Physics of Particles and Fields, known as the Standard Model. It is expressed in the language of Relativistic Quantum Field Theory, whose first successful contribution had been QED (Quantum Electrodynamics) in 1946-48. The interactions it describes are generated by Quantum (local) Gauge Theories, as applied to the Lie Groups

$$SU(3)_{color} \times SU(2)^{I}_{left} \times U(1)^{Y}. \tag{2.1}$$

I and Y stand respectively for the *weak* isospin and hypercharge. These are thus Yang-Mills interactions, mediated by 4-vector potentials, transforming like the adjoint representations of these groups, plus the inhomogeneous term needed to restore gauge invariance; in geometrical language, these are Principal Fiber Bundles, with Minkowski spacetime as base manifolds and the above three groups as structure groups. The potentials are connection 1-forms, the corresponding field strengths are the curvature 2-forms. This geometrical formulation became popular in 1974-76, when it was first used to produce non-perturbative solutions (the analog to a Schwarzschild or Kerr solution in Einstein's GR) such as "instantons" or magnetic monopoles (the latter in broken gauge theories – see the coming sections). It is a testimony to the strong links between mathematics and physics, or to the fitness of the mathematical description of the physical world. Back in 1872, Felix Klein (supported by Sophus Lie) had launched the "Erlangen Program", aiming at a "more natural" description of mathematics, through the use of group theory, coupled with differential geometry (transformation groups, algebraic topology, etc..). Surprisingly, physics is now entirely *à la* Erlangen.. Around 1956, it did not look as if the Strong and Weak Interactions would turn out to be geometrical. General Relativity (1915) was the first theory to geometrize physics (at the classical level, to date) but the nuclear interactions seemed very far from such a description. True, Hermann Weyl, inspired by GR, had reformulated QED in 1929 as a local gauge theory of the quantum phase of charged (i.e. complex) matter fields (after first trying mistakenly in 1919 to relate the electromagnetic field to scale invariance). In 1953, C.N. Yang and R.L. Mills generalized the treatment to non-Abelian Lie groups. It was first a tool looking for applications - which were soon to be found.

B. THE ELECTROWEAK UNIFICATION

First, the Weak Interactions turned out (1957) to involve vector and axial-vector currents, resembling the electromagnetic current - an indication of the action of a Lie group and of a vector potential. The W^+, W^- were indeed discovered in 1982. This was a beautiful vindication of "off-mass-shell" physics: from the value of the beta decay coupling constant (the Fermi coupling, $G_F = 10^{-5} m_N^{-2}$), measured in a reaction with a total available energy of slightly more than $1MeV$ (the mass of the neutron, minus those of the proton, electron and neutrino - if any), it was assumed that this really involved a coupling of the order of the charge on the electron, and that the value of G_F results from the presence of a propagator for an off-mass-shell W^-. The W^- is emitted by the neutron and decays into $e + \nu$. The emission of the W^- violates the conservation of energy by a factor of 85,000; it has to take place within a time such that $(\Delta t \times \Delta E) < h/2\pi, \Delta E = M_W c^2$, i.e. $\Delta t = .3 \times 10^{-26} sec$. This implies

$$\frac{G_F}{\sqrt{2}} = \frac{e^2}{8M_W^2 \, sin^2\theta_w}\,, \tag{2.2}$$

with a precise prediction $M_W = 85GeV$, for $sin^2\theta_w = .225$. θ_w is the "Weinberg angle", defining the mix between $SU(2)$ and $U(1)$.

Thus, the 1957 apparent similarity between weak and electromagnetic currents was more than just a vague resemblance: the electro-weak gauge involves the $3+1$ parameters of $SU(2)^I_L \times U(1)^Y$, with W^+, W^0, W^- as connections for $SU(2)$ and a Y^0 for $U(1)$. The symmetry is broken "spontaneously" (we shall review the mechanism later on) and the two electrically neutral potentials, W^0, Y^0, mix and recombine. One combination corresponds to an unbroken direction of the gauge symmetry; its connection field thus remains massless – the Maxwell field $A = sin\theta_w W^0 + cos\theta_w Y^0$. The orthogonal combination $Z = -cos\theta_w W^0 + sin\theta_w Y^0$ gauges a broken symmetry and is thus massive ($91GeV$). The weak neutral currents which couple to the Z^0 were indeed discovered in 1973; they play an important role in astrophysics, since the Kamiokande neutrino detector is based on observing the elastic scattering of neutrinos off electrons $\nu_e^0 + e \rightarrow \nu_e^0 + e$, in which one can thus observe the direction from which the neutrinos arrive: this very much enhances the confidence in the identification of solar neutrinos and was also very important in the detection of the neutrinos from Supernova 1987A.

C. HADRONS, QUARKS AND QCD; THE STANDARD MODEL GAUGE GROUP

The understanding of the structure of hadronic matter starts with our discovery of ("flavor" in the present nomenclature) $SU(3)$ (the "Eightfold Way") in 1961, soon followed by our identification of a substructure (1962). This became the quark model (1963), yielding beautiful predictions, though these were derived from a non-relativistic treatment. Quantum Chromodynamics (QCD), the $SU(3)_{color}$ gauge, was then adopted in 1974-5, as a description of the Strong Interactions at the fundamental level (between quarks in nucleons, etc..). First, "Asymptotic Freedom" (AF) was found as an intrinsic feature of Yang-Mills (unbroken) gauge theories: the coupling of the gauge potentials to matter (and to themselves, since they also carry the color charges, in contradistinction to the electromaglnetic field, which is itself electrically neutral) weakens and vanishes at very large momenta (the "UV" limit). This explains why the quarks, which are so tightly bound inside a nucleon that they cannot be scattered out, nevertheless appear "floating", almost unbound, inside the nucleon (i.e. at very short distances) when probed by large momentum transfer photons - or in "impulse approximation" calculations of the non-relativistic quark model. This is the UV end, and QCD can be used here perturbatively. Extrapolating to the opposite, IR end, one could then conjecture that at some "large" distance (such as $1fm$, the "size" of a nucleon) the same couplings become so large that they diverge, and the quarks or gluons are trapped forever.

This is known as "color confinement". After 20 years, it has yet to be proven (beyond Wilson's 1974 demonstration, using a lattice approximation). QCD has been very successful at the UV end, for instance in understanding "scaling", as displayed in deep-inelastic electron-nucleon scattering (scaling results from the pointlike character of the interaction of photons with the quarks, inside the nucleon). No progress has been made, to-date, in rederiving from QCD the beautiful (experimentally validated) detailed predictions of the "naive" quark model. Since 1990, I have personally been involved in such an attempt, with very tantalizing results; however, up to now, they are mostly qualitative [9]. I hope we can develop the idea into a useful quantitative dynamical tool.

Summarizing the gauge interactions of the SM, we have the Lagrangian,

$$\begin{aligned}\mathcal{L} = &-\frac{1}{4g_3^2}\,F^a_{\mu\nu}F^{a\mu\nu} - \frac{1}{4g_1^2}\left[F^i_{\mu\nu}F^{i\mu\nu} + \frac{1}{tan^2\theta_w}\,F^0_{\mu\nu}F^{0\mu\nu}\right]\\ &-\left[D_\mu\Phi_m\right]^2 - \left[-\mu^2|\Phi^2_m| + \lambda|\Phi^2_m|^2\right] - \overline{\Psi}(\gamma^\mu D_\mu + \mathcal{M})\Psi \\ &a = 1,\ldots 8,\quad i = 1,\ldots 3,\quad m = +,o\,;\end{aligned} \tag{2.3}$$

Here g_3 is the QCD strong coupling, g_1 the weak $SU(2)$ coupling. g_2 of $U(1)$ has been replaced by $g_1 tan\theta_w$. The a are the $SU(3)_{color}$ indices in the octet representation; the matter fields Ψ include the quarks, assigned to the **3** representation (and the antiquarks in **3***) with the leptons ($SU(3)_{color}$ scalars) and anti- leptons; $\mathcal{M}$ is the bare mass matrix, and all internal and Lorentz-spinor indices have been left out. The i are $SU(2)$ indices, in the adjoint $I_L^W = 1$ triplet representation. $U(1)^Y$ is gauged by the "0" component of the field strengths. All matter fields are either spinors or scalars under the electroweak $SU(2)_L^I$ (see table I). The covariant derivatives ensure parallel-transport under the full SM reducible group, acting upon the relevant representations. The Φ is a two (complex) components ($m = +,o$), (i.e. a charged and a neutral) field, with its ($m = -,\overline{o}$) conjugate, $\overline{\Phi}$; it transforms as a scalar under $SU(3)_{color}$ and as an isospinor $I_L^W = 1/2, Y^W = -1$ under the electroweak gauge, i.e. the same quantum numbers as the ν_L^0, e_L^- lepton isodoublet in table 1. In the context of the reducible gauge group $SU(2)_L^{weak} \times U(1)^{weak}$, Φ is one more type of matter field, with a further "incidental" role in the "spontaneous" breakdown of the electroweak symmetry. However, in an alternative **irreducible** formulation, involving the gauging of the (internal) supergroup $SU(2/1)$, the $\Phi, \overline{\Phi}$ form the "odd" part of the gauge supermultiplet [10]. This complex field is known in particle physics as the "Higgs field". We shall describe this spontaneous symmetry breakdown mechanism (SSB) in sections F,G - starting with the Landau 1938 theory of phase transitions, which inspired it.

D. MATTER FIELDS AND INDICATIONS OF GUT ACTION; SUPERSYMMETRY

The SM organizes matter in 3 generations of 15 fields each - a classification with a not very economical basic set.. Note that the number of generations is now known to stop at three, as long as the generation's neutrino is lighter than some $40GeV$ (to the extent that neutrinos might yet be massive). This number represents a triumph for astrophysics, since nucleosynthesis calculations relating to the Big Bang already put an upper bound of 7, then 4 generations, from the deuterium production [11]. In 1988, both CERN and SLAC, studying Z^0 decays, put the bound at the 3 existing generations, (as defined by their "light" neutrinos, "light" meaning here masses smaller than some 45 GeV) [12].

Each generation includes a weak-isospin doublet of left-chiral quarks, plus their right-chiral partners (i.e. two weak-isospin singlets), each in three "colors" (i.e. as an $SU(3)_{color}$ defining representation **3**), altogether 12 chiral quark fields; in addition, there are three leptons per generation: a left-chiral neutrino and its related left- chiral charged lepton (making up a second weak-isospin doublet) and the right-chiral partner of that charged lepton, i.e. a weak-isospin singlet. The total is thus 15 fields per generation, altogether 45 "fundamental" matter fields (see table t1).

The $SU(2)_L^{I(weak)}, U(1)^{Y(weak)}$ quantum numbers obey an isospin-hypercharge rule (yielding the electric charges) similar to that of the Strong Interactions' Gell-Mann/Nishijima rule of 1953:

$$Q_e = I^3_{L,weak} + Y_{weak}/2 \tag{2.4}$$

As a result, the left- and right-chiral fermions have completely different quantum numbers and couple very differently to the $SU(2), U(1)$ gauges. Primordially, they are totally different fields and should be massless, like neutrinos. They pair-off as a result of unknown interactions which provide them with their UV masses – presumably interactions mediated by scalar fields with Yukawa couplings. These break the chiral symmetry, very little for the 1st generation fields, much more for the 2nd and 3rd. In addition, QCD amplifies these masses for the quarks, with a stronger breaking of chiral symmetry, making up effective "constituent" quarks (e.g. $300MeV$ instead of $5 - 9.5MeV$ for the u,d quarks). In my derivation, QCD effectively induces another (strongly interacting) scalar field, which interferes additively with the above Yukawa scalar fields. At the same time, the diagonalized states (in any case) come out somewhat "mixed'. The best known example is that of the "Cabibbo angle", causing the "original" *d',s'* quark fields (or "current quarks") to be given - in terms of the Strong Interactions' "constituent quark" eigenstates *d,s* - as $d' = cos(\theta_C)d + sin(\theta_C)s, s' = -sin(\theta_C)d + cos(\theta_C)s$, with $\theta_C = 15^o$. This explains the related beta-decays of the neutron (udd) and the $\Lambda^0(uds)$, effectively representing the decay of the d' in (udd'). For three generations, there is the possibility of four such mixing angles (or parameters), including a possible complex phase in the b, t system, which might explain the observed $\epsilon = 0.002$ CP violation, in $K_L^0 \to 2\pi$. The 3×3 array of parameters is known as the Cabibbo-Kobayashi-Maskawa (CKM) matrix.

The composition of a generation is not halfhazard - the sum of the electric charges has to vanish, for the theory to be renormalizable. Indeed, $6 \times (2/3) + (6 \times (-1/3) + (0) + 2 \times (-1) = 0$. This is essential for the cancellation of divergent "quantum anomalies" due to triangle diagrams such as in $\pi^0 \to 2\gamma$, in which the pseudo-scalar meson couples (with a γ^5) to a charged fermion, which then runs in a triangle and emits two photons. The total effect of the set of charged fields in one generation cancels, due to the vanishing of their added charges. This *anomaly* constraint is an indication of the existence of a larger algebraic structure, in which the SM reducible gauge group (2.1) is embedded, a GUT (formerly read as "grand unification theory", now changed to "gauge unification theory"). A second such hint comes from the uniqueness of the electric charge quantum. Why should'nt there be particles with 2.43567... times the charge on the electron? The unique quantum implies a non-linear relation such as could be given by an encompassing simple (irreducible) group.

A third indication of a GUT is provided by the Renormalization Group (RG) calculations. The g_1, g_2 (or g_1, e) become equal (up to an algebraic factor) above about $100GeV$; RG calculations show that g_3 rf QCD becomes equal to the first two, at about $10^{16}GeV$. This energy is therefore designated as M_{GUT}. The implication is that above that energy, there is a GUT symmetry regime, a larger and simple gauge group (i.e. irreducible) such as $SU(5)$, $SO(10)$ or $E(8)$ - or under a supersymmetrized version of such a group. The GUT or superGUT contain the SM reducible gauge group $SU(3) \times SU(2) \times U(1)$ and reduces to it as a result of yet another spontaneous symmetry breakdown mechanism, acting at M_{GUT}. In the language of condensed matter physics, this is a *phase transition.* In the immediate post Big Bang evolution of the universe, one conjectures that when the temperature cooled down from $10^{19}GeV$, it went through such a phase transition, at $10^{16}GeV$, bringing it to a state in which the QCD parameters separated from the rest.

We should note that having two SSB systems, one at $100GeV$ and the other at $10^{15}GeV$ is an unstable result, using Relativistic Quantum Field Theory (the lower system tends to rise to the level of the higher one through radiative corrections). *Supersymmetry* was suggested as a possible explanation: the Φ and W, Y bosons would then have chiral fermionic partners, massless in the chiral symmetry limit; both these partners and the gauge and Higgs bosons themselves would then acquire masses lower than $1TeV$, as a result of either chiral symmetry breaking, supersymmetry breaking or both - a stable solution. The only alternative explanation which has been suggested to-date is "technicolor" - regarding the lower Higgs boson as a composite object, rather than as an elementary field; this removes the mass-raising radiative corrections altogether. The SSB boson would then be made of "techniquarks" as a "pseudo-Goldstone/Higgs boson", just as the pion is now considered to represent a quark-antiquark composite - even though it serves as SSB Goldstone boson for "flavor" chiral $SU(2)$ symmetry. Both solutions predict a wealth of new particles at $.5 - 1TeV$ and the now defunct SSC accelerator project was designed to search for them and identify the solution selected by nature (present evidence appears to favor supersymmetry).

There is a fourth, much weaker, "piece of evidence" for this picture of a GUT broken at $10^{16}GeV$, this time from Cosmology [13]. One assumes that the Big Bang (at a "temperature" of $10^{19}GeV$) produced equal amounts of particles and antiparticles. Presently, however, there is an overwhelming dominance of matter in the universe: had many galaxies been made of antimatter, we would have observed, in the cosmic background radiation (CBR), a strong component of gamma rays, produced by annihilation in the contact zones. Assuming a GUT broken at $10^{16}GeV$, with decays involving CP violation (extending by the above $\epsilon = .002$) due to a CKM phase, and matter/antimatter annihilation stopping only when all that is left belongs to one "side" only (the side favored by ϵ), one finds

$$\frac{N_B}{N_\gamma} = 10^{-9} - 10^{-10} \tag{2.5}$$

where (N_B stands for the density of baryons and N_γ for that of the photons) this figure fits observational estimates, based on the measurements of the $3K$ CBR.

Generation1		Generation2		Generation3		$SU(3)$ (QCD)	$SU(2) \times U(1)$ quantum numbers			
Field	Mass	Field	Mass	Field	Mass	Color	e-charge	I_W^L		Y_W
								I	I_3	
u_R	$5MeV$	c_R	$1.2GeV$	t_R	$172GeV$	3red 3green 3blue	2/3	0	0	4/3
u_L	$5MeV$	c_L	$1.2GeV$	t_L	$172GeV$	3red 3green 3blue	2/3	1/2	+1/2	1/3
d'_L	$9MeV$	s'_L	$190MeV$	b_L	$4.2GeV$	3red 3green 3blue	-1/3	1/2	-1/2	1/3
d'_R	$9MeV$	s'_R	$190MeV$	b_R	$4.2GeV$	3red 3green 3blue	-1/3	0	0	-2/3
ν_L^e	0?	ν_L^μ	0?	ν_L^τ	0?	1	0	1/2	+1/2	-1
e_L^-	.5MeV	μ_L^-	106MeV	τ_L^-	1,784MeV	1	-1	1/2	-1/2	-1
e_R^-	.5MeV	μ_R^-	106MeV	τ_R^-	1,784MeV	1	-1	0	0	-2

Table 1: Matter Fields, in the Standard Model.

E. THE TOP QUARK

The heavier quarks (c,b,t) were all first detected in paired systems, i.e. mesons, $\Psi = \bar{c}c(3,097MeV)$, $\Upsilon = \bar{b}b(9,460MeV)$ and now $\bar{t}t(348GeV)$. As a matter of fact, one also measures the ratio $R = \frac{\sigma(e^+e^- \to hadrons)}{\sigma(e^+e^- \to \mu^+\mu^-)}$, which is proportional to the sum of the squares of the electric charges of the quarks "available" at the given energy. Thus R starts at 5/3 (only u,d quarks, in ω^0, ρ^0); then climbs to 2 (s quarks in ϕ^0, at $1.02GeV$). At the Ψ^0 mass $(3.1GeV)$ i.e. $\bar{c}c$, $R = 10/3$ and at $\Upsilon(9.4GeV)$ it gets to 11/3 and stays fixed throughout a very much longer interval, before climbing up to what is thought to be the "final" value of $R = 5$. The top was found [14] by a team of 350 authors (reporting in a 50 pages report) at FNAL in April 1994, just before these lectures were delivered (see Figure 2). Note that a $1.8TeV$ (center of mass) energy was needed, since the system has to produce many extra accompanying particles, aside from the top (the same will be true in the future experiments searching for the Higgs meson - see in the sequel). In the top identification, $p+\overline{p}$ annihilate - presumably into at least three gluons; the latter then produce various hadron clusters, plus two which can be identified as having resulted from one gluon making a $t+\overline{t}$ pair, characterized by their large angle split. The presence of the heaviest quark is detected through the identification of the decay

$$t \rightarrow b + W^{-}, W^{-} \rightarrow e^{-} + \overline{\nu}_{e}^{0}, \tag{2.6}$$

and of its charge-conjugate for the $\overline{t}$. The b and $\overline{b}$ (each "cruising" respectively together with two other quarks/antiquarks, presumably of the first generation) are identified through their weak-interaction cascading decays.

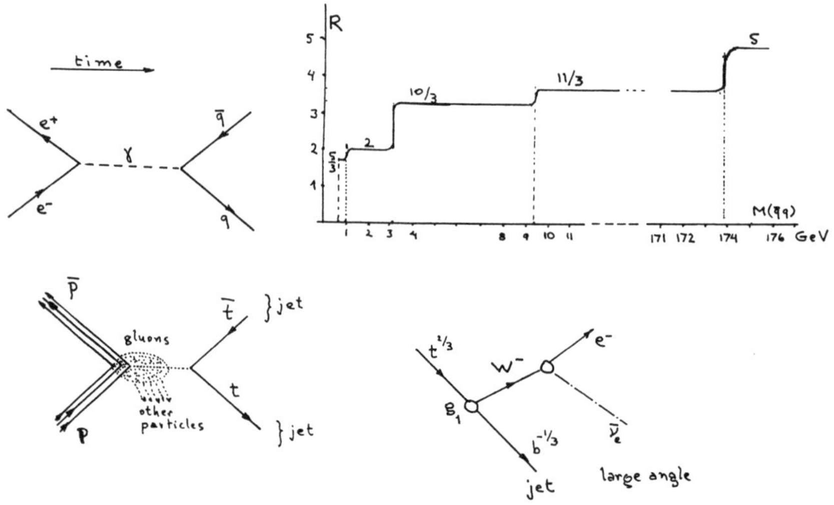

Figure 2: Production of the Top Quark

F. SPONTANEOUS SYMMETRY BREAKDOWN - THE LANDAU (1938) THEORY

"The Higgs Mechanism", was introduced in particle physics by Peter Higgs - and independently by five or six other authors, around 1964-67; as a matter of fact, this was also the problem suggested to me for my thesis in 1960 by Salam*. It is a direct adaptation of the 1938 Landau theory of phase-transitions, which I shall now describe.

---

* I preferred to stick to the search for a "global" symmetry (it then meant a symmetry relating π and K couplings), which I had started on my own; Salam's reaction was "you are embarking on a highly speculative search; if you insist, then do it, but first learn your group theory in depth - beyond the group theory I know and taught you!"

Our model is taken from the behavior of paramagnetic materials. Above the *"critical temperature"* T_c, there is no magnetization. Temperature is the reflection of the kinetic energy of the average molecule (here, of a metallic atom). In this phase, this kinetic energy is larger than the mutual magnetic alignment energy between neighboring atoms. Underneath T_c, the kinetic energy is weaker, and neighboring atoms - with their rotating electrons turning them effectively into tiny electromagnets - now manage to overcome the stochastic "jumping" due to the temperature and to align parallelly with each other. Landau wrote for the "Free Energy",

$$F = A(T)M^2 + B(T)M^4, \quad A = a(T - T_c), \tag{2.7}$$

to which Ginzburg later added a term proportional to the squared gradient of M, the magnetization 3-vector. The powers of M in (2.7) are all even, to make scalars and to fit the parity of the energy. Now for $T > T_c, A > 0$; the minimal free energy

$$F_{min}(T > T_c) = 0, \quad M_{min}(T > T_c) = 0, \tag{2.8a}$$

since both A, B, are positive. However, when the temperature drops beneath the critical, i.e. $T < T_c$, the coefficient of M^2 is negative, i.e. $A = -|A|$. The minimal free energy is found by making $F' = \partial F/\partial M^2 = 0$, yielding $-|A| + 2BM^2 = 0$. Thus

$$F_{min}(T < T_c) = -\frac{|A|^2}{2B} + \frac{B|A|^2}{4B^2} = -\frac{|A|^2}{4B^2}, \quad M_{min}(T < T_c) = \pm(\frac{|A|}{2B})^{1/2}. \tag{2.8b}$$

See figure 3.

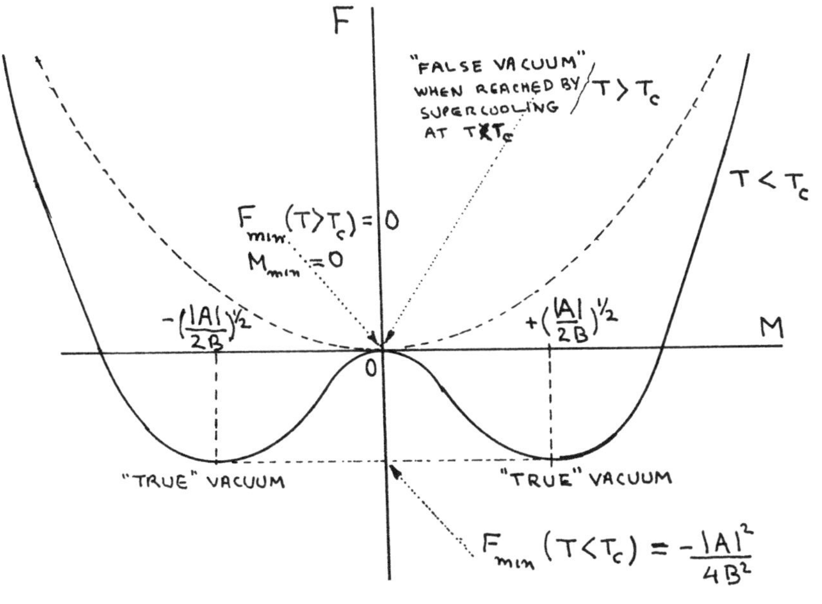

Figure 3: Magnetization minima, above and below the critical temperature.

Below the critical temperature T_c, there are two allowed *"true"* vacuum states, respectively at positive and negative values of M. And yet, the $M = 0$ vacuum state corresponding to temperatures higher than T_c can sometimes be maintained even when $T < T_c$, through an "evasion" of the phase transition, in a delicate treatment; for instance, in the transition corresponding to liquification (from a gas phase), one may preserve the gaseous phase underneath the evaporation point, by *"supercooling"* (the same is true for the liquid/solid transition). In such cases, the "$M = 0$" vacuum is unstable and is known as a *false* vacuum; a small perturbation will cause it to vanish and be replaced by either of the "true" vacua of (2.8b). In quantum phases separated by a potential wall, this will happen, the wall notwithstanding - through the *"tunneling"* of the vacuum state from its temporary "false" level, down to a "true" one, meanwhile emitting away the extra *vacuum-energy*, in a process known as *"reheating"*. In our exposition of "Inflationary Cosmology", this concept of an unstable *false vacuum* will play the key role: a vacuum energy, in the language of classical General Relativity, is simply a local cosmological constant, the λ of equations (1.3-1.5). It can then trigger an exponential expansion (a de Sitter phase), which will last as long as the vacuum energy is there.

Another characteristic of this kind of phase transition is the creation of *"domains"*. In the magnetic material, the alignment between neighboring atoms will be different in different regions of the material. The alignment occurs between neighboring atoms in some random direction, corresponding to their positions in the kinetic energy "dance" at the instant when the magnetic energy wins, over that stochastic jumping. Should we want to make a magnet from this piece of material, we shall still have to bring in an outside strong magnetic field, which will rotate the various domains with their different orientations into one single overall parallel alignment. Before this action, the domains point in different directions, with a *"domain wall"* separating any two such domains. Such walls represent high-density energy concentrations, due to the "tension" between neighboring magnetization vectors. Geometrically, they are given by *"topological defects"*. Domains and walls - or rather their absence when observing the universe - represented yet another paradox in fitting the physics of the SM with cosmology.

G. THE HIGGS MECHANISM AND THE DOMAINS/MONOPOLES PARADOXES

In the adaptation of the Landau theory to particle physics, the original motivation leading to the identification of the presence of SSB came from having to deal with gauge ("vector potential") bosons which were found to be massive, even though the relevant mass terms broke the local gauge symmetry. This was already true for the ρ^+, ρ^0, ρ^- triplet phenomenologically found to gauge (strong) isospin $SU(2)$ and, as a matter of fact, for the entire $\mathbf{8}+\mathbf{1}$ of (strong, i.e. flavor) $SU(3)$ which we identified in 1961, including, aside from the ρ triplet, ϕ^0, ω^0 isosinglets, gauging linear combinations of (strong) hypercharge and baryonic charge and the $K^0, \overline{K^0}$ isodoublet. For these phenomenological fields, a solution was found by regarding them as fulfilling the gauge role "incidentally", as a result of the fact that their quantum numbers made them appear as S-matrix poles in dispersion relations involving the weak and electromagnetic currents (see Gell-Mann's 1962 article in our 1964 book, "The Eightfold Way"). But such arguments could not be raised for the gauge mesons of the weak and electromagnetic currents themselves! And yet gauge-violating masses for the W, Z mesons would also deny the theory its chance for renormalizability.

In the (2.3) Lagrangian, the kinetic energy term of the Φ and its conjugate include an interaction (within the covariant derivative) between the gauge vector-bosons and the $\Phi, \overline{\Phi}$, which becomes quadratic in both types of fields in this term. This made it possible to think of the mass as a "soft" mass, resulting from the fact that (the $\Phi_1^0 = (\Phi^0 + \overline{\Phi}^0)/\sqrt{2}$, i.e. the real part)

$$< 0|\Phi_1^0|0 >= v \neq 0 \tag{2.9}$$

(modeled after M_{min} in (2.8b), as against (2.8a)) giving for the vector mesons a mass proportional to v. This also finds its analog in the "mass gap" in condensed matter physics. The only vector field in the set $\mathbf{W}, Y$ which does not get a mass is that linear combination of W^0, Y^0 which is not involved in the Higgs field's covariant derivative; since the component in (2.9) is electrically neutral, this is the electromagnetic field. 't Hooft proved in 1971 that such a mass does not hurt renormalizibility. The Higgs field potential in (2.3) follows Landau's (2.7) precisely, with the (Lorentz-scalar) Φ carrying a non-trivial representation of the internal gauge group, instead of the magnetization's space-vector behavior. The charged components of Φ are shown to provide for effective longitudinal components for the charged W mesons - absent in massless but essential for massive mesons. The same is true of the imaginary component Φ_2^0, which gives mass to the Z^0. As a result, only the real Φ_1^0 (above the vacuum energy v of (2.9)), has to exist - at least as an effective massive field (it gets its mass from the quartic term in the potential). The SM predicts (θ_w is the Weinberg angle - see section 2B) equation (2.2) and the values,

$$M(W^+) = M(W^-) = v\ sin\theta_w; \quad M(Z) = v\ cos\theta_w; \quad M(\gamma) = 0. \tag{2.10}$$

and does not fix the mass of Φ_1^0. Experimentally, $sin^2\theta_w = .225$. Our $SU(2/1)$ theory predicts $sin^2\theta_w = .25$ and $M(\Phi) = 2M(W)$, i.e. a Higgs mass of $170GeV$. This value for the Weinberg angle fits the region of $2 - 5TeV$, according to a recent calculation applying the renormalization group [16]; using this fact and the recently obtained value of the mass of the top quark, we find RG corrections to the quartic coupling, bringing down the prediction for the Higgs mass to around $130 - 150GeV$.

Now to the paradoxes. If the history of the universe did pass through a SSB of some GUT at $10^{15} - 10^{16}GeV$, one would have expected to observe "domains" in the universe, i.e. regions with different values of the relevant v_{GUT} and of $v_{electroweak}$ itself, reflected as different values of the fine-structure constant (the charge on the electron), the Fermi constant, θ_w, etc. Nothing of the sort is seen; on the contrary, one can check, e.g. that the fine-structure constant is the same at a red-shift of 5, from the lines in quasar spectra.

The solutions to a Yang-Mills equation for a simple group $\mathcal{G}$, with a SSB Higgs field spanning a group quotient $\mathcal{G}/\mathcal{H}$, include magnetic monopoles. Their mass should be extremely large and their interaction relatively strong. They should be produced in the walls separating different domains. The search for monopoles in cosmic radiation, etc.. has been on for many years and the experimentally established upper bounds on their flux are several orders of magnitude below the theory's predictions. Where are the monopoles? These were the two paradoxes presented by the application of the particle physics SM to cosmology.

3. Lecture III: The Inflationary Model

A. RESOLUTION OF THE PARADOXES BY THE INFLATIONARY MECHANISM; OUTLINE

Inflationary Cosmology is the name given to a hypothesis, first suggested by A.H. Guth in 1981 [1] as a solution to the paradoxes we listed: (1) Homogeneity without causal relationships (horizons); (2) Apparent quasi-flatness requiring tremendous fine-tuning; (3) Observations showing no domain structure; (4) Absence of magnetic monopoles. The theory, in addition, provides (5) a mechanism for the initiation and dynamics of the Big Bang, as a natural physical process; and (6) a mechanism for the creation of all matter in the universe. A.D. Linde [3] developed point (5), leading to the rebirth of Steady-State Cosmology - on a much grander scale, as compared to the Bondi/Gold/Hoyle 1948 theory, discarded after the discovery of the CBR, generally accepted as evidence for the Big Bang. P.J. Steinhardt also contributed various features, including [2] the most plausible mechanism for the *termination* of that *inflationary phase* in the history of "our" universe. We start with a general outline of the model (see Figure 4).

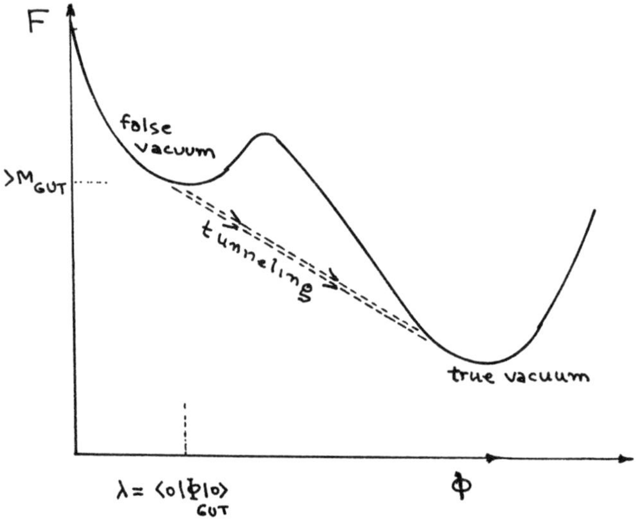

Figure 4: False and True Vacua,
Tunneling through the Separating Potential Barrier.

The Inflationary model consists in hypothesizing such an inflationary phase, lasting during the first 10^{-35} seconds in the history of the observable universe. During this stage, the universe grows in size (the scale function $S(t) = e^{Ht}$), at least by a factor of the order of 10^{35}, i.e. about 10^{100} in volume. It is assumed that this growth is that of a de Sitter model (see section IA), caused (or triggered) by the local occurence of a false vacuum (see IIF) with an energy somewhere between $10^{16} - 10^{19} GeV$ (see IIG), corresponding to Compton wavelengths of $10^{-33} - 10^{-30} cm$, i.e. an energy-density of $10^{105} - 10^{110} erg/cm^3$. The energy-density stays fixed throughout the inflationary stage, without violating energy conservation, because the total energy stays null all the time (the negative gravitational binding energy cancels the inertial Mc^2 - see IC). This universal vacuum energy (about $10^{85} - 10^{90}$ g, for a "universe" of the size of an orange) is next released as "reheating", when the false vacuum manages to tunnel through to the true vacuum. The former (false) vacuum energy now generates matter; the microscopic mechanism would involve quark and lepton pair creation, by vacuum gravitons. All the matter in the observed universe, which is of the order of 10^{90} particles (assuming the photons represent the results of fermion annihilations - see IIB,D) is created at that stage. As a matter of fact, the end-of-inflation mini-universe could have included much more than that "orange" - the extra matter corresponding precisely to what lies behind and beyond our present 10^{10} light-years horizon. The inflationary stage is terminated by a mechanism which has been described as a *graceful exit*, being replaced by a Friedmann "linear" expansion with $\Omega \sim 1$ and leading after 10^{10} years to the present observable universe. The earliest part of that Friedmann stage includes reheating and the creation of matter, then annihilations - yielding the $N_B/N_\gamma = 10^{-8}$ ratio (see section IIB), followed by a recombination phase, with the separation of the radiation from the matter, that radiation thus forming the CBR, with its inhomogeneities as revealed by the 1993 COBE satellite observations.

The model resolves the paradoxes in the following way: 1) Acausal homogeneity: it becomes causal, because in the de Sitter phase, distances (as defined by the scale function S) grow exponentially, while the horizon is given by the constant (1.4) inverse Hubble function (see section ID) $cH^{-1} = S/\dot{S}$. Thus the ratio favors the distance scale exponentially! All the matter entering our horizon was thus "put" there during the inflationary stage. It was therefore causally connected, prior to the Friedmann stage, during the crucial formative instants. 2) Flatness and Fine-Tuning. During the inflation, $k = 0, \Omega = 1$. As a result, Ω has stayed close to unity since, i.e. a quasi-flat geometry. There is then no need for fine-tuning, the Friedmann evolution equations are stable. Another conclusion: look for the missing ("dark") matter, or for a sizable cosmological constant, capable of "making up" for that missing matter, if the latter is not found. 3) Domains: no room for them in our entire observable universe, all of it representing just a tiny fraction of one domain in that phase transition. 4) Monopoles: they are generated on "walls" which are very far (perhaps at 10^{100} light-years, rather than at $10^9 - 10^{10}$). They might thus "arrive here" in 10^{90} years, except that with such a larger universe, they would be very diluted and their flux negligible and probably undetectable even then..

The new features: 5) The "creation" of a universe is triggered by the presence of a localized vacuum energy, in a "false vacuum" quantum state. This may result from a vacuum fluctuation of the gravitational field, or of the $10^{16} GeV$ Higgs field (now designated as the *inflaton*) - or by some other mechanism, such as a collision between two cosmic rays of unusually high energy (the "normal" cosmic rays stop roughly at $10^{11} GeV$), an event which might take place once every $10^{10^{10}}$ years...[16]. It even becomes conceivable to create a universe in the laboratory..[17]. More plausibly, it has been suggested that adequate "seeds" for new universes might be provided by just those domain walls whose absence in the observations was one of the paradoxes which brought about the idea of inflation. Such a wall (a topological "defect") represents tremendous tensions and could be the source of a very large vacuum energy [18]. An important philosophical implication is that the universe is "expanding" or "reproducing" *eternally*. We shall return to this point and exhibit some novel conceptual "re-educational" features, resulting from this combination of horizons with eternity.

6) The creation of matter (and the seeds of new spatial expansion) is a discrete natural process, occuring at some irregular intervals. No violation of energy conservation is involved.

In the remaining sections we shall discuss the issue of the "graceful exit", list some possible tests of the theory (including the fit with the COBE results) and discuss the related issues in *metaphysics* about which we were just hinting.

B. "EXTENDED" INFLATION AND THE "GRACEFUL EXIT".

Once the idea of an inflationary stage was conceived, the hardest part in working out the details turned out to consist in finding the mechanism which stops it and replaces it by a Friedmann expansion. Physically, the transition starts with the *true* vacuum appearing as bubbles within the *false* vacuum's spatial extension. These t.v. bubbles then have to start joining up, until they end up replacing the f.v. everywhere. The difficulties are caused by the fact that the universe meanwhile continues to expand, so that the bubbles stand little chance of catching up with each other, in the joining process. The dimensionless "bubble spreading rate" $\epsilon = \tau / H^4$ grows with the tunneling factor τ but is also strongly dependent on the Hubble constant H.

"Old" inflation (Guth's original theory) had no "exit" mechanism; "new" inflation, suggested by Linde and by Albrecht and Steinhardt in 1982, assumed a "slow roll" to the true vacuum, with no separating potential barrier to tunnel through. This is a "spinodal" first-order phase transition, instead of the more common nucleation and merging. These ideas had to be discarded because of the necessity for extensive fine-tuning of the potential, to have the transition go at the right rate, while also having observationally acceptable density fluctuations, after the transition. "Chaotic" inflation was a suggestion of chaotic vacuum fluctuations as an alternative to the GUT SSB involving a Higgs field, as the vacuum-energy trigger of inflation. However, it added nothing to the exit problem. Finally, "extended inflation" [2] has provided the most plausible solution, to date. It involves a slow-down in the inflationary expansion, happening at the same time as the tunneling and joining of the t.v. bubbles. In explaining de Sitter cosmology in section IA, we saw (1.4) that this model involves a Hubble *constant* (literally: 'unfortunately', we use the same term in the Friedmann models, in which H is time-dependent and is "constant" only with respect to spatial dependence). In a de Sitter model, H's value is proportional to the square-root of Newton's constant G. Thus, a gravitational theory with a "varying Newton constant" (another semantical nonsense..) can cause a slowdown of the inflationary expansion; such a theory generally involves a scalar field - whether a la Jordan/Dirac, Brans-Dicke [19] or according to a more modern version, involving SSB of local scaling or conformal invariance. This kind of scalar field is known as a *dilaton*, and was suggested in the Seventies by Englert et al. [20]; in such theories, the Einstein Lagrangian is written as $(\sigma(x))^2 R$, with $\sigma(x)$ the dilaton field; its dimension being -1, its square has the same dimensions as G^{-1}; the modified Lagrangian is now a "soft" (no dimensional constants) $d = -4$ and could fit in a renormalizable theory [21]. Its vacuum expectation value $< 0|\sigma(x)|0 > \neq 0 \sim G^{-1/2}$, just as the Fermi constant, as a result of equations (2.2), (2.9), (2.10), arises through $G_F^{-1} \sim < 0|\Phi_1^0(x)|0 >$. Note that there are strong experimental upper bounds on a Brans-Dicke force nowadays; however, such a component might well have existed during the inflationary stage, finally "freezing" at the present value of Newton's constant, which is known not to change by more than 10^{-11} per year.

The Inflationary Lagrangian is now

$$\begin{aligned}\mathcal{L} = f(\sigma) < 0|(\sigma)^{-2}|0 > R - \frac{1}{2}(\partial_\mu\sigma)^2 - \\ - \frac{1}{2}(\partial_\mu\Xi)^2 - V(\Xi) - \frac{1}{4}F.F - \\ - \frac{1}{2}(\partial_\mu\Phi)^2 - V(\Phi)\end{aligned} \tag{3.1}$$

where σ, Ξ, Φ are scalar fields respectively breaking spontaneously the local invariances of the scale, GUT and electroweak gauges (should supersymmetry be the answer to the hierarchy puzzle, there might be additional fields and SSB structure), the V are the corresponding quartic potentials and F is the GUT field-strength. $f(\sigma)$ grows during the inflationary stage, thus causing a de facto decrease of the Newton constant (the inverse of the vacuum expectation value in the first term). The SSB action of σ results in its acquiring mass; so do any extra components of the gravitational gauge field beyond those gauging the Lorentz group, i.e. either $SU(2,2)/SL(2,C)$ for conformal symmetry, or, for the linear case $\overline{GL(4,R)}/SL(2,C)$ [21].

C. OBSERVATIONAL TESTS.

The predictions of Extended Inflation fit the COBE observations [22]. Very roughly, the inhomogeneities measured by this satellite should reflect the fluctuations of the Ξ field during inflation, as compared to the overall scale of the (orange-size) universe; the latter is determined by the σ field, whose action stops the exponential expansion. The inhomogeneities should go as

$$\frac{\delta\rho}{\rho} \sim \frac{\left[M_{GUT}(<\Xi>)\right]^2}{\left[M_{Planck}(<\sigma>)\right]^2} \tag{3.2}$$

which fits a $10^{-5} - 10^{-6}$ level. Other observational predictions concern gravitational radiation. The merging of bubbles, during the exit from inflation, should provide a gravitational equivalent to the CBR, i.e. GCBR. This should appear, at $10m$ wavelengths (the red-shifted size of the cH^{-1} horizon during the inflationary stage) as a very narrow peak, $10^5 - 10^6$ times higher than the waves emitted by the Ξ fluctuations [23]. There is a fair chance that such a peak might be detected by the LIGO experiment, now in preparation.

D. SURREALITY - ANOTHER FORCED MODIFICATION OF OUR CONCEPT OF REALITY.

We now reach the 'philosophical' implications.

First - the Big Bang becomes a parochial feature. Every time, somewhere in the universe, there is a large vacuum energy density (i.e. $\lambda > 10^{16}GeV$), the conditions are appropriate for the *creation of a "universe"*.

Secondly - there is *Eternity*. We have returned to a Steady-State model, except that the creation of matter and of spacetime comes in bursts, rather than continuously, as in the 1948 theory.

Thirdly, there are those *singularity* theorems. In one example, developed by Farhi and Guth [18], quantum mechanics intervenes so as to avoid singularities, even though these are "compulsory", at the classical level. The Hawking-Penrose theorems force any world-line to have a singularity, either in its past or in its future. For example, the collapsing black hole of Figure 1 will reach its singularity in its future, when it will get to $r = 0$ (never, in the time-variable of the distant observer, but very quickly for the infalling one). On the other hand, the white hole (in the same figure) has come *out of* an $r = 0$ singularity. This happened "prior" to $t = -\infty$ in the distant observer's system, but not so long ago, in that of an outcoming one. Here is a manner in which quantum mechanics could be used to evade singularities: start by following the course of the future black hole, but rather than going with it to the singularity awaiting it in its future, *tunnel away*, over to the white hole's world-line; since this line has already fulfilled the theorem by having had its singularity in its past, you have managed to skip singularities altogether - thanks to quantum tunneling..

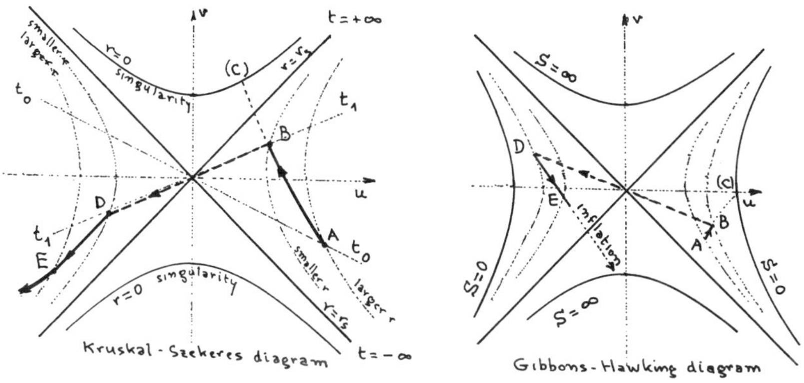

Figure 5: Birth of a new Universe
(in Kruskal-Szekeres and in Gibbons-Hawking coordinates)

Fourthly, there is the conceptual shock of Surreality. Let us go on with our above example of a world-line which avoids singulatities. Figure 5, involving an adaptation of the Kruskal-Szekeres diagram, by Gibbons and Hawking [24] to the de Sitter case) describes such a sequence, for the case of a universe created at A by the presence of an appropriate vacuum fluctuation: it starts as a tiny black hole, collapsing to B, then evading the singularity awaiting it at C, by tunneling to D, then expanding. In the system of the outside observer, who stayed in the neighborhood of A, this is the end of the story: as far as his measurements will go, there will forever remain a tiny black hole at B, which will just have stopped its collapse (i.e. be at D). However, in the system of the infalling particle at A, it falls to B, changes the state to an expanding one at D, then undergoes inflation at D-E (during $10^{-35} sec$), changes into a Friedmann model at E - and 10^{10} years later there are, on an obscure planet in that new universe, some physicists reading a paper about inflation..

NOTA BENE: IN THE SYSTEM OF OUR OBSERVER AT A, ALL OF THIS (beyond D) WILL NEVER HAPPEN! The new universe will never exist in the time-frame of that observer. Not even for $t = \infty$. And yet, we cannot deny the "EXISTENCE" of the new universe. The two universes are not disconnected: there is one world line which comes from A in the "old" universe and continues in the new one. Already after the discovery of Special Relativity in 1905, it became clear that our intuitive view of time is erroneous, a pragmatic notion fitting the circumstances of biological evolution. What we then had to drop was our notion of absolute simulataneity, replacing it by a relative one. With horizons and eternal cosmology, it turns out that there is a "reality" - in the sense of the Einstein-Podolski-Rosen question "is there an underlying reality?" - outside of our notion of time. The new universe has no existence between $t = -\infty$ and $t = +\infty$, in our notion of time - and yet it will exist and last for eons. This change calls for the same type of "conceptual reeducation" as in the 1905 case.

I have tried to construct a "replacement" - a variable on which we can fall back and try to preserve the essentials of our notion of time. The angle variable in the plane of the Kruskal-Szekeres diagram, seems to provide such a conceptual crutch, when mapped on the line namely, $\tau = \sum_i^{\oplus} tan^{-1}(tanht_i)$. One can then have a (τ, t) diagram, with the different universes growing out of each other in a tree diagram (Figure 6).

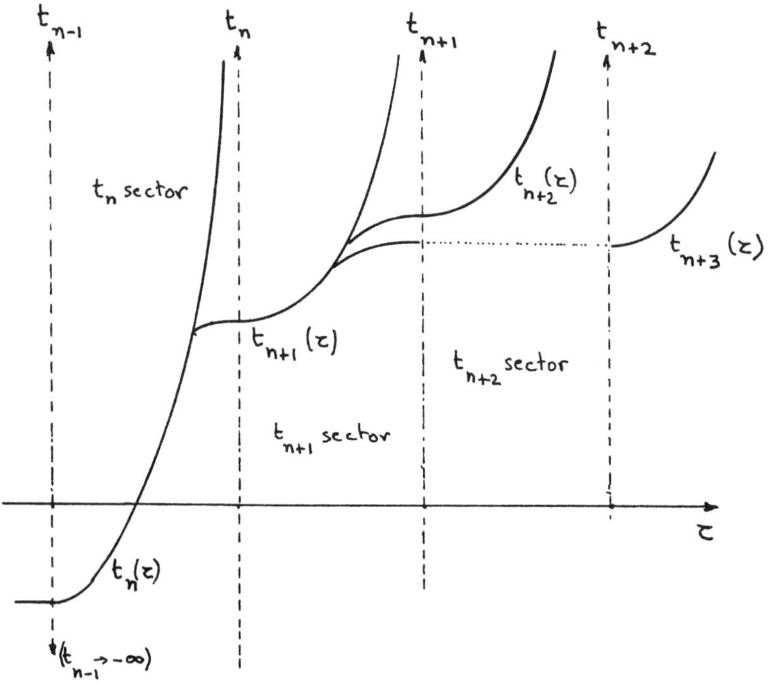

Figure 6: "Extended Time" and the Extended Time Tree" of Universal History

These lectures represent no more than an introduction, to the ideas and physics generated by Guth's original suggestion of an inflationary stage in cosmogony. I hope some of the readers will go further and partake in this exciting exploration of a GREATER UNIVERSE. Guth and Linde have each written a book on the subject; there is also a collection of reprints of key papers [25]; I also list recent reviews, updating the material [26].

ACKNOWLEDGEMENTS

I would like to thank Prof. R. Orava and M. Chaichian, for the hospitality of SEFT (Helsinki) and Prof. Richard Kerner for the hospitality of the Laboratoire de Gravitation et Cosmologie Relativistes (Paris), where these lectures were written up.

REFERENCES

1. A.H. Guth, *Phys. Rev.* **D23** (1981) 347.
2. D. La and P.J. Steinhardt, *Phys. Rev. Lett.* **62** (1989) 376.
3. A.D. Linde, *Phys. Lett.* **129B** (1983) 177 and **175B** (1986) 395.
4. Y. Ne'eman, to be pub. in *Found. of Phys. Lett.*
5. S. Coleman and S. Hughes, *Phys. Lett.* **B309** (1993) 246.
6. L. Susskind, *Phys. Rev. Lett.* **71** (1993) 2367; L. Susskind, L. Thorlacius and J. Uglum, *Phys. Rev.* **D48** (1993) 3742; G.'t Hooft, *Nucl. Phys.* **B335** (1990) 138. S. Coleman, *Nucl. Phys.* **B307** (1988) 867.
7. M.D. Kruskal, *Phys. Rev.* **119** (1960) 1743; G. Szekeres, *Pub. Math. Debrecen* **7** (1960) 285.
8. I.D. Novikov, *Astr. Zh.* **41** (1964) 1075; Y. Ne'eman, *Ap.J.* **141** (1965) 1303; Y. Ne'eman and G. Tauber, *Ap.J.* **150** (1967) 755.
9. Dj. Sijacki and Y. Ne'eman, *Phys. Lett.* **B247** (1990) 571, bf B276 (1992) 173.
10. Y. Ne'eman, *Phys. Lett.* **B81** (1979) 190; D.B. Fairlie, *Phys. Lett.* **B82** (1979) 97.
11. G. Steigman, D.N. Schramm and J. Gunn, *Phys. Lett.* **B66** (1977) 202. The limit was first put at 7 species of light neutrinos, then improved and brought down to 4.
12. B. Adeva et al.,*Phys. Lett.* **B248** (1990) 464; D. Decamp et al., *Phys. Lett.* **B231** (1989) 519.
13. For a review of the Baryon Asymmetry issue, see E.W. Kolb and M.S. Turner, *Ann. Rev. Nucl. Part. Sci.* **33** (1983) 645.
14. I would like to thank CDF for making this information available prior to publication and Prof. R. Schwitters for his help in making the material available to me.
15. D.S. Hwang, C.Y. Lee and Y. Ne'eman, "BRST Quantization of SU(2/1) Electroweak Theory in the Superconnection Approach - and the Higgs Meson Mass", TAU preprint.
16. The possibility of two energetic cosmic rays colliding was treated by P. Hut and M.Rees, *Nature* **302**(1983) 508, but with a different motivation, checking whether SSC energies might not generate a Bang.. To attain M_{GUT} energies, it is further necessary to evaluate the tail end of the cosmic rays spectrum (which stops at $10^{11} GeV$ for more "practical" purposes).
17. E. Farhi, A.H. Guth and J. Guven, *Nucl. Phys.* **B339** (1990) 417; W. Fischler, D. Morgan and J. Polchinski, *Phys. Rev.* **D42**(1990) 4042.
18. A. Linde and D. Linde, "Topological Defects as Seeds for Eternal Inflation" (preprint SU-ITP-94-3); A. Vilenkin, "Topological Inflation" (preprint hep-th/2085)

19. C. Brans and R.H. Dicke, *Phys. Rev.* **24** (1961) 924. For the method of generalizing Einsteinian gravity, see J. P. Duruisseau and R. Kerner, *Class. Quantum Grav.* **3** (1886) 817.
20. F. Englert, E. Gunzig, C. Truffin and R. Windey, *Phys. Let.* **57B** (1975) 73; F. Englert, C. Truffin and R. Gastmans, *Nucl. Phys.* **B117** (1976) 407.
21. Y. Ne'eman and Dj. Sijacki, *Phys. Lett.* **B200** (1988) 488; Y. Ne'eman and C.Y. Lee, *Phys. Lett.* **B233** (1989) 286; C.Y. Lee and Y. Ne'eman, *Phys. Lett.* **B242** (1990) 59.
22. G.F. Smoot et al., *Ap. J.* **396** (1992) L1.
23. M.S. Turner and F. Wilczek, *Phys. Rev. Lett.* **65** (1990) 3080; A. Kosovsky, M.S. Turner and R. Watkins, report FNL91-333 (1992); G.F. Smoot and P.J. Steinhardt, *Class. Quantum Grav.* **10S** (1993). 19.
24. G.W. Gibbons and S. Hawking, *Phys. Rev.* **D15**(1977) 2738.
25. R.B. Abbotts and S.-Y. Pi, *Inflationary Cosmology*, World Scientific Pub., Singapore (1987).
26. A.H. Guth, *Proc. Natl. Acad. Sci. USA* **90** (1993) 4871. P.J. Steinhardt, *Class. Quantum Grav.* **10** (1993) S33. A. Linde, in *Gravitation and Modern Cosmology*, A. Zichichi, V. de Sabbata and N. Sanchez, eds., Plenum Press, New York (1990) pp. 101-120.

FORMATION OF GALAXIES IN A DARK MATTER DOMINATED UNIVERSE - A NEW APPROACH

P. G. MACEDO and J. P. M. de CARVALHO
*Grupo de Cosmologia do C.A.U.P.**
Rua do Campo Alegre, 823
4100 Porto - Portugal

ABSTRACT. Conditions for the growth of small density perturbations in a self-gravitating fluid mixture of two components are studied using a dynamical systems approach.

It is shown that besides the existence of exponentially growing and decaying modes, which are present for values of the perturbation wave-number smaller than a critical value, two other, pure oscillatory, modes exist at all scales. Moreover it is shown that the growing mode always affects both components of the fluid and not only one of them. However, it is found that the growth of the density perturbations is different in the two components and depends on the scale of the pertubations.

It is shown that, at another critical scale, a resonance between the baryonic and the dark perturbations occurs. For dark matter particles having a mass $\sim$ 30 eV, this critical scale at the time of recombination is of the same order of magnitude of typical galaxies.

1. Introduction

The study of the mechanisms responsible for the formation of structures in the universe was started by the pioneer work of Jeans, (1902, 1928). Since then, the mechanism proposed by him has been extensively applied as a criterion of stability in several models of galaxy formation.

However, by the recombination epoch, believed to be the one after which baryonic perturbations, which led to the formation of galaxies, could begin to grow, the critical scale predicted by the classical Jeans theory is not at all related to the galactic scale. Instead, it is well known that just before recombination the critical Jeans Mass is $\sim 10^{16-17} M_{\odot}$, (the mass of rich clusters or even superclusters of galaxies), and immediately after recombination the Jeans Mass drops abruptly, more than 10 orders of magnitude, to a value $\sim 10^{5-6} M_{\odot}$, which is related to the mass of globular clusters, (Weinberg 1972, Zel'dovich & Novikov 1983).

The origin of the discrepancy between the Jeans Mass, before and after recombination, and the mass of a typical galaxy, a few $10^{11} M_{\odot}$, has not been, up till now, well understood.

On the other hand, the observation of flat galactic rotation curves at large galactic radii as well as high galaxy velocity dispersions in clusters has led to the missing mass

---

* A Member of the European Cosmology Network

M. M. Shapiro et al. (eds.), Currents in High-Energy Astrophysics, 281–290.

problem and to the dark matter conjecture to solve it. This fact has, in particular, led to the hypothesis that dark matter halos exist around galaxies.

The existence of dark matter implies that when studying the dynamics and the formation of structure in the universe one can not use the simple Jeans stability criterion for a one component gas, but has to study how does the gravitational instability arise in a mixture of at least two components. This formulation of the problem leads to a different concept of critical Jeans Mass, as well as to a better understanding of how typical galactic scales arise.

Since it is generally accepted that dark matter is constituted by WIMPs (Weakly Interacting Massive Particles), we shall consider the case for which the two components (a baryonic and a dark matter (DM) one, hereafter component B and D respectively) interact only gravitationally. The DM component is assumed to be made of neutrino-like particles, though one should point out that the formalism used is quite general and applies to any type of a two-component mixture.

In Section 2. we establish the linear coupled equations which describe the dynamical system for the density perturbations in the two-component fluid we wish to study. In Section 3. we briefly describe a qualitative analysis of the dynamical system. In Section 4. we describe a resonance between the baryonic and the dark components which is, in our opinion, responsible for the formation of structure at the typical galactic scales. Finally in Section 5. we discuss the cosmological implications of the results obtained.

For numerical purposes we shall consider that the onset of gravitational instability occurs in a flat ($\Omega = 1$) universe at recombination epoch, after the decoupling between radiation and baryonic matter, when the radiation temperature, the same as for baryonic matter, is $\sim$ 3000 K, (Kolb & Turner 1990). We shall assume that $\Omega = \Omega_B + \Omega_D$ and use for Ω_B a value in agreement with the limits imposed by standard nucleosynthesis. Assuming $0.5 \leq \mathrm{h} \leq 1$ such a value is ~ 0.05, where h is the Hubble constant in units of 100 Km/s/Mpc, i.e. $\mathrm{h} = \mathrm{H_0}/(100\ \mathrm{Km/s/Mpc})$.

2. Dynamical equations for a two component fluid

The dynamics of a non-relativistic self-gravitating fluid can be described by the usual set of hydrodynamical equations: continuity equation, Euler equation and Poisson's equation.

We shall use a perturbative analysis (de Carvalho and Macedo 1994, hereafter Paper I) and, to first order in the perturbation, write the dynamical variables as

$$\rho_j = \rho_{j_0} + \delta\rho_j \qquad \boldsymbol{v}_j = \boldsymbol{v}_{j_0} + \delta\boldsymbol{v}_j \qquad p_j = p_{j_0} + \delta p_j \qquad \Phi_T = \Phi_{T_0} + \delta\Phi_T$$

where Φ is the total gravitational potential, ρ_j, $\boldsymbol{v}_j$ and p_j are respectively the density, velocity and pressure fields for fluid elements of component j, (the index j taking the values D or B) and the index "0" indicates the unperturbed value, and δ the first order perturbation.

It's common to consider the zero order solution to be an infinite fluid at rest, ($\boldsymbol{v}_0 = 0$), for which the density and pressure are the same everywhere, (unperturbed fluid). We must point out, as it is well known, that this solution does not exist in a material universe, since it would imply $\rho_0 = 0$. However this problem is removed when one considers the more realistic case of a fluid embedded in the expanding universe and therefore one can start perturbing this static and uniform configuration.

Fourier analysing the solution of the hydrodynamical equations in the spatial variables, they can be written to first order in the perturbations of the fluid mixture, after some manipulation, as follows:

$$\begin{cases} \ddot{\Delta}_D + \left(c_D^2 k^2 - W_D\right) \Delta_D - W_B \Delta_B &= 0 \\ \ddot{\Delta}_B + \left(c_B^2 k^2 - W_B\right) \Delta_B - W_D \Delta_D &= 0 \end{cases} \tag{1}$$

where k is the wave number of the perturbation and W_j is given by $W_j = 4\pi G \rho_{j_0}$; Δ_j is the density contrast $\delta\rho_j/\rho_{j_0}$ and c_j the adiabatic sound speed in the component j. The double dot stands for second time derivative.

One can write the system of differential equations (1) as the following autonomous dynamical system of four first order differential equations which we write in a matrix form as:

$$\dot{\boldsymbol{x}} = \boldsymbol{A}\,\boldsymbol{x} \tag{2}$$

where $\boldsymbol{x} \equiv (x_1, x_2, x_3, x_4)^{\mathrm{T}} \equiv (\dot{\Delta}_D, \Delta_D, \dot{\Delta}_B, \Delta_B)^{\mathrm{T}}$, the superscript T standing for transpose and $\boldsymbol{A}$ is given by:

$$\boldsymbol{A} \equiv \boldsymbol{A}(k) = \begin{pmatrix} 0 & W_D - c_D^2 k^2 & 0 & W_B \\ 1 & 0 & 0 & 0 \\ 0 & W_D & 0 & W_B - c_B^2 k^2 \\ 0 & 0 & 1 & 0 \end{pmatrix} \tag{3}$$

We shall be particularly interested in the second and fourth components, $x_2(t)$ and $x_4(t)$, of the general solution $\boldsymbol{x}(t)$, which represent the dynamical behaviour of the density perturbations in components D and B respectively.

3. Qualitative study of the dynamical system

When the matrix $\boldsymbol{A}(k)$ has 4 different eigenvalues (which occurs for all values of k except the critical point $k = k_M = \sqrt{k_B^2 + k_D^2}$ where k_B and k_D are the Jeans wave numbers of components B and D respectively) there are 4 linearly independent eigenvectors associated to the eigenvalues. We shall call k_M the Jeans wave number of the mixture.

In this case the general solution $\boldsymbol{x}(t)$ of the system (2) is well known (Arnold 1973), and can be written in the form of a linear combination of particular solutions $\mathrm{e}^{\lambda_j t}\,\boldsymbol{\xi}_j$, i.e.:

$$\boldsymbol{x}(t) = \sum_{j=1}^{4} \alpha_j\, \mathrm{e}^{\lambda_j t}\, \boldsymbol{\xi}_j \tag{4}$$

where λ_j and $\boldsymbol{\xi}_j$ are respectively the eigenvalues of the matrix $\boldsymbol{A}$ and its associated eigenvectors and α_j are the coefficients which are determined by the initial conditions.

The 4 eigenvalues λ_j of $\boldsymbol{A}(k)$ are given by

$$\begin{cases} \lambda_1 = -\lambda_2 = \frac{1}{\sqrt{2}}\sqrt{f + \sqrt{h^2 + 4\,W_B W_D}} \\ \lambda_3 = -\lambda_4 = \frac{1}{\sqrt{2}}\sqrt{f - \sqrt{h^2 + 4\,W_B W_D}} \end{cases} \tag{5}$$

and their associated eigenvectors $\boldsymbol{\xi}_j$ can be written in the following form:

$$\begin{cases} \boldsymbol{\xi}_1 \equiv (\beta_1\,\lambda_1,\ \beta_1,\ \lambda_1,\ 1)^{\mathrm{T}} \\ \boldsymbol{\xi}_2 \equiv (\beta_2\,\lambda_2,\ \beta_2,\ \lambda_2,\ 1)^{\mathrm{T}} \\ \boldsymbol{\xi}_3 \equiv (\beta_3\,\lambda_3,\ \beta_3,\ \lambda_3,\ 1)^{\mathrm{T}} \\ \boldsymbol{\xi}_4 \equiv (\beta_4\,\lambda_4,\ \beta_4,\ \lambda_4,\ 1)^{\mathrm{T}} \end{cases} \tag{6}$$

where β_j are the following functions of k

$$\begin{cases} \beta_1 = \beta_2 = \frac{1}{2W_D}\left(h + \sqrt{h^2 + 4\,W_B W_D}\right) \\ \beta_3 = \beta_4 = \frac{1}{2W_D}\left(h - \sqrt{h^2 + 4\,W_B W_D}\right) \end{cases} \tag{7}$$

In this relationships the k-dependent functions f and h are given by:

$$f(k) = W_B + W_D - k^2\left(c_B^2 + c_D^2\right) \tag{8}$$

$$h(k) = W_D - W_B + k^2\left(c_B^2 - c_D^2\right) \tag{9}$$

In order to study the behaviour with scale of the dynamical system at the onset of the structure formation process, one needs to analyse the k-dependence of the eigenvalues λ_j, the eigenvectors $\boldsymbol{\xi}_j$ and the coefficients α_j, which appear in Eq. (4). From Eq. (5) one can conclude that λ_3 and λ_4 are pure imaginary conjugate eigenvalues, for all $k \neq 0$, while λ_1 and λ_2 have two different regimes according to whether k is greater or less than the critical value k_M. For $k > k_M$ the eigenvalues λ_1 and λ_2 are pure imaginary conjugate and for $k < k_M$ they are real ones with $\lambda_1 > 0$ and $\lambda_2 < 0$.

This means that for $k > k_M$ one has two pairs of oscillatory modes with constant amplitude. The perturbations in these modes, propagating like acoustic waves in the mixture, avoid any growing of the density contrast.

On the other hand for $k < k_M$, although oscillations do not disapppear, due to the imaginary values of λ_3 and λ_4, the real and positive eigenvalue λ_1 originates a growing mode and the real and negative eigenvalue λ_2 a decaying one.

Since $\lambda_2 = -\lambda_1$ and $\lambda_4 = -\lambda_3$ we shall only be interested in the dependence with scale of the real eigenvalue λ_1 and of the imaginary part of λ_3.

The eigenvalue λ_1 is an increasing function of scale. Since it is real and positive it is related to the growth of the perturbations, i.e. $1/\lambda_1$ is the characteristic time of collapse.

The eigenvalue λ_3 is a decreasing function of scale. Since λ_3 is the frequency associated to the acoustic modes one can see, from its dependence on scale, that although oscillations are present in all scales, for large scales acoustic waves give way to collapse, in the sense that their frequency approaches zero, i.e. their period tends to infinity.

From relations (7), one can see that β_j are real-valued functions for all $k > 0$, with $\beta_1 > 0$ and $\beta_3 < 0$. β_1 is the ratio of the density contrasts, Δ_D/Δ_B, in the collapse mode, i.e. it gives us the relative composition, or relative percentage of components B and D in the growing mode. Therefore the condition $\beta_1 > 0$ means that both components participate in the collapse of the mixture (as well as in the decaying mode, which being transient will not interest us). On the other hand, since β_3 is the same ratio but in the oscillating modes, the condition $\beta_3 < 0$ means that, in the acoustic modes, the two components oscillate with opposite phases.

From this, it is clear that the change of relative growth of perturbations with scale can be seen analysing the behaviour of β_1 with k (or $l = 2\pi/k$).

The observed behaviour of β_1 with scale means that the relative growth of the perturbations in the two components changes crucially when the scale is of the order of l_D (the Jeans scale length of component D). For scales well below l_D the baryonic perturbation grows faster than the dark one and above l_D the two perturbations grow almost at the same rate, i.e. for the largest scales the composition of the perturbation, in the collapsing mode, is identical to the one in the unperturbed state.

In the next section, it will be shown that there is a critical scale $l_c \sim l_D$ where a resonance between the two components occurs.

4. The structure formation resonance

The dynamical system (2) represents two gravitationally coupled oscillators whose natural frequencies ω_j, (i.e. the ones the components would have if uncoupled), are given by

$$\omega_j = \left(c_j^2 k^2 - W_j\right)^{1/2} . \tag{10}$$

It is easily seen that the wave-number k_c for which the two natural frequencies of oscillation of the two components coincide is given by:

$$k_c = \left(\frac{W_D - W_B}{c_D^2 - c_B^2}\right)^{1/2} \tag{11}$$

It is reasonable to expect a resonance to occur, at least in some of the mode amplitudes of the perturbations, for a value of k near k_c. In fact, it will be shown below that such a resonance does indeed occur.

Note that ω_B and ω_D can only be equal if $W_D - W_B$ and $c_D - c_B$ have the same sign. This is a necessary condition for the existence of a resonance in the system. When applied to our universe, for which the observational evidences seem to indicate it to be dominated by dark matter, (i.e. $W_D > W_B$), this means that such a resonance can not occur in a cold dark matter (CDM) dominated universe.

The existence of the resonance is easily verified analysing the mode amplitudes and their dependence on scale.

After some straightforward manipulation, the second and fourth components of the general solution (4) can be written as:

$$\Delta_{D}(\tau) = \Delta_{D_0} \left[\zeta_1 \left(e^{\lambda_1 \tau} + e^{-\lambda_1 \tau}\right) + \zeta_2 \cos(i\lambda_3 \tau)\right] \tag{12}$$

and

$$\Delta_{B}(\tau) = \Delta_{B_0} \left[\zeta_3 \left(e^{\lambda_1 \tau} + e^{-\lambda_1 \tau}\right) + \zeta_4 \cos(i\lambda_3 \tau)\right] \tag{13}$$

where $\tau = t - t_0$ and $i = \sqrt{-1}$. The coefficients ζ_j give the mode amplitudes for the perturbations in both components: in component D, ζ_1 is the common amplitude of the growing and decaying modes and ζ_2 is the amplitude for the oscillating mode; in component B, the common amplitude of the growing and decaying modes is ζ_3 and ζ_4 is the one for the acoustic mode. These coefficients are given by the following expressions:

$$\begin{aligned}
\zeta_1 &= \tfrac{\beta_1}{2} \tfrac{1-Q_0\beta_3}{\beta_1-\beta_3} \\
\zeta_2 &= \beta_3 \tfrac{Q_0\beta_1-1}{\beta_1-\beta_3} \\
\zeta_3 &= \tfrac{1}{2} \tfrac{Q_0^{-1}-\beta_3}{\beta_1-\beta_3} \\
\zeta_4 &= \tfrac{\beta_1-Q_0^{-1}}{\beta_1-\beta_3}
\end{aligned} \tag{14}$$

where Q_0 is the ratio between the density contrast in the two components at $t = t_0$, i.e. $Q_0 = \Delta_{B_0}/\Delta_{D_0}$.

All these coefficients present a peculiar behaviour for a scale near l_c, but one should first of all point out that the pairs ζ_1 , ζ_2 and ζ_3 , ζ_4 are related by:

$$2\,\zeta_1 + \zeta_2 = 2\,\zeta_3 + \zeta_4 = 1, \tag{15}$$

i.e. a maximum of ζ_1 corresponds to a minimum of ζ_2 and a maximum of ζ_3 corresponds to a minimum of ζ_4.

This physicaly means that in the resonant region of the perturbation mass spectrum most of the material, which for other scales was oscillating in the acoustic modes, does for this scale participate in the collapse, therefore enhancing its efficiency.

We shall therefore analyse the functions ζ_1 and ζ_3. Looking at their derivatives with respect to k, one can see that their maxima occurs for values of k given by:

$$k_{\text{max}}(\zeta_1) = \left[k_c^2 - \frac{2\,W_D}{Q_0\left(c_D^2 - c_B^2\right)}\right]^{1/2} \tag{16}$$

for ζ_1, and

$$k_{\text{max}}(\zeta_3) = \left[k_c^2 + \frac{2\,W_B}{q\left(c_D^2 - c_B^2\right)}\right]^{1/2} \tag{17}$$

for ζ_3, where $q = 1/Q_0 = \Delta_{D_0}/\Delta_{B_0}$.

Looking at Eqs. (11) and (16), one can see that ζ_1 will have a maximum in the range $0 < k < k_M$, only if $q < \frac{1}{2}\left(1 - \frac{W_B}{W_D}\right)$. On the other hand, from eqs. (11) and (17), one can see that ζ_3 will have a maximum in the same range, only if q satisfies the condition $q > 2\left(\frac{c_D^2}{c_B^2} - \frac{W_D}{W_B}\frac{c_B^2}{c_D^2}\right)^{-1}$.

It is reasonable to assume that the composition of the initial perturbation in each of the components is the same as the composition of the unperturbed state (i. e. $q = W_D/W_B$) . It is easy to see that in this case, ζ_1 will never have a maximum in the desired range, while ζ_3 can have one if the following condition is satisfied:

$$\frac{c_D^2}{c_B^2} > \frac{W_B}{W_D} + \sqrt{\frac{{W_B}^2}{{W_D}^2} + \frac{W_D}{W_B}}. \tag{18}$$

From this analysis one concludes that, in a universe like ours, the resonance will only affect the baryonic component and therefore we shall, from now on study only the behaviour of the amplitude ζ_3 which is the one associated to the growing mode in the baryonic component. Substituting the value of $k_{\max}(\zeta_3)$ given by eq.(17) in eq.(14), one obtains for the maximum value $\zeta_3^{\max}$, of the function ζ_3:

$$\zeta_3^{\max} = \frac{1}{4}\left(1 + \sqrt{1 + \frac{W_D}{W_B}q^2}\right) \tag{19}$$

which depends only on the unperturbed densities of the components and on the initial conditions.

In the case of a universe dominated by dark matter ($W_D/W_B \gg 1$) and for the inital conditions specified above, one has $\zeta_3^{\max} \simeq \frac{1}{4}\,(W_D/W_B)^{3/2}$. From this expression it is clear that the height of the resonant peak increases as the ratio W_D/W_B increases. This means that the more the Universe is dominated by dark matter the more the effect of the structure formation resonance in the baryonic matter is important.

With initial conditions $\boldsymbol{x}_0 \equiv (0, 10^{-3}\,\Omega_D, 0, 10^{-3}\,\Omega_B)^{\mathrm{T}}$, we represent in Fig. 1. the dependence of ζ_3 on scale, where its resonant behaviour for $l \sim l_c$ is clear.

The importance of the resonance is also measured by the width of its band (Feynman 1966); a larger resonant band corresponding to a weaker resonance. The farther apart the sound speeds c_D and c_B in the two components are, the smaller the resonant band width is, indicating a stronger and sharper resonance (see Macedo and de Carvalho 1994, hereafter Paper II).

The scale l_c and the associated mass scale M_c coincides with the scale where the peaks in ζ_3 and ζ_4 occur, which leads us to relate their occurrence with a resonance between the baryonic and dark perturbations. This resonance is linked to the fact that, at this scale l_c, there is an enhancement of the number of particles in both components, (but particularly in the baryonic one), which participate in the collapse mode, instead of participating in the acoustic modes as it happens for other scales.

The behaviour of the coefficients ζ_j, particularly ζ_3, clearly favours the growth of baryonic perturbations with a scale-length near l_c relative to all other scales.

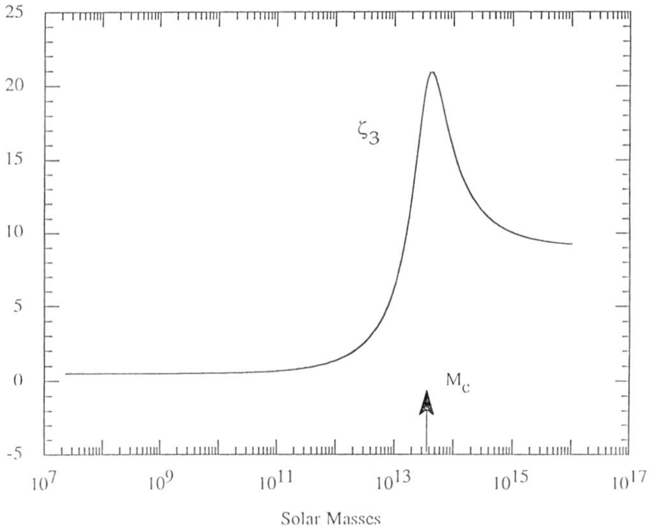

Fig. 1. Coefficient ζ_3 and its dependence on scale. The vertical arrow indicates the total mass M_c inside a spherical region of diameter l_c. Note the clear change in behaviour of this coefficient around this scale

5. The resonance and its effect on the formation of galaxies

We believe that the resonance we have just described is responsible for the formation, at the recombination epoch, of massive structures at the galactic scale. This resonance provides an explanation for why do galaxies, with the typical masses one observes, form around the recombination epoch if the Jeans mass for baryons either before or just after this epoch is either much larger or smaller than this typical mass. The typical galactic mass should, according to this work correspond to the mass of the perturbations which, at that epoch, would have a scale near the resonant scale above mentioned. One would therefore have no need to assume that the mass spectrum of perturbations at the time of recombination was already populated in a similar way to the one we observe now. Instead, one can say that the typical galactic mass spectrum would result simply from the preferred scale set up by this resonance.

In fact, as we have shown (for details see Paper I and Paper II), in a two-component fluid, a resonance between the two components of the fluid must always occur provided that the component with a greater density is the one with a greater sound speed, see eq.(11). One can therefore put some constraints on the nature of the dark matter if galaxies are to be formed by such a resonant effect. Assuming that the density parameter of the universe is $\Omega = 1$ and adopting a value for the density of the baryonic component in agreement with standard light-element nucleosynthesis (Walker et al. 1991), one is immediately led to the conclusion that the universe is dominated by dark matter or $W_D > W_B$.

In such a case, if $c_B > c_D$ (case of CDM), one concludes from eq.(11) that no resonance between the two components is possible and therefore, the galactic scale would not be a preferred one, as it seems to be. One is therefore led to believe that most, if not all, of the dark matter present in the universe has a *hot nature*, i.e. it is constituted by light neutrino-like particles as implied by $c_D \gg c_B$.

Since the scale at which the resonance peak occurs depends on c_D, see eq.(17), which in turn is a function of the value of the mass of HDM particles m_D, one can estimate the mass for which the resonance peak occurs at scales characteristic of the ones of typical galaxies. The total mass M inside a spherical region of radius $l/2 = \pi/k$ is given by

$$M = \frac{4}{3}\pi^4 \frac{\rho}{k^3}, \tag{20}$$

where k is the wave number and $\rho = \rho_B + \rho_D$ is the total density in that region.

In order to obtain bounds for the mass of dark matter particles we shall assume that galaxies form by means of the above described resonance and therefore have mass scales which correspond to the resonant scale for ζ_3. The range of galactic masses which one observes in the universe will therefore imply a specific range for the mass of dark matter particles.

Substituting in eq.(20) the value of the wave number k by its resonant value $k_{\max}(\zeta_3)$, given by eq.(17), one can establish the following relation between M and m_D, where M is given in solar masses and m_D in eV:

$$m_D = \frac{210\,\pi^{5/6}}{\rho^{1/6}\,M^{1/3}} \tag{21}$$

In deriving the above equation, we have assumed that $W_D \gg W_B$ and that $c_D \gg c_B$. The sound speeds c_j at recombination, appearing in eq.(17) can be given by $c_{j_{rec}} = [(5/3)K_{Bol}T_{j_{rec}}/m_j]^{1/2}$, where K_{Bol} is the Boltzman constant and $T_{j_{rec}}$ and m_j are respectively the temperature at recombination and mass of the constituting particles of component j. One should emphasize that from the decoupling, between the neutrino-like particles and radiation, until the epoch at which these particles become non-relativistic, which for $m_D \sim 30\,\mathrm{eV}$ occurs at a red-shift $z_{NR} \sim 6 \times 10^4$, (Doroshkevich et al. 1988), their temperature T_D is given by $\mathrm{T}_D = (4/11)^{1/3}\,\mathrm{T}_\gamma$, where T_γ is the radiation temperature. Since then, it decays with a^{-2}, where a is the scale factor of the universe (Gao & Ruffini 1981; Padmanabhan 1993). One can therefore estimate that, at recombination, such a dark component D will have a temperature $\mathrm{T}_{D_{rec}} \sim 40\,\mathrm{K}$.

Assuming that a typical galaxy, formed at recombination epoch, has a baryonic mass of a few $10^{11}\,\mathrm{M}_\odot$, wich corresponds to a total mass $M \equiv 10^{13}\mathrm{M}_\odot$, one can estimate that the mass of the dark matter particles is of the order of $30\,\mathrm{eV}$. This result agrees quite well with the Sciama's Decaying Dark Matter Hypothesis (Sciama 1990a, 1990b) and the cosmological upper bounds for neutrino's masses (Peebles 1993).

One should point out that this study assumes galaxies to have started their formation during recombination. However, this can well be not the case and they could have started their formation over a wider time interval starting from recombination.

In that case, the unperturbed densities ρ_j and the sound speeds c_j in the two components would vary during such period, allowing the scale of the resonance to be shifted. This might in turn provide an explanation for the wide mass range for galaxies. One must also emphasize that this study has left a lot of open questions, in particular, it will be interesting to study what happens in a 3-component case like the dark matter mixed models. This is an open question which we intend to address in future work.

On the other hand, one should also mention that although resonances between periodic coupled oscillators are common in all branches of physics, this type of resonance occuring between (non periodic) collapsing modes are, to our knowledge, unheard of, and we suspect that in other branches of physics such as in plasmas, where one has a 2-component fluid whose components are coupled by the electromagnetic interaction, similar resonances may perhaps occur, giving rise to instabilities. We think that such open question is also worth investigating.

Acknowledgements: We thank JNICT for financial support and Centro de Astrofísica da Universidade do Porto (CAUP) for the facilities provided.

6. References

Arnold V. I. 1973, Ordinary Differential Equations, MIT Press, Cambridge MA
De Carvalho J.P.M., Macedo P.G., (1994), Submited to A.A. (Paper I)
Macedo P.G., De Carvalho J.P.M., (1994), Submited to A.A. (Paper II)
Doroshkevich A.G., Klypin A.A., Khlopov M.Yu 1988, Sov. Astron., 32(2), 127
Feynman R.P., Leighton R.B., Sands M., (1966), The Feynman Lectures on Physics, Vol.I, Addison-Wesley Pub. Comp., Massachusetts
Gao J. G., Ruffini R. 1981, Phys. Lett., 100B, 47
Grishchuk L. P., Zel'dovich Ya. B. 1981, Sov. Astron., 25(3)
Jeans J. 1902, Phil. Trans. R. Soc., 199A, 49
Jeans J. 1928, Astronomy and Cosmogony, Cambridge University Press, Cambridge
Kolb E., Turner M. 1990, Early Universe, Addison-Wesley, Massachusetts
Padmanabhan T. 1993, Structure Formation in the Universe, Cambridge University Press, Cambridge
Peebles P. J. E., (1993), Principles of Physical Cosmology, Princeton Series in Physics, Princeton University Press, Princeton
Polyachenko V. L., Fridman A. M. 1981, Sov. Phys. JETP, 54(1)
Sciama D. W. 1990a, MNRAS 246, 191
Sciama D. W. 1990b, Nat 348, 617
Walker T. P., Steigman G., Schramm D. N., Olive K. A., Kang H., (1991), Ap.J., 376, 51
Weinberg S. 1972, Gravitation and Cosmology, J. Wiley, New York
Zel'dovich Ya. B., Novikov I. D. 1983, Relativistic Astrophysics (Vol. 2), The University of Chicago Press, Chicago

NEW WAYS IN COSMOLOGY:

PART I: FRIEDMANN-LEMAITRE MODEL DERIVED FROM THE LYMAN ALPHA FOREST IN QUASAR SPECTRA

H.J. BLOME[1], W. PRIESTER[2], J. HOELL[3]

[1] Deutsche Forschungsanstalt für Luft- und Raumfahrt (DLR) D-51140 Köln, Germany
[2] Institut für Astrophysik, Universität Bonn and Max-Planck-Institut für Radioastronomie, Auf dem Hügel 71, D-53121 Bonn, Germany
[3] Deutsche Agentur für Raumfahrtangelegenheiten (DARA), D-53227 Bonn, Germany

ABSTRACT. Einstein's cosmological term λ_o and the matter density parameter $\Omega_{M,0}$ of a Friedmann-Lemaître model are derived using the line separations in redshift Δz and the number counts dN/dz of the Ly-α forest lines in high resolution spectra of distant quasars (redshift: $1.8 < z < 4.5$). The analysis is based on the complete Friedmann equations (i.e. including the cosmological constant Λ) with two different methods: The bubble wall model and the cloud model.

The first is based on a characteristic pattern $\Delta z[z]$ in the spectra which can be related to a bubble wall structure in the galaxy distribution. The second method assumes a homogeneous distribution of clouds in a comoving system.

Both methods yield the same Friedmann-Lemaître model: $\Omega_{M,0} = 0.014$ and $\lambda_0 = 1.080$. A comparison with the density parameter $\Omega_{B,0} \cdot h_0^2 = 0.0125$ of the primordial nucleosynthesis results in a Hubble parameter $78 < H_0 < 110$ km/(s·Mpc) with the optimum $H_0 = 90$ km/(s·Mpc) and an age of the universe of $t_0 = 30 \cdot 10^9$ years. This is sufficient for galaxy formation at $z \approx 5$ in a low-density universe.

M. M. Shapiro et al. (eds.), Currents in High-Energy Astrophysics, 291–300.

1. Introduction

When Einstein in 1931 realized that the cosmological constant Λ could not be determined from the then available observations he recommended to drop the Λ-term for reasons of logical economy (logische Oekonomie) until astronomical observation forces us to introduce it again (Einstein (1931)(1956)). Recently, Steven Weinberg (1993) took issue with Einstein's statement regarding Λ: "The experience of 75 years in theoretical physics has taught us that any term in our basic equations which is not forbidden by symmetry or other fundamental principles does indeed exist in our theories." It has to be determined from experiment or observation.

The determination of the value of the cosmological constant is one of the current problems in cosmology. A value greater than zero has major impacts on the age of the universe, on the cosmic expansion rate, on the action of gravity on very large scales, and on the possible existence of nonbaryonic dark matter.

In the frame of Einstein's field equations, Λ corresponds to an inherent space curvature. In the Friedmann cosmology of an homogeneous and isotropic universe the inverse square root ($\Lambda^{-0.5}$) defines a cosmic unit of length. Its value corresponds to the curvature radius R of all infinitely expanding Friedmann-Lemaître models with spherical metric at the epoch when the matter density parameter $\Omega_M[t]$ passes through its maximum value in the course of evolution. This epoch ist vital for galaxy formation in a primordial H-, He-gas (Priester et al. 1994 [PHB 1994]). The decoupling of the cloud from the Hubble expansion requires a matter density parameter $\Omega_M > 4$ and a total density parameter $\Omega_{total} = \Omega_M + \lambda \geq 6$ at that epoch (Hoell, Priester (1994)). This is provided by the best-fit model derived by the Friedmann regression analysis of the Ly-α forest in 21 quasar spectra with $1.8 < z < 4.5$ (Liebscher et al. (1992)[LPH 1992 a, b])(Fig.1) with

$$\Omega_{M,0} = \frac{\varrho_{M,0}}{\varrho_{c,0}} = 0.014 \pm 0.002 \qquad (1)$$

and

$$\lambda_0 = \Lambda \frac{c^2}{3H_0^2} = 1.080 \pm 0.006 \qquad (2)$$

Here, $\varrho_{M,0}$ is the present matter density and $\varrho_{c,o} = 3H_0^2/8\pi G$ is the critical density, which depends on the Hubble expansion rate $H_0 = \dot{R}[t_0]/R_0$ only. The cosmological term can further be interpreted as $\lambda_0 = \varrho_\Lambda/\varrho_{c,0}$ with the Λ-equivalent density $\varrho_\Lambda = \Lambda c^2/8\pi G$. It obeys the unusual equation of state: pressure $p_\Lambda = -\varrho_\Lambda c^2$. (Hoell, Liebscher, Priester(1994)[HLP], Priester, Hoell, Blome (1994)). The equation of state follows when the Λ-term is shifted to the right hand side of Einstein's equation. It then can be interpreted as the ground level of all matter fields within the energy-momentum tensor or - in short - as the true quantum vacuum as distinguished from the primordial quantum vacuum out of which the quarks and leptons are believed to have originated by a phase transition. This problem will be considered in part II of our lectures. The transition from the primordial to the true quantum vacuum is one of the paramount problems in cosmology and quantum field theory.

2. The Lyman-alpha forest

The numerous absorption lines observed in the spectra of all high-redshift quasars below the Ly-α emission wavelength are generally regarded as caused by hydrogen clouds or filaments mostly in the expected large numbers of dwarf galaxies or in between the larger galaxies with a smaller portion of the clouds in these galaxies themselves. We restrict the wavelength range between the Ly-α emission and the Ly-β emission in order to avoid confusion by the Ly-β absorption lines.

There are two different methods how the observed line numbers can be analysed with cosmological models. They are a) the bubble wall model and b) the cloud model. Here we intend to show that both methods yield the same Friedmann-Lemaître model: $\Omega_{M,0}$ (equ.1) and λ_0 (equ.2), if the data are analysed using the complete Friedmann equation, which includes the Λ-term. It also shows that there is no need to require strong evolutionary effects in the absorption cross-sections of the clouds. This was the case in the investigations of previous workers who had tried the analysis with the abbreviated Friedmann equation where Λ was postulated to be zero. A brief history was given by Murdoch et al. (1986) and by Lu et al. (1991), see also Peebles (1993). In the following paragraphs we present a brief account of the theoretical approach for both methods using the complete Friedmann equation. The detailed theory including the listings of the data was given by Liebscher, Priester, Hoell (1992)[LPH 1992 a, b].

3. The Friedmann regression for the bubble wall model

The absorbers are assumed to consist of hydrogen clouds or filaments in or between galaxies situated in the walls of a bubble structure. This is based on the observation that the spectra of quasars within the redshifts $1.8 < z < 4.5$ show a characteristic pattern where the absorption lines are separated in the statistical average by typically $6Å < \Delta\lambda < 9Å$. If the void structure in the distribution of galaxies in our cosmological neighbourhood ($z < 0.04$) is a universal phenomenon and not a local pecularity then it should be observable as a bubble wall pattern in the Ly-α forest. A close inspection of high resolution spectra (e.g. Pettini et al. 1990) shows that the broader lines can be caused by more than one filament within one bubble wall, which differ in relative peculiar motion along the line of sight. The Friedmann regression analysis [LPH 1992 a, b] relates the observed redshift intervals $\Delta z[z] = \Delta\lambda/1216Å$ between the absorption lines to the bubble size at the redshift z (see equ. (16) in [LPH 1992 a] The bubbles are assumed to expand with the Hubble flow (Fig.2)

$$\Delta z[z] = R_0 \cdot \Delta\chi \cdot H[z] \cdot c^{-1} \tag{3}$$

Here $R_0 \cdot \Delta\chi$ is the typical bubble diameter at the present time t_0. $\Delta\chi$ is the comoving dimensionless inner diameter of the bubbles, correspondending to the size of voids in the galaxy distribution.

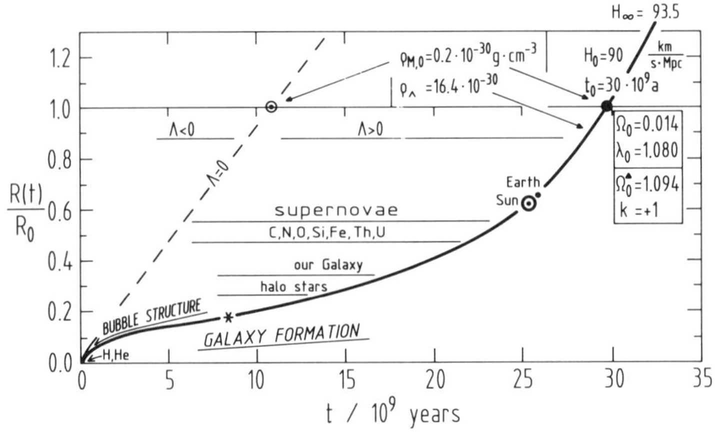

Fig.1. Expansionrate of our best-fit model with $t_0 = 30 \cdot 10^9$ *years. Friedmann-Lemaître model with spherical metric and* $R_0 = 36 \cdot 10^9$ *lightyears. At* $t = 7 \cdot 10^9$ *a is the curvature radius* $R = 1/\sqrt{\Lambda} = 6 \cdot 10^9$ *lightyears.*

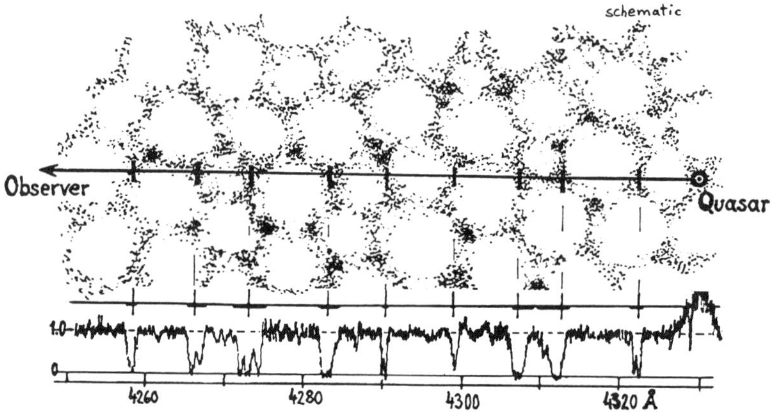

Fig.2. A universal bubble structure in the spatial distribution of galaxies and corresponding Ly-α absorption lines (schematic).

The Hubble expansion rate H[z] follows from the Friedmann equation consisting of the Λ-term, the curvature term and the matter density term. The radiation term can here be neglected as it is dominant only within the first 10^6 years.

$$H^2[z] = \frac{\Lambda c^2}{3} - \frac{kc^2}{R^2[z]} + \frac{8\pi G}{3} \cdot \varrho_{M,0} \cdot \left(\frac{R_0}{R[z]}\right)^3 \tag{4}$$

or

$$H^2[z] = H_0^2\left(\lambda_0 - (\lambda_0 + \Omega_{M,0} - 1)(1+z)^2 + \Omega_{M,0}(1+z)^3\right) \tag{5}$$

The combination of eqs. (3) and (5) yields the Friedmann regression form [LPH 1992 a, b] for the observed Δz:

$$(\Delta z[z])^2 = a_0 + a_2(1+z)^2 + a_3(1+z)^3 \tag{6}$$

with

$$\Omega_{M,0} = \frac{a_3}{(a_0 + a_2 + a_3)} \tag{7}$$

$$\lambda_0 = \frac{a_0}{(a_0 + a_2 + a_3)} \tag{8}$$

$$R_0 \cdot \Delta\chi = \frac{c}{H_0}\sqrt{a_0 + a_2 + a_3} \quad . \tag{9}$$

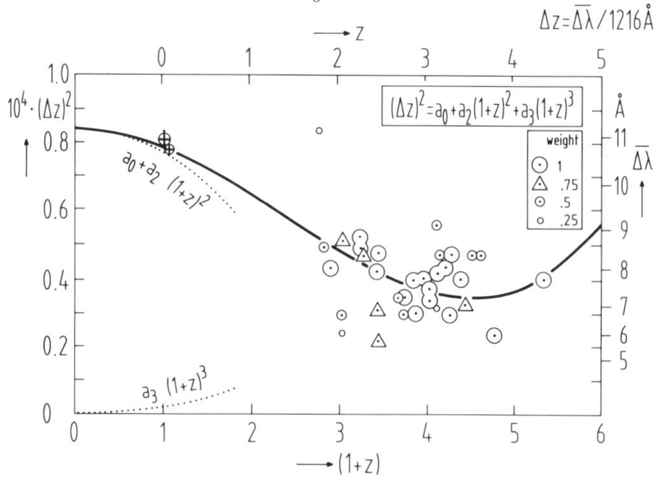

Fig.3. Friedmann regression analysis of the observed Δz in the spectra of 21 quasars as function of the redshift (1+z).

The data from 1360 lines in the spectra of 21 quasars are plotted in Fig.3. The optimum values for the coefficients obtained in our analysis are $10^4\ a_0 = 0.8433$, $10^4\ a_2 = -0.0736$, $10^4\ a_3 = 0.0111$. The results with the (1σ)-errors are given in eqs. (1) and (2) and in Fig.4. Compare also the error discussions in [LPH 1992 a, b], [HLP 1994] and [PHB 1994]. The astonishing small errors result from the fact that the Friedmann equation (5) does not contain a linear term in (1+z). This makes the regression very robust against the scatter in the Δz-data (Fig.4).

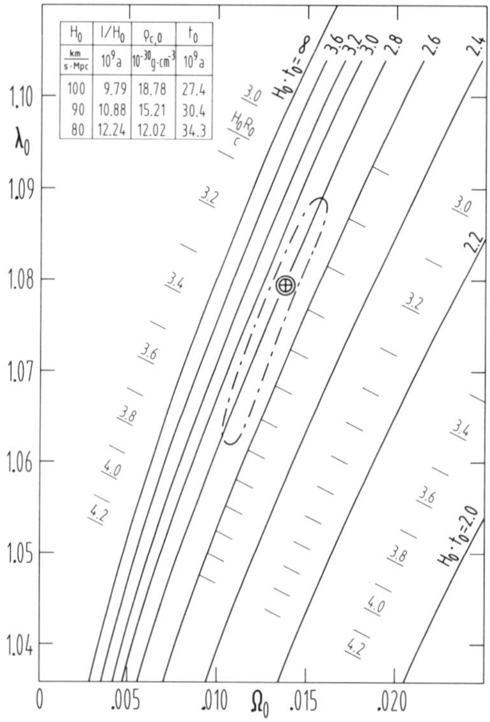

Fig.4. The circle marks the value of the .cosmological term λ_0 and of the density parameter $\Omega_{M,0}$ resulting from the Friedmann regression analysis. The dash-dotted ellipse gives the (1σ)-error curve. The solid lines represent the age of the universe t_0 (in units of H_0^{-1}). The perpendicular marks give the present curvature radius R_0 (in units of c/H_0).

We further point out the important result that the regression predicts the typical size ($\Delta z = 0.009$) of the nearby bubbles for $z \rightarrow 0$ which agrees with the observed void diameters of the galaxy distribution.

4. Regression formula for the cloud model

This approach is based on a homogeneous distribution of hydrogen clouds with a constant average number density Φ_0 in a comoving volume and with an invariant physical size. This implies clouds which are decoupled from the Hubble expansion similar to the galaxies. They could mostly be envisaged as clouds in dwarf galaxies with almost no star formation in the redshift interval $1.8 < z < 5$. The previous investigators calculated the number dN/dz of clouds intercepted along the line of sight (in an interval dl) in the frame of a cosmology based on the restricted Friedmann equation, where the constant Λ was assumed to be zero (Peterson 1978). As it was not possible to represent the data directly, it was concluded that this showed overwhelming evidence for a strong evolution in the physical parameters and in the number densities of the clouds (Murdoch et al. (1986), Lu et al. (1991)). Here we intend to demonstrate that this conclusion rests entirely on the assumption of a $\Lambda \equiv 0$ cosmology and that it is possible to fit the observed numbers on the basis of a Friedmann-Lemaître cosmology ($\Lambda > 0$) without evolutionary effects.

The theory for the analysis of the observed number on lines dN in a redshift interval dz was given by Peterson (1978), see also Lu et al. (1991) and [HLP 1994] Here we repeat the essential steps. The number of clouds dN intercepted along the line of sight dl is given by

$$dN = \sigma \, \Phi \, dl \tag{10}$$

with the here invariant physical cross section $\sigma = \sigma_0$ of the clouds and the number density Φ of the clouds with

$$\Phi = \Phi_0 \left(\frac{R_0}{R} \right)^3 = \Phi_0 \, (1+z)^3. \tag{11}$$

Φ refers to the number of clouds which produce absorption lines with an intrinsic equivalent width EW larger than an observationally fixed value. On the line of sight we have

$$dl = R[t] \, d\chi = - c \, dt \, , \tag{12}$$

where R[t] is the scale factor at time t and $d\chi$ the comoving radial coordinate of the Friedmann metric. From the redshift relation $1+z = R_0/R$ follows that

$$\frac{dz}{dt} = -\frac{R_0}{R[t]} \cdot H[t] = -(1+z) H[z] \tag{13}$$

with H[z] the Hubble expansion rate at the absorption time of a cloud which is observed with redshift z. Eqs. (12) and (13) yield

$$\frac{dl}{dz} = \frac{c}{(1+z) H[z]}. \tag{14}$$

With eqs. (10) and (11) we have

$$\frac{dN}{dz} = \sigma_0 \cdot \Phi_0 \cdot (1+z)^2 \cdot \frac{c}{H[z]} \tag{15}$$

Thus the observed numbers dN/dz should be represented by the basic formula

$$\frac{dN}{dz} = \sigma_0 \cdot \Phi_0 \cdot \frac{c(1+z)^2}{H_0\sqrt{\lambda_0-(\lambda_0+\Omega_0-1)(1+z)^2+\Omega_0(1+z)^3}} \qquad \textbf{(16)}$$

A satisfactory representation of the numbers observed by Lu et al. (B) and Röser (A) is obtained with eq. (16) with $\lambda_0 = 1.080$ and $\Omega_0 = 0.014$ with $\sigma_0 \cdot \Phi_0 c \cdot H_0^{-1} = 3.1$ for (B) and 3.72 for (A). These results are in agreement with the results of the Friedmann regression of the bubble wall approach. The numbers from Röser with EW $\geq$ 0.32 Å are systematically 1.2 times larger than the numbers from Lu et al. with EW $\geq$ 0.36 Å. They depend on the chosen lower limit of the intrinsic (i.e. rest frame) equivalent width (Fig.5).

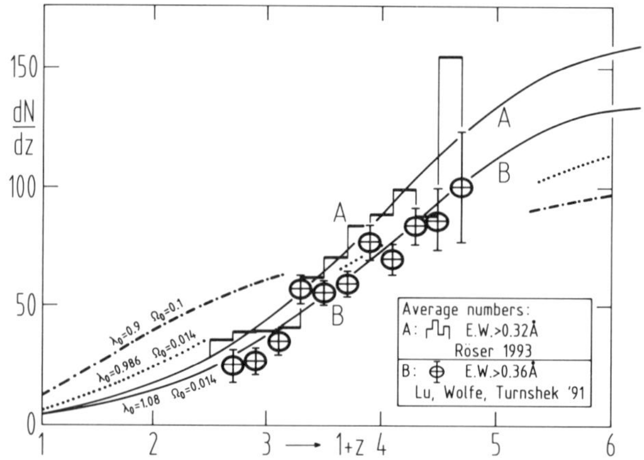

Fig.5. Line counts dN/dz as function of redshift (1+z). Data from Lu et al. (1991) (B) and Röser (1993) (A) with lower limits of the intrinsic equivalent width (0.36 Å for (B) and 0.32 Å for (A).

It can easily be seen that here also the errors of the regression are relatively small. In order to demonstrate this we have included a dotted curve in Fig. 5 which represents the corresponding Friedmann-Lemaître model with euclidean metric: $\lambda_0 + \Omega_0 = 1$ with $\lambda_0 = 0.986$. The dash-dotted curve shows another model with euclidean metric ($\lambda_0 = 0.9$, $\Omega_0 = 0.1$) favored by Fukugita et al. (1990, 1993). It was derived from number versus magnitude counts of galaxies and from the visibility of gravitational lenses if structure and dust in the lensing galaxies are taken into account. It is obvious that these two euclidean models are incapable of fitting the Ly-α data. The previous investigators have fitted the data with an approximation formula

$$\frac{dN}{dz} \approx (1+z)^{\gamma} \qquad \textbf{(17)}$$

and concluded strong evolution in the clouds, when they obtained $\gamma > 1$. A criticism of this method is given in [HLP 1994] together with a discussion of the possible limits on the evolution of the amount of absorbing hydrogen mass in the clouds. The fact that the observed dN/dz are averages over a large distance dl along the line of sight explains easily why both methods yield the same Friedmann-Lemaître model. The dl corresponds in size to a crossing of 20 bubbles and walls. Thus it is possible to assume that the clouds are located in the walls of the bubble structure. A detailed discussion can be found in [HLP 1994].

5. The best-fit Friedmann-Lemaître model

Table 1. The parameters of the best fit model

matter density parameter	$\Omega_{M,0}$	$= 0.014 \pm 0.002$
cosmological term	λ_0	$= 1.080 \pm 0.006$
Hubble number	H_0	$= 90$ km/(s Mpc)
matter density	$\varrho_{M,0}$	$= 0.21 \cdot 10^{-30}$ g·cm^{-3}
radiation density	$\varrho_{R,0}$	$= 0.47 \cdot 10^{-33}$ g·cm^{-3}
Einstein's constant	Λ	$= 3.1 \cdot 10^{-56}$ cm^{-2}
cosmic unit of length	$R_\Lambda = \Lambda^{-0.5}$	$= 6 \cdot 10^9$ ly
age of the universe	t_0	$= 30 \cdot 10^9$ years
curvature radius	$R_0 = 3.3\ c/H_0$	$= 36 \cdot 10^9$ ly
volume of the universe	$V_0 = 2\pi^2 \cdot R_0^3$	$= 7.8 \cdot 10^{86}$ cm^3
mass of the universe	$M = \varrho_{M,0} \cdot V_0$	$= 1.6 \cdot 10^{56}$ g
number of baryons	N	$= 1.0 \cdot 10^{80}$
specific entropy	$S = n_\gamma / n_b$	$= 3.2 \cdot 10^9$

The two methods of the regression analysis of the Ly-α forest yield the present cosmological term λ_0 and the present matter density parameter $\Omega_{M,0}$ independent of the Hubble parameter H_0. The value of H_0 and its permissible range can be derived by comparing $\Omega_{M,0} = 0.014 \pm 0.002$ of all gravitating particles (including possible nonbaryonic particles) with the density parameter of the total number of baryons obtained by the primordial nucleosynthesis (Walker et al. (1991)): $\Omega_{B,0} \cdot h_0^2 = 0.0125 \pm 0.0025$ with $h_0 = H_0/100$ km/(s·Mpc). The close agreement between both numbers points to H_0 in the range $78 < H_0 < 110$ km/(s·Mpc) with an optimum value of $H_0 = 90$ km/(s·Mpc). It shows that a significant contribution by nonbaryonic particles would require $H_0 > 110$ km/(s·Mpc). $\Omega_{B,0} \cdot h_0^2$ corresponds to a total baryonic density $\varrho_{B,0} = (0.24 \pm 0.05) \cdot 10^{-30}$ g·cm^{-3} independent of H_0. The term "total number of baryons" comprises the visible and the dark baryonic matter. The latter can be envisaged as consisting of "dead" stars (e.g. neutron stars), brown dwarfs etc. and nonluminous clouds. In our old universe most of the early generations, in particular the early halo stars, have burned up their thermonuclear fuel a long time ago. As Persic and Salucci (1992) have shown, only a small portion of the baryons is contained in visible objects (galaxies, gas, dust). They found $\Omega_{b,0} \approx 0.003$ for the visible baryons. Thus between 80 and 90 percent are hidden in dark objects. The total baryonic matter seems sufficient to explain the flat rotation curves of galaxies and the virial masses. All this is in accordance with our low-density universe where the expansion has been dominated by the Λ-term or the corresponding true quantum vacuum with an energy density $\varrho_\Lambda c^2 = \Lambda c^4/8\pi G$ for the last $15 \cdot 10^9$ years. Einstein's cosmological constant depends on the square of the Hubble radius. Using the total range of H_0 derived above we obtain

$$(c/H_0)^2 = (1.06 \pm 0.35) \cdot 10^{56}\ \mathrm{cm}^2 \tag{18}$$

and

$$\Lambda = 3\lambda_0 \cdot H_0^2 \cdot c^{-2} = (3.1 \pm 1.2) \cdot 10^{-56}\ \mathrm{cm}^{-2}. \tag{19}$$

In an old Friedmann-Lemaître universe the present value of the Hubble parameter H_0 is already close to its final value H_∞. It approaches its asymptotic value as:

$$H_\infty = c\sqrt{\frac{\Lambda}{3}} = H_0\sqrt{\lambda_0}\ . \tag{20}$$

The term λ_0 approaches 1 in the distant future. That shows that the Hubble radius c/H_0 does not change significantly anymore, despite the fact that the curvature radius R[t] approaches infinity for $t \rightarrow \infty$.

The present age of our universe is

$$t_0 = \frac{1}{H_0}\int_0^1 \frac{dx}{\sqrt{\Omega_{M,0}\,x^{-1} + \lambda_0 x^2 - (\lambda_0 + \Omega_{M,0} - 1)}} = 2.8 \cdot \frac{1}{H_0}\ . \tag{21}$$

This corresponds to $30 \cdot 10^9$ years for $H_0 = 90$ km/(s·Mpc). The present curvature radius is

$$R_0 = c \cdot H_0^{-1}\,(\Omega_{M,0} + \lambda_0 - 1)^{-0.5} = 3.3 \cdot c/H_0\ . \tag{22}$$

$H_0 = 90$ km/(s·Mpc) yields $R_0 = 3.4 \cdot 10^{28}$ cm $= 36 \cdot 10^9$ ly.
The inverse square root of Λ defines a cosmic unit of length $R_\Lambda = \Lambda^{-0.5}$. It corresponds to the curvature radius at the time of slow expansion, when the clouds decouple from the Hubble expansion. We obtain $R_\Lambda = 6 \cdot 10^9$ ly, corresponding to a redshift z= 5.

NEW WAYS IN COSMOLOGY:

PART II: ALTERNATIVE MODELS FOR THE VERY EARLY UNIVERSE

H.J. BLOME[1], W. PRIESTER[2], J.HOELL[3]

[1)] *Deutsche Forschungsanstalt für Luft- und Raumfahrt (DLR) D-51140 Köln, Germany*
[2)] *Institut für Astrophysik, Universität Bonn and Max-Planck-Institut für Radioastronomie, Auf dem Hügel 71, D 53121 Bonn, Germany*
[3)] *Deutsche Agentur für Raumfahrtangelegenheiten (DARA), D-53227 Bonn, Germany*

ABSTRACT: The past evolution of the Friedmann-Lemaître model, derived from the Ly-α forest ($\Omega_{M,0} = 0.014$, $\lambda_0 = 1.080$), is discussed depending on the Friedmann time t and on the normalized curvature radius $x = R[t]/R_0$. For $x < 10^{-12}$ all particles are relativistic. Their density is proportional to x^{-4}. The curvature radius is calculated for $t = 10^{-32}$ s, the time when the primordial quarks and leptons possibly originated from a phase transition of the Higgs scalar field or - in a pragmatic view - of the primordial quantum vacuum.

The advantages and disadvantages of four scenarios for the very early universe are discussed: 1.) Big Bang, 2.) Inflation, 3.) Big Bounce, 4.) Cosmic-Egg. It is shown that all scenarios can be connected with the curvature radius obtained from the closed Friedmann-Lemaître model with spherical space metric. This demonstrates that inflation does not require euclidean metric.

The explanation for the transition from the high-energy density of the primordial quantum vacuum to the true quantum vacuum, represented by the Λ-term with an equivalent density $\varrho_\Lambda = 16.4 \cdot 10^{-30}\ \mathrm{g \cdot cm^{-3}}$, is one of the outstanding problems of cosmology.

M. M. Shapiro et al. (eds.), Currents in High-Energy Astrophysics, 301–312.

6. The evolution of the universe

Any discussion of the very early universe for Friedmann-times $t < 10^{-10}$ s is necessarily speculative as it is impossible - at least at the present time - to obtain proofs by experimental or observational results. Here we intend to follow a very pragmatic way by extrapolating our Friedmann-Lemaître model ($\Omega_{M,0} = 0.014$, $\lambda_0 = 1.080$, $H_0 = 90$ km/(s·Mpc) (see Table 1 in part I)) into the very early times, where at $t = 10^{-32}$ s the primordial quarks and leptons are believed to have originated in the course of a phase transition. We shall demonstrate that the model can easily be connected with different alternatives of inflationary scenarios (A. Guth (1981), A. Linde (1981)), the cosmic-egg model (M. Israelit, N. Rosen (1989)) and the Big-Bounce model (H.J. Blome, W. Priester (1991)).

At first we look at the history of the evolution in the standard picture which is accepted by a majority of astrophysicists (Fig. 6). In our low-density universe matter dominates over the radiation for $t > 10^7$ years (Fig. 7) corresponding to $T = 10^3$ K or to a normalized curvature radius $x = R/R_0 = 3{\cdot}10^{-3}$. The "recombination" to neutral hydrogen took place at about $T = 3600$ K corresponding to a redshift $1+z = R_0/R = 1200$. This was accomplished in the time interval between $6{\cdot}10^5$ and $8{\cdot}10^5$ years. That is long before matter began to dominate the cosmic evolution.

The annihilation of the antibaryons began at $T \approx 10^{13}$ K. Thus the freezing out of the stable baryons was accomplished at $x = R/R_0 = 10^{-12}$. From then on we can follow the evolution of the matter density parameter $\Omega_M[x]$. Before that time all particles were relativistic, and the evolution followed according to the radiation density parameter $\omega[x]$. This parameter dominates the evolution from $t = 10^{-32}$ s to 10^{14} s. For later times it can be neglected as then $\Omega_M[x]$ dominates the evolution. In our present universe we have the contribution of the background photons with $\omega_0 = 3{\cdot}10^{-5}$. If the neutrinos have a vanishing rest mass (≈ 0) we would expect a total $\omega_0 \approx 5{\cdot}10^{-5}$.

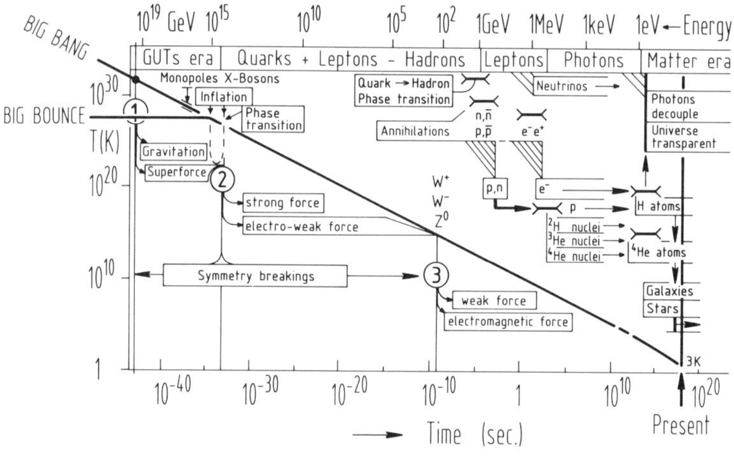

Fig.6. Schematic evolution of the universe showing the temperature T as function of time from the Planck-time $t_{PL} = 5{\cdot}10^{-44}$ s to our present time t_0. The upper part shows the history of particles from the primordial quarks to the formation of stars. The lower part demonstrates the symmetry breakings of the fundamental forces.

We further note that our best-fit model does not allow any significant contribution by non-baryonic particles. Thus their possible contribution was taken to be zero.

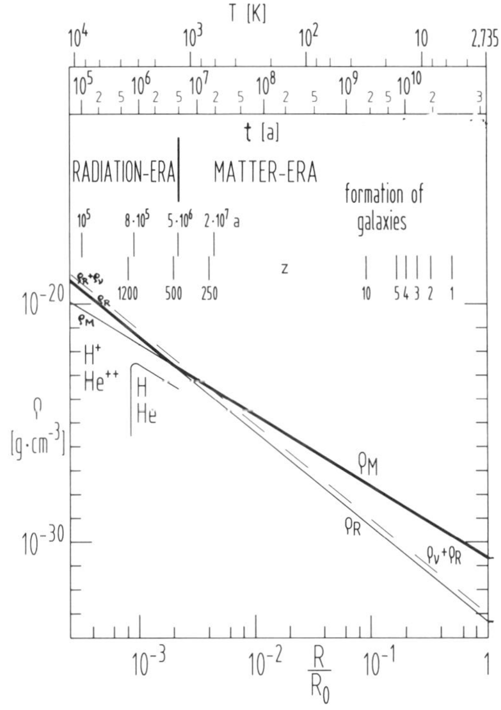

Fig. 7. Evolution of the matter density $\varrho_M[x]$ and the radiation density $\varrho_R[x]$ in the time interval from $t = 10^5$ years to $t_0 = 30.10^{10}$ years. The abscissa is the normalized curvature radius $x = R/R_0$.

The evolution of the density parameters and of the cosmological term can be obtained from the complete Friedmann equation which includes the radiation density parameter $\omega = \varrho_R[t]/\varrho_c[t]$. This term was neglected in eq. (5) of part I. The complete Friedmann equation reads with $x = R[t]/R_0$

$$H^2[x] = H_0^2 \left(\lambda_0 - (\lambda_0+\Omega_{M,0}+\omega_0-1)\ x^{-2} + \Omega_{M,0}\ x^{-3} + \omega_0\ x^{-4}\right) . \qquad (23)$$

From the definition of

$$\begin{aligned} \Omega_M[x] &= \varrho_M[x] \ / \ \varrho_c[x] \\ \varrho_c[x] &= 3H^2[x] \ / \ 8\pi G \\ \lambda[x] &= \Lambda c^2 \ / \ 3H^2[x] \\ \omega[x] &= \varrho_R[x] \ / \ \varrho_c[x] \end{aligned} \qquad (24)$$

follows the evolution of the parameters using the conservation law

$$\Omega[x] = \frac{\Omega_{M,0}\; H_0^2}{x^3 \quad H^2[x]} \qquad \textit{for } x > 10^{-12} \tag{25}$$

$$\omega[x] = \frac{\omega_0 \;\; H_0^2}{x^4 \;\; H^2[x]} \;, \tag{26}$$

$$\lambda[x] = \lambda_0 \, \frac{H_0^2}{H^2[x]} \quad . \tag{27}$$

We note that for instance at x = 10^{-12} the curvature radius is already R = $3.4 \cdot 10^{16}$ cm = 0.034 ly and the Hubble expansion rate is H[x] $\approx 10^{22} \cdot H_0 \approx 3 \cdot 10^4\ s^{-1}$. From eq. (25) follows that the cosmological term λ[x] at that time is extremely small ($\lambda \approx 10^{-44}$).

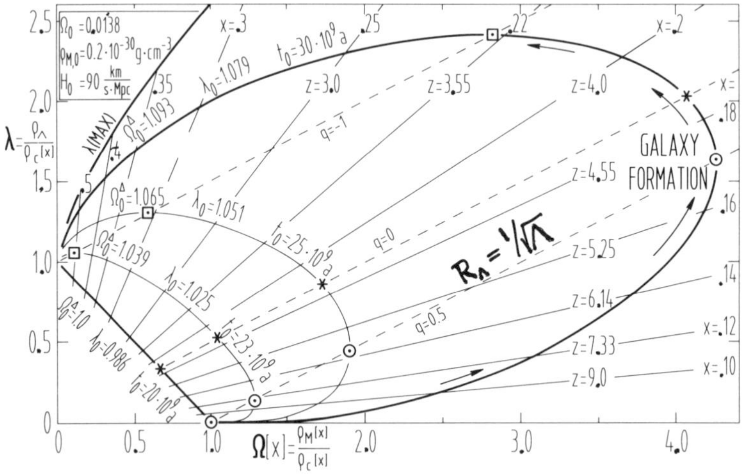

Fig. 8. Evolutionary tracks of 4 Friedmann-Lemaître models $\Omega_M[x]$ versus λ[x] with x = $R[t]/R_0$ = 1/(1 + z). The present density parameter is $\Omega_M[1]$ = 0.014 for all four models. The ages t_0 are calculated with H_0 = 90 km/(s·Mpc).

Fig. 8 shows the evolution of $\Omega_M[x]$ versus λ[x] for $10^{-12} \leq x \leq 1$. All Friedmann models begin with $\Omega_M = 1$ and λ = 0 because $\varrho_c[0] = \infty$ and $H[0] = \infty$ for x = t = 0. Ω[x] is still 1.00000 for x = 10^{-12} when the stable baryons freeze out. In Fig. 8 the evolution of four models is given. Important is the one with $\Omega_M > 4$ at a redshift of z = 5 corresponding to x = 0.167. A large Ω_M is required for galaxy formation in a hydrogen gas during the slow phase (z ≈ 5) (Hoell, Priester (1994)). It is further noteworthy that in these models with spherical space metric the curvature radius is always $R_\Lambda = 1/\sqrt{\Lambda}$ at the epoch, when the density parameter passes through its maximum. These points lie on a straight line for which the deceleration parameter is $q = 0.5$.

The extrapolation of our model towards the very early times yields a curvature radius $R \approx 600$ cm at $t = 10^{-32}$ s and $\omega \approx 1 + 10^{-48}$. These values can easily be obtained from all models with an early inflationary phase when the phase transition is completed at $t = 10^{-32}$ s after an increase of the volume by a factor in the range of 10^{84} to 10^{93} (Fig. 9). There is no need to require an euclidean metric with $\Omega_M + \omega \equiv 1$ or with $\Omega_M + \omega + \lambda \equiv 1$. This result is supported by the investigation of Madsen and George Ellis (1988).

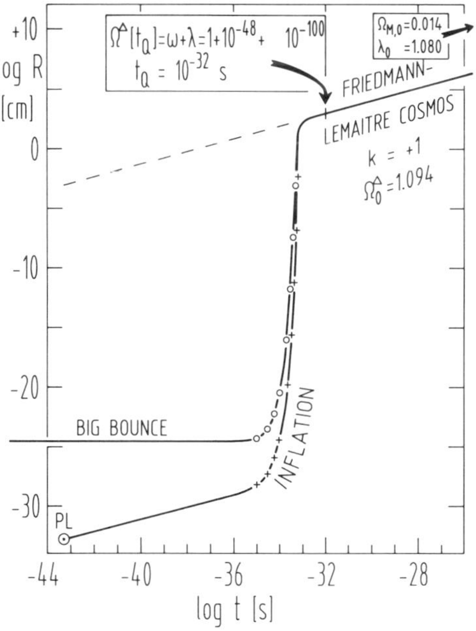

Fig. 9. Curvature radius R as function of time in the very early universe $10^{-44} < t < 10^{-26}$ s leading to the present model with $\Omega_{M,0} = 0.014$ and $\lambda_0 = 1.080$. The inflation model starts from a Planck bubble with $R_{PL} = 1.6 \cdot 10^{-33}$ cm at $t = 5.4 \cdot 10^{-44}$ s. The dashed line gives the extrapolation towards the singularity of a Big Bang model. The Big Bounce model avoids the singularity at $t = 0$.

7. Scenarios for the very early universe

In this paragraph we intend to follow again our pragmatic approach and discuss the advantages and disadvantages of different scenarios.

A singularity is unavoidable in the Friedmann cosmology for $t \to 0$ if the condition for the energy dominance $\varepsilon + 3p \geq 0$ is fulfilled at all times (Hawking, Ellis (1968), Penrose (1965), Kundt (1968)). It is, however, this condition which is likely to be violated in the very early universe if the quantum vacuum or the Higgs scalar field dominates the dynamics of the expansion. The equation of state is then pressure $p = - \varepsilon$ (Zeldovich (1968)).

Furthermore, the usual concept of separating spacetime and energy may be violated during the very early universe (Einstein (1956)), when the curvature of space varies very significantly on the scale of a Planck-time

$$t_{PL} = (\hbar G/c^5)^{0.5} = 5 \cdot 10^{-44} \text{ s} .$$

At this time quantum-gravitational effects become dominant. The typical size of a mini universe emerging from the spacetime foam (Misner et al. (1973)) is assumed to be of the order of a Planck-length $L_{PL} = (\hbar G/c^3)^{0.5} = 1.6 \cdot 10^{-33}$ cm, and the energy density is postulated to be of the order of the Planck density $\varepsilon_{PL} = \varrho_{PL} \cdot c^2$ with $\varrho_{PL} = c^5/\hbar G^2 = 5.2 \cdot 10^{93}$ g·cm^{-3}.

Attempts to describe the quantum creation of the universe as quantum tunneling from "nothing" (i.e. no classical spacetime and matter) have been made e.g. by Starobinskii (1980), Atkatz, Pagels (1982), Vilenkin (1988).

The singularity of the Big Bang implies at $t = 0$ an infinitely large energy density and temperature and in addition an infinitely large Hubble expansion rate. These conditions seem to be highly unphysical. Therefore it became common practice in the so-called standard cosmology to begin the Friedmann universe at the Planck-time without further explanation for the conditions at $t < t_{PL}$. The Planck time is the assumed threshold time of classical cosmology, defined by the validity of Einstein's field equations and the quantum field theory. One justification could be seen in the fact that the back extrapolation of the present Friedmann-Lemaître universe yields an energy density $\varepsilon_F = \varrho_F c^2$ at the Planck time which is of the same order of magnitude as the Planck density

$$\varrho_F[t_{PL}] = 3 / (8\pi G \cdot t_{PL}^2) = 3 / (8\pi) \cdot \varrho_{PL} .$$

On the other hand the curvature radius $R_F[t_{PL}]$ of the present universe at the Planck time is

$$R_F[t_{PL}] = 2 \cdot 10^{-3} \text{ cm} = 10^{30} L_{PL} .$$

This is an enormous excess over the postulated mini universe with a size of a Planck bubble. In order to eliminate this discrepancy, a mechanism of exponential expansion is required to reconcile the classical model with its possible quantum origin. Already Gliner (1966) pointed out that an exponential expansion can be realized when the universe is dominated by a substrate with positive energy density but negative pressure $p = - \varepsilon$. This would be provided by a dominating quantum vacuum.

The inflationary scenario (Guth (1981), Linde (1983)) achieves the rapid expansion at the epoch $t = 10^{-35}$ s when the constant energy density of the Higgs scalar field with $\varrho_{Hi} \approx 10^{76}$ g·cm^{-3} surpasses the decreasing energy density of the Friedmann expansion which started from a Planck bubble with ϱ_{PL} at t_{PL}. As the pressure of the Higgs fields obeys the equation of state $p_{Hi} = - \varrho_{Hi} c^2$ the exponential inflation takes place. At $t = 10^{-33}$ s the curvature radius R has increased by a factor of 10^{31} matching now the curvature of our Friedmann-Lemaître universe at that moment. Thus a phase transition is required which transforms the Higgs energy into the primordial elementary particles (quarks and leptons). This stops the exponential expansion. At 10^{-32} s the transformation must have been fully accomplished for the merger into the closed Friedmann-Lemaître model with spherical space metric. This demonstrates that the inflation does not and cannot predict a euclidean metric for our universe (Fig.9). Since we have a closed model with spherical space metric we must conclude that the beginning at the Planck time also started from a closed Planck bubble. This contains a problem, which is not accounted for in the inflationary scenario, in particular the missing explanation for the origin of the exceedingly large energy density with $\varrho = \varrho_{PL}$ and the extremely large expansion rate $H[t_{PL}] = 1/(2t_{PL}) = 10^{43}\ s^{-1}$ at the Planck time.

Generally it is believed that the origin of our universe began with a fluctuation of the quantum vacuum. A fluctuation which provides a Planck density in a Planck bubble would disappear again within $5 \cdot 10^{-44}$ s according to Heisenberg's uncertainty principle. In order to overcome this serious obstacle an extremely large expansion rate between two adjacent bubbles with distance L_{PL} is required. This could be achieved by a Hubble expansion in a Friedmann model with

$$H[t_{PL}] = c\ /\ (2\ L_{PL}) = 1\ /\ (2\ t_{PL})\ .$$

This corresponds to a separation with the speed of light.

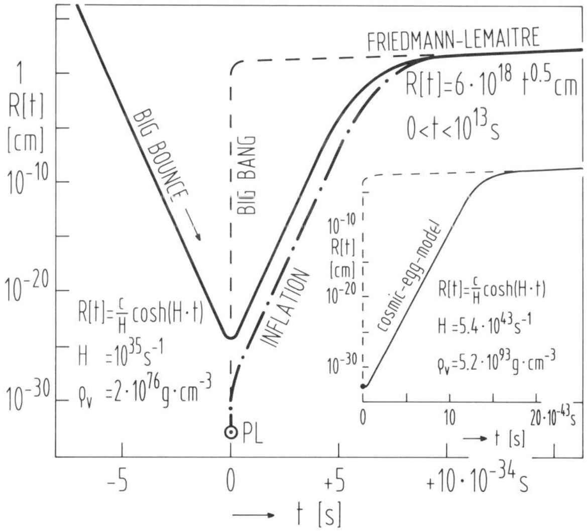

Fig. 10. Four scenarios for the very early universe: Curvature radius R as function of time (linear scale $|t| < 16 \cdot 10^{-34}$ s). Big Bang and the inflation model start here from a Planck bubble (PL) filled with relativistic matter and the scalar Higgs field. At $t = 10^{-35}$ s begins the inflation (dash-dotted-line). The cosmic-egg model begins also from a Planck bubble, but filled with "prematter" ($p = -\varrho c^2$). The Big Bounce is based on a quantum vacuum with $\varrho_V = 10^{-17}\ \varrho_{PL}$. The de-Sitter-solution provides the minimum curvature radius $R_{Min} = 2 \cdot 10^8\ L_{PL}$.

Unsolved is further the question of the underlying space structure on which the fluctuation took place. Here we face the paramount problem of the origin or creation of the universe for which no satisfactory explanation exists, at least in our opinion. This points again to the singularity with the sudden creation of space and energy. Is there a possibility within existing theories like General Relativity and Quantum Field Theory to avoid the singularity? At this point I like to quote Yakov Zeldovich's vision on page 630 in his Selected Works, Vol. II (1993) from his paper dedicated to the 60th birthday of John Archibald Wheeler (1970):

"It is possible that the observed energy (or, to be more precise, entropy) of the universe and its observed isotropy are both connected to pair creation (near the universe's "initial singularity"). It is quite probable that the quantization of the metric (near the initial singularity) was unavoidable and affected its subsequent state. But here we encounter also an aspect new in principle: in cosmology - as distinguished from all other fields of science - one requires ideas concerning the choice of the initial state, and not simply ideas concerning the evolution of a given state by given laws."

Big Bounce (Blome, Priester (1991)): An initial singularity could have been avoided within the realm of validity of General Relativity with the following line of arguments. The lowest state of the quantized fields defines the energy level of the quantum vacuum. This picture is supported by quantum field theory, in which real matter (elementary particles) represents the excited states of the space-filling matter fields. Thus, the vacuum is the more fundamental entity, which can exist independent of real particles. Therefore, it seems a natural hypothesis that the quantum vacuum was the primordial stage before elementary particles came into being (Priester, Blome (1987)). This hypothesis implies the decoupling of the origin of matter (quarks, leptons, photons etc.) from the creation of spacetime.

Our approach ("Big Bounce") is based on the assumption of a primordially homogeneous and isotropic spacetime, in which all fields exist in their ground state. Because of the connection with our present closed Friedmann-Lemaître universe we must also require the primordial space metric being spherical. Furthermore, the model must merge into the Friedmann-Lemaître model most favorably at the Friedmann time $t = 10^{-32}$ s in analogy to the inflationary scenario.

In order to achieve this, we must require a phase transition out of the primordial quantum vacuum which provides the origin of the quark and leptons at $t = 10^{-32}$ s, very similar to the phase trasition in the inflationary scenario. Here, however, we do not need to postulate the Planck density at the Planck time. In contrast to that we solve the Friedmann equations in the early times for a universe which does not yet contain real matter but is filled with a constant energy density of the quantum vacuum. The energy density which provides the merger at $t = 10^{-32}$ s is represented by an equivalent density $\varrho_V = 2 \cdot 10^{76}$ $g \cdot cm^{-3}$ with $p_V = - \varrho_V c^2$. The solution of the Friedmann equation for $-\infty < t < 10^{-32}$ s is of the de-Sitter-type for a closed universe with the curvature radius R given by

$$R[t] = \frac{c}{H} \cosh(Ht) \tag{28}$$

with

$$H = \sqrt{\frac{8\pi G}{3} \varrho_V} = 10^{35} s^{-1} \qquad [29]$$

This expansion rate H is constant during the time interval from minus infinity to the time of the phase transition. The curvature radius passes through a minimum $R_{min} = 3 \cdot 10^{-25}$ cm corresponding to $2 \cdot 10^8$ L_{PL}. This is the minimum radius which could occur at any time. Thus we identify it with the Friedmann time $t = 0$. We have a universe without a singularity and with a maximum energy density ($\varepsilon = 2 \cdot 10^{97}$ $erg \cdot cm^{-3}$ corresponding to $\varrho_V = 2 \cdot 10^{76}$ $g \cdot cm^{-3}$) provided by the primordial quantum vacuum. This energy is far below the Planck energy ($\varepsilon = 5 \cdot 10^{114}$ $erg \cdot cm^{-3}$). Thus we can assume to remain in the validity range of GRT and QFT. In this solution the universe lasts from "eternity to eternity" (Fig. 10).

Finally we would like to point out that the Big Bounce model should not be confused with the Eddington-Lemaître bounce models which are special solutions of the Friedmann equation without the early phase transition. These bounce models can nowadays be excluded since the observed maximum redshifts (z=4.9) are not anymore compatible with the present matter density $\varrho_{M,0}$.

The cosmic-egg model (M. Israelit, N. Rosen (1989)): This singularity-free model emerges from a small bubble ("cosmic egg") with a size of nearly Planck length filled with "prematter" of Planck density, characterized by the equation of state $p = - \varrho c^2$ agreeing with the one of the quantum vacuum. The cosmos springs forth from the quantum epoch and starts at the Planck time with a de-Sitter-solution of the field equations for a spherical space metric. This model was parallel and independently developed with regard to our Big Bounce model (1987, 1991).

In Fig. 11 we point out the essential differences of the two models. Because the cosmic-egg models begins with Planck density, the cosinus-hyperbolicus function produces an immediate exponential expansion. Thus the merger into the Friedmann-Lemaître model must occur already at $t = 10^{-42}$ s. This causes a critical situation with the GUT's, the Grand Unified Theories, which predict the production of a huge number of magnetic monopoles with masses of 10^{17} GeV/c^2 at $t = 10^{-37}$ s. If the monopoles would be produced as expected they would cause an early collapse of the universe. This situation is avoided in the inflationary scenarios and in the Big Bounce model.

Mark Israelit and Nathan Rosen used for the Friedmann model $H_0 = 46.4$ km/(s·Mpc) with $\Omega_{M,0} = 1.16$, $\Lambda = 0$ and $R_0 = 52.8 \cdot 10^9$ ly in contrast to our choice of three Friedmann-Lemaître models in Fig. 11 with $\Omega_{total} = \Omega_{M,0} + \lambda_0 = 1.28$, 1.09 and 1.03, $H_0 = 75$ km/(s·Mpc) and $R_0 = 25$, 44 and $75 \cdot 10^9$ ly. This leads to a sufficiently good agreement between both models at $t = t^{-32}$ s. It shows that both approaches beginning with a de-Sitter solution provide a reasonable way to avoid the initial singularity.

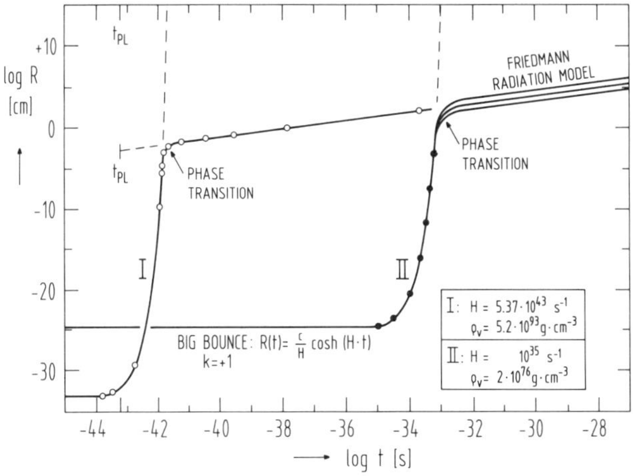

Fig. 11. Curvature radius R[t] as function of time. The cosmic-egg model (I) is marked by open circles. The Big Bounce model (II) is marked by black dots. Both models merge into Friedmann models.

Conclusions: At the beginning of the Friedmann expansion at $t = 10^{-32}$ s two important conditions must have been provided for a cosmos which was able to develop into our present universe: the creation of the primordial matter (quarks, leptons) and the extremely large expansion rate at that time. This could have been provided by the extremely large density of a primordial quantum vacuum. Its connection to the true quantum vacuum represented by the Λ-term in our universe is one of the unsolved paramount problems of cosmology. It is , however, very satisfying, that the numerical value of Λ and its corresponding equivalent density $\varrho_\Lambda = 16.4 \cdot 10^{-30}$ $g \cdot cm^{-3}$ can now be obtained from the analysis of the Ly-α forest in high-resolution spectra of quasars. The Λ-equivalent density in our present universe is about 80 times bigger than the baryonic matter density. This implies that the Hubble expansion was small about $20 \cdot 10^9$ years ago. This was an essential prerequisite for galaxy formation in a hydrogen gas.

We would like to conclude with a quotation from Yakov Zeldovich's autobiography (1984) with his vision on the future development of theoretical physics: "The idea of an all-encompassing unified physical theory seems brighter, and geometry is playing an increasingly great role. Perhaps, in the highest sense, not literally, Einstein will prove to be right, and his theory, which reduces the forces of gravity to geometry, will turn out to be the model of the unified theory. It is possible that precisely cosmology will turn out to be the testing ground for the verification of new theories."

References:

Atkatz, D., Pagels, H.: 1980.
Phys. Rev. D 25, 2065.
Blome, H.J., Priester, W.: 1985.
Astroph. Space Sci. 117, 327.
Blome, H.J., Priester, W.: 1991.
Big Bounce in the early universe. Astron. Astrophys. 250, 43-49.
Einstein, A.: 1931.
Sitzungsber. Preuss. Akad. Wiss. 96, 235.
Einstein, A.: 1956.
Grundzüge der Relativitätstheorie. Vieweg, Braunschweig 1956, p. 84.
Felten, J.E., Isaacman, R.: 1986.
Rev. Mod. Phys. 58, 689.
Fukugita, M.,Takahara, F., Yamashita, K., Yoshii, Y.: 1990.
Astroph. J. 361, L 1.
Fukugita, M., Peebles, P.J.E.: 1993.
Princeton Preprint.
Gliner, E.B.: 1966.
Sov. Phys. JETP 22, 378.
Guth, A.: 1981.
Phys. Rev. D 23, 347.
Hawking, S.W., Ellis, G.F.R.: 1968.
Astrophys. J. 152, 25.
Hoell, J., Liebscher, D.E., Priester, W.: 1994.
Confirmation of the Friedmann-Lemaître universe. Astron. Nachr. 315, 89-96. [HLP].
Hoell, J., Priester, W.: 1994.
Galaxy formation in a Friedmann-Lemaître model. Panchromatic view of galaxies. (G. Hensler, Ch. Theis, J. Gallagher eds.). Editions Frontières, p. 29-33.
Israelit, M., Rosen, N.: 1989.
Astroph. J. 342, 627.
Kundt, W.: 1968.
Springer Tracts in Mod. Phys. 47, 111.
Liebscher, D.E., Priester, W., Hoell, J.: 1992.
A new method to test the model of the universe. Astron. Astrophys. 261, 377. [LPH 1992 a].
Liebscher, D.E., Priester, W., Hoell, J.: 1992.
Lyman α forest and the evolution of the universe. Astron. Nachr. 313, 265 [LPH 1992 b].
Linde, A.D.: 1983.
Phys. Lett. 129 B, 177.
Lu, L., Wolfe, A.M., Turnshek, D.A.: 1991.
Astroph. J. 367, 19.
Madsen, M.S., Ellis, G.F.R.: 1988.
MNRAS 234, 67
Misner, C.W., Thorne, K.S., Wheeler, J.A.: 1973.
Gravitation. Freeman, San Francisco. 1279 pp.

Murdoch, H.S., Hunstead, R.W., Pettini, M., Blades, J.C.: 1986.
Astroph. J. 309, 19.
Peebles, P.J.E.: 1993.
Principles of Physical Cosmology. Princeton University Press, 718 pp.
Penrose, R.: 1965.
Proc. Roy. Soc. A 284, 159.
Persic, M., Salucci, P.: 1992.
MNRAS 258, 14p-18p.
Peterson, B.A.: 1978.
IAU Symp. 79, Dordrecht.
Pettini, M., Hunstead, R.W., Smith, L.J., Mar, D.P.: 1990.
MNRAS 246, 554.
Priester, W.:, Blome, H.J.: 1987.
"Big Bang" oder "Big Bounce"? Sterne und Weltraum, 26, 83- 89 and 140-144.
Priester, W., Hoell, J., Blome, H.J.: 1989.
Physik. Bl. 45, 51
Priester, W., Hoell, J., Blome, H.J.: 1994.
The scale of the universe. Preprint April 29, 1994. [PHB].
Röser, H.J.: 1993.
Max-Planck-Inst. f. Astronomy, Heidelberg. Habil. Thesis (in preparation).
Starobinskii, A.A.: 1980.
Phys. Lett. 91 B, 99.
Vilenkin, A.: 1988.
Phys. Rev. D 37, 888.
Walker, T.P., Steigman, G., Schramm, D., Olive, K., Kang, H.S.: 1991.
Astrophys. J. 376, 1.
Weinberg, S.: 1989.
Rev. Mod. Phys. 61, 1.
Weinberg, S.: 1993.
Dreams of a Final Theory. Pantheon Books, New York.
Zeldovich, Y.B.: 1968.
Sov. Phys. Usp. 11, 381.
Zeldovich, Y.B.: 1979.
Large Scale Structlure of the Universe. Dordrecht; Reidel 409-421.
Zeldovich, Y.B.: 1984.
An Autobiographical Afterword. Selected Works Vol. II (J.P. Ostriker ed.) Princeton Univ. Press (1993) p. 632-644.
Zeldovich, Y.B.: 1993.
Selected Works Vol. II, p. 630 (Ostriker, J.P. ed.) Princeton Univ. Press, 644 pp.

PARTICIPANTS

Philip D. ALLEN
Department of Physics
University of Leeds
Leeds LS2 9JT
United Kingdom
Tel: 05-32-333-825
Fax: 05-32-333-846
E-mail: PHY5PDA@SUN.LEEDS.AC.UK

Matthew G. BARING
Code 665
NASA Goddard Space Flt. Ctr.
Greenbelt, MD 20771
USA
Fax: 001-301-286-1682
E-mail: BARING@LHEAVX.GSFC.NASA.GOV

Antonio Renato Perissinotto BIRAL
Deparamento de Paios Cosmicos
Instituto de Fisica "Gleb Wataghin"
Universidade Estadual de Campinas
Caixa Postal 6165/DRC
13081 - Campinas - SP
Brazil
Fax: 55-192-39-3127
E-mail: BIRAL@IFI.UNICMP.BR

David BIRD
Department of Physics
University of Utah
Salt Lake City, Utah 84112
USA
Fax: 001-801-581-4801
E-mail: DBIRD@COSMIC.PHYSICS.UTAH.EDU

Hans-Joachim BLOME
DLR
German Aerospace Research Institute
MD-NR
Linder Höhe
51140 Köln
Germany
Tel: 49-220-360-12968
Fax: 49-220-367-668

Enrique BRONCHALO
Dept. de Fisica
Universidad de Alcala de Henares
28871 Alcala de Henares
Madrid, Spain
Fax: 34-1-889-49-53

Edmund CLAYTON
Dept. of Physics & Astronomy
Louisiana State University
Baton Rouge, LA 70803-4001
USA
Fax: 001-504-388-1222
E-mail: CLAYTON@PHEPDS.DNET.NASA.GOV

Kai DAUMILLER
Inst. fur Kernphysik I
Kernforschungszentrum Karlsruhe
Postfach 3640
76021 Karlsruhe
Germany
Fax: 49-7247-82-3548
E-mail: DAUMILLER@IK1MV2.KFK.DE

Stanley P. DAVIS
Code 668
Bldg. 2 Room 61
Goddard Space Flight Center
Greenbelt, MD 20771
USA
Tel: 001-301-286-9852
Fax: 001-301-286-1681 or 1682
E-mail: DAVIS@GROSSC.GSFC.NASA.GOV (GROSSC::DAVIS)

Ocker C. DE JAGER
Potchefstroom University
Physics Dept.
Priv. Bag X6001
2520 Potchefstroom,
South Africa
Fax: 27 148 99 2421
E-mail: FSKOCDJ@PUKNET.PUK.AC.ZA

Vladimir DOGIEL
Department of Theoretical Physics
P.N. Lebedev Inst.
Russian Academy of Sciences
Leninsky Prospect 53
117924 CSP, Moscow
Russia
Fax: 7-095-938-2251
E-mail: DOGIEL@FIAN.FREE.NET

Lev DORMAN
Cosmic-Ray Department
IZMIRAN 142092
Troitsk, Moscow Region
Russia
Tel: 7-095-334-0924
Fax: 7-095-334-0124
E-mail: DORMAN@COSRAY.IZMIRAN.
TROITSK.SU. - OR -
LAP@WISE1.TAU.AC.IL

Ronald EKERS
Australian Nat'l Telescope Facility
P.O. Box 76
Epping, NSW 2121
Australia
Tel: 61-2-372-4300
Fax: 61-2-372-4310
E-mail: REKERS@ATNF.CSIRO.AU

Anne-Marie ELO
Physics Department
University of Oulu
90570 Oulu
Finland
Fax: 358-81-553-1287
E-mail: FYS-AME@FINOU.OULU.FI

Victoria FONSECA
Dpto. Fisica Atomica
Facultad Ciencias Fisicas
Universidad Complutense
28040 Madrid
Spain
Fax: 34-1-394-5193
E-mail: FONSECA@EUCMDX.GAE.
UCM.ES

Raj GANDHI
Mehta Inst. of Physics & Mathematics
10 Kasturba Gandhi Marg
Allahabad, 211001
India
E-mail: RAJ@MRI.ERNET.IN

Lewis GEER
Department of Physics
Washington University, Box 1105
1 Brookings Drive
St. Louis, MO 63130 USA
Tel: 001-314-935-6285
Fax: 001-314-935-6219
E-mail: LEWIS@COSRAY.WUSTL.EDU

Vitaly L. GINZBURG
P.N. Lebedev Physical Institute
Russian Academy of Sciences
Leninsky Prospect 53
117924 GSP Moscow
Russia
Fax: 007-095-938-2251

Fatma GÖK
Department of Physics
Middle East Technical University
66531 Ankara,
Turkey
Fax: 90-312-210-1281
E-mail: FATMA@NEWTON.PHYSICS.
METU.EDU.TR

Oleg GOLDSHMIDT
School of Physics and Astronomy
Tel Aviv University
69978 Tel Aviv,
Israel
Tel: 972-3-640-8895
Fax: 972-3-642-9306
E-mail: GOLDSH@CCSG.TAU.AC.IL or OLEG@WISE.TAU.AC.IL

Mohini GUPTA
Department of Physics
University of Bombay-Kalina Campus
Sanata Cruz (E)
Bombay 400098,
India
Tel: 91-22-202-8104
Fax: 91-22-266-2735
E-mail: MOHINI@FREENET.FSU.EDU

Simon HART
Department of Physics
University of Leeds
Leeds LS2 9JT
United Kingdom
Tel: 05-32-333-825
Fax: 05-32-333-846
E-mail: PHY6SPH@SUN.LEEDS.AC.UK

German HERMANN
MPI fur Kernphysik
Saupfercheckweg 1
69117 Heidelberg
Germany
Tel: 49-6221-516-510
Fax: 49-6221-516-540
E-mail: HERMANN@ENULL.MPI-HD.MPG.DE

Yunsil HO
Department of Physics
Columbia University
P.O. Box 137
Irvington, NY 10533
USA
Tel: 001- 914-591-8100
Fax: 001-914-591-8120
E-mail: HO@NEVIS.COLUMBIA.EDU

Minghuey A. HUANG
Department of Physics
University of Utah
Salt Lake City, UT 84112
USA
Tel: 001-801-581-5504
Fax: 001-801-581-4801
E-mail: HUANG@COSMIC.PHYSICS.UTAH.EDU

Gottfried KANBACH
MPI fur Extraterr. Physik
Giessenbachstrasse
85740 Garching bei Munchen
Germany
Tel: 49-89-3299-3544
Fax: 49-89-3299-3569
E-mail: GOK@MPE-GARCHING.MPG.DE

Georg B. KHRISTIANSEN
Inst. of Nuclear Physics
Moscow State University
117234 Moscow,
Russia
Fax: 007-095-939-1158

Nick D. KYLAFIS
University of Crete
Physics Department
714 09 Heraklion, Crete
Greece
Tel: 30-81-239-757, X155
Fax: 30-81-239-735
E-mail: KYLAFIS@IESL.FORTH.GR

Axel LINDNER
II. Inst. fur Exp. Physik
University of Hamburg
Luruper Chaussee 149
22761 Hamburg
Germany
Tel: 49-40-8998-2176
Fax: 49-40-8998-2950
E-mail: LINDNER@HP-CLUSTER.DESY.DE

Ignazio LONGO
Dipartimento di Fisica
Universita di Bologna
via Irnerio 46
40126 Bologna
Italy
Fax: 39-51-247-244

Paulo Gali MACEDO
Centro de Astrofisica da U.P.
R. do Campo Alegre-823
4200 Porto
Portugal
Tel: 351-1-600-1672
Fax: 351-2-600-3654
E-mail: PGMACEDO@NCC.UP.PT

Eric MANNEL
Columbia University (Nevis Lab)
P.O. Box 137
Irvington, NY 10533
USA
Fax: 001-914-591-8120
E-mail: MANNEL@NEVIS.COLUMBIA.
EDU

John C. MATHER
Code 685
Goddard Space Flight Center
Greenbelt, MD 20771
USA
Tel: 001-301-286-8720
Fax: 001-301-286-1617
E-mail: (15382::MATHER)
MATHER@STARS.GSFC.NASA.GOV

Pieter MEINTJES
Department of Physics
Unit for Space Research
University of Potchefstroom
Potchefstroom 2520
South Africa
E-mail: PTM@MPE-GARCHING.MPG.DE

Christian NAUNDORF
Department of Astronomy
MPI - Univ. of Bonn
Auf dem Hugel 69
53175 Bonn
Germany
Fax: 49-228-525-229
E-mail: P675CEN@MPIFR-
BONN.MPG.DE

Yuval NE'EMAN
Department of Physics
University of Tel Aviv
Tel Aviv
Israel

Martina NIEMEYER
MPI fur Radioastronomy
Auf dem Hugel 69
53175 Bonn,
Germany
Tel: 49-228-525-285
Fax: 49-228-525-229
E-mail: MNIEMEYER@MPIFR-
BONN.MPG.DE

Masaki NISHIZAWA
Institute for Cosmic Ray Research
3-2-1 Midori-cho, Tanash
Tokyo 188,
Japan
Tel: 81-424-69-9530
Fax: 81-424-62-3096
E-mail: NISIZAWA@EMSUN3.ICRR.U-
TOKYO.AC.JP

Paolo PAPINI
INFN - Florence
Largo Enrico Fermi 2
55100 Firenze
Italy
E-mail: PAPINI@VAXFI.FI.INFN.IT

Wolfgang PRIESTER
Inst. für Astrophysik
MPI fur Radioastronomie
Auf dem Hügel 71
53121 Bonn
Germany
Tel: 49-228-73-3671
Fax: 49-228-73-3672

Clem PRYKE
Department of Physics
University of Leeds
LS2 9JT Leeds
United Kingdom
Tel: 05-32-333-825
Fax: 05-32-333-846
E-mail: PHY5CLP@PHYS-IRC.NOVELL.LEEDS.AC.UK

Lev PUSTIL'NIK
Israel Cosmic Ray Research
Israel Jordan Valley College
Str. Sheisaf 11/6, P.O. Box 3024
129000 Quatsrin
Israel
Tel: 972-6-962-849
E-mail: LAP@WISE1.TAU.AC.IL

Olaf REIMER
Physics Dept.
University of Siegen
Adolf-Reichwein Strasse
57068 Siegen,
Germany
Fax: 49-271-740-4377
E-mail: REIMER@SICIP1.PHYSIK.UNI-SIEGEN.DE

Alastair RODGERS
Department of Physics
University of Leeds
Leeds LS2 9JT
United Kingdom
Tel: 05-32-333-825
Fax: 05-32-333-846
E-mail: PHY5AJR@PHYS-IRC.NOVELL.LEEDS.AC.UK

Vincenzo SALAMONE
IFCAI-CNR
Istituto di Fisica Cosmica
Piazza Verdi, 6
90100 Palermo
Italy
Tel: 39-091-605-5111

Dirk SCHMELE
II. Inst. fur Exp. Physik
University of Hamburg
Luruper Chaussee 149
22761 Hamburg
Germany
Tel: 49-40-8998-2194
Fax: 49-40-8998-2950
E-mail: SCHMELE@DESYVAX.DESY.DE

Alexander SERBER
Astrophy. & Cosmic Plasma Physics
Inst. of Applied Physics
46 Uljanov Str.
603600 Nizhny-Novogrod,
Russia
Fax: 007-8312-36-2061
E-mail: SERBER@APPL.NNOV.SU

Maurice M. SHAPIRO
205 Yoakum Pkwy.
#1514
Alexandria, VA 22304
USA
Tel: 001-703-370-1985
Fax: 001-703-370-1985

Rein SILBERBERG
Code 7654, Space Physics
Cosmic and Gamma Ray Astrophysics
Naval Research Laboratory
Washington, DC 20375
USA
Fax: 001-202-761-6473

Chandra SIVARAM
Department of Physics
Indian Inst. of Astrophysics
Srjapur Road, Koramangala
560034 Bangalore
India

Hennie SMIT
Physics Department
Potchefstroom University
Priv. Bag X6001
2520 Potchefstroom,
South Africa
Fax: 27-148-99-2421
E-mail: FSKHJS@PUKNET.PUK.AC.ZA

Arthur E. SMITH
Department of Chemistry
MSB-203C
University of Alabama in Huntsville
Huntsville, AL 35899 USA
Fax: 001-205-895-6207
E-mail: SMITHA@SSL.DNET.NASA.GOV
(8858::SMITHA)

F. GRAHAM SMITH
Jodrell Bank (NRAL)
Macclesfield
Cheshire, SK11 9DL,
United Kingdom
Fax: 44-61-273-5867
E-mail: FGS@JB.MAN.AC.UK

Todor STANEV
Bartol Research Institute
University of Delaware
Newark, DE 19716
USA
Tel: 001-302-831-1495
Fax: 001-302-831-1843
E-mail: STANEV@BARTOL.UDEL.EDU
(BARTOL::STANEV)

Magdalena TEODORESCU
Ministry of Communications
CP 22-90
Bd. Libertatu
74556 Bucharest
Romania
Tel: 40-1-674-6681
Fax: 40-1-312-4591

Klaus THIELHEIM
Inst. fur Kernphysik
Universitat Kiel
Otto Hahn Platz 1
24118 Kiel
Germany
Tel: 49-431-880-3216
Fax: 49-431-880-3216

Enis TUNCER
Dept. of Physics
Middle East Technical University
06531 Ankara,
Turkey
Tel: 90-312-240-3109
Fax: 90-312-210-1281
E-mail: TUNCER@NEWTON.PHYSICS.
METU.EDU.TR

Joseph UNGER
Inst. fur Kernphysik I
Kernforschungszentrum Karlsruhe
P.O. Box 3640
76021 Karlsruhe,
Germany
Fax: 49-07-247-82-3548
E-mail: UNGER@IK3DC1.KFK.DE

John P. WEFEL
Department of Physics and Astronomy
Louisiana State University
202 Nicholson Hall
Baton Rouge, LA 70803-4001 USA
Tel: 001-504-388-8696
Fax: 001-504-388-1222
E-mail: (Decnet: 7894::WEFEL)
WEFEL@PHEPDS.DNET.NASA.GOV

Yvonne WETZEL
Inst. fur Kernphysik I
Kernforschungszentrum Karlsruhe
Postfach 3640
76021 Karlsruhe,
Germany
Fax: 49-72-4782-3548
E-mail: YVONNE@IK1MV2.KFK.DE

Barbara WIEBEL
Fachbereich Physik
University of Wuppertal
Gausstrasse 20
42097 Wuppertal,
Germany
Tel: 49-202-439-2829
Fax: 49-202-439-2811
E-mail:WIEBEL@WPDS5.PHYSIK.UNI-WUPPERTAL.DE (14029::WIEBEL)

Nathan YANASAK
Department of Physics
University of Utah
201 James Fletcher Bldg.
Salt Lake City, UT 84112 USA
Tel: 001-801-581-5917
Fax: 001-801-581-4801
E-mail: YANASAK@EINSTEIN.PHYSICS.UTAH.EDU

Maria ZIELINSKA
Department of Physics
University of Lodz
u1. Uniwersytecka 5
Pomarska
90-950 Lodz 1
Poland
E-mail: MZ@ZPK2.U.LODZ.PL

Ralf ZINK
MPI fur Kernphysik
Saupfercheckweg 1
Postfach 10 39 80
69128 Heidelberg
Germany
Fax: 49-6221-516-540
E-mail: ZINK@EU1.MPI-HD.MPG.DE

SUBJECT INDEX

AUTHOR INDEX